The Handbook of Environmental Chemistry

Editor-in-Chief: O. Hutzinger

Volume 5 Water Pollution
Part L

The Handbook of Environmental Chemistry

Recently Published and Forthcoming Volumes

Environmental Specimen Banking
Volume Editors: S. A. Wise and P. P. R. Becker
Vol. 3/S, 2006

Polymers: Chances and Risks
Volume Editors: P. Eyerer, M. Weller and C. Hübner
Vol. 3/V, 2006

Persistent Organic Pollutants in the Great Lakes
Volume Editor: R. A. Hites
Vol. 5/N, 2006

Antifouling Paint Biocides
Volume Editor: I. Konstantinou
Vol. 5/O, 2006

Estuaries
Volume Editor: P. J. Wangersky
Vol. 5/H, 2006

The Caspian Sea Environment
Volume Editors: A. Kostianoy and A. Kosarev
Vol. 5/P, 2005

Marine Organic Matter: Biomarkers, Isotopes and DNA
Volume Editor: J. K. Volkman
Vol. 2/N, 2005

Environmental Photochemistry Part II
Volume Editors: P. Boule, D. Bahnemann and P. Robertson
Vol. 2/M, 2005

Air Quality in Airplane Cabins and Similar Enclosed Spaces
Volume Editor: M. B. Hocking
Vol. 4/H, 2005

Environmental Effects of Marine Finfish Aquaculture
Volume Editor: B. T. Hargrave
Vol. 5/M, 2005

The Mediterranean Sea
Volume Editor: A. Saliot
Vol. 5/K, 2005

Environmental Impact Assessment of Recycled Wastes on Surface and Ground Waters
Engineering Modeling and Sustainability
Volume Editor: T. A. Kassim
Vol. 5/F (3 Vols.), 2005

Oxidants and Antioxidant Defense Systems
Volume Editor: T. Grune
Vol. 2/O, 2005

The Rhine

Volume Editor:
Thomas P. Knepper

With contributions by
M. H. Bik · H.-J. Brauch · P. Diehl · R. Forster · T. Gerke
J. Hartkopf · A. Jeuken · W. Jülich · F. Karrenbrock · T. P. Knepper
R. W. P. M. Laane · F. T. Lange · M. Leemann · K. Lindner · J. Lowis
J. Müller · M. Peschka · M. Petri · C. K. Schmidt · P. Seel · R. Steen
J. van Steenwijk · P. Stoks · P. de Voogt · H. Weil · R.-D. Wilken
H.-G. Willemsen · S. Wisser

Springer

Environmental chemistry is a rather young and interdisciplinary field of science. Its aim is a complete description of the environment and of transformations occurring on a local or global scale. Environmental chemistry also gives an account of the impact of man's activities on the natural environment by describing observed changes.
The Handbook of Environmental Chemistry provides the compilation of today's knowledge. Contributions are written by leading experts with practical experience in their fields. The Handbook will grow with the increase in our scientific understanding and should provide a valuable source not only for scientists, but also for environmental managers and decision-makers.
The Handbook of Environmental Chemistry is published in a series of five volumes:

Volume 1: The Natural Environment and the Biogeochemical Cycles
Volume 2: Reactions and Processes
Volume 3: Anthropogenic Compounds
Volume 4: Air Pollution
Volume 5: Water Pollution

The series Volume 1 The Natural Environment and the Biogeochemical Cycles describes the natural environment and gives an account of the global cycles for elements and classes of natural compounds.
The series Volume 2 Reactions and Processes is an account of physical transport, and chemical and biological transformations of chemicals in the environment.
The series Volume 3 Anthropogenic Compounds describes synthetic compounds, and compound classes as well as elements and naturally occurring chemical entities which are mobilized by man's activities.
The series Volume 4 Air Pollution and Volume 5 Water Pollution deal with the description of civilization's effects on the atmosphere and hydrosphere.
Within the individual series articles do not appear in a predetermined sequence. Instead, we invite contributors as our knowledge matures enough to warrant a handbook article.
Suggestions for new topics from the scientific community to members of the Advisory Board or to the Publisher are very welcome.

Library of Congress Control Number: 2005937701

ISSN 1433-6863
ISBN-10 3-540-29393-0 Springer Berlin Heidelberg New York
ISBN-13 978-3-540-29393-4 Springer Berlin Heidelberg New York
DOI 10.1007/b11739

Springer is a part of Springer Science+Business Media

springer.com

Printed in Germany

Cover design: E. Kirchner, Springer-Verlag
Typesetting and Production: LE-TEX Jelonek, Schmidt & Vöckler GbR, Leipzig

Printed on acid-free paper 02/3141 YL – 5 4 3 2 1 0

Prof. Dr. D. Mackay
Department of Chemical Engineering
and Applied Chemistry
University of Toronto
Toronto, ON, M5S 1A4, Canada

Prof. Dr. A. H. Neilson
Swedish Environmental Research Institute
P.O. Box 21060
10031 Stockholm, Sweden
ahsdair@ivl.se

Prof. Dr. J. Paasivirta
Department of Chemistry
University of Jyväskylä
Survontie 9
P.O. Box 35
40351 Jyväskylä, Finland

Prof. Dr. Dr. H. Parlar
Institut für Lebensmitteltechnologie
und Analytische Chemie
Technische Universität München
85350 Freising-Weihenstephan, Germany

Prof. Dr. S. H. Safe
Department of Veterinary
Physiology and Pharmacology
College of Veterinary Medicine
Texas A & M University
College Station, TX 77843-4466, USA
ssafe@cvm.tamu.edu

Prof. P. J. Wangersky
University of Victoria
Centre for Earth and Ocean Research
P.O. Box 1700
Victoria, BC, V8W 3P6, Canada
wangers@telus.net

The Handbook of Environmental Chemistry Also Available Electronically

For all customers who have a standing order to The Handbook of Environmental Chemistry, we offer the electronic version via SpringerLink free of charge. Please contact your librarian who can receive a password or free access to the full articles by registering at:

springerlink.com

If you do not have a subscription, you can still view the tables of contents of the volumes and the abstract of each article by going to the SpringerLink Homepage, clicking on "Browse by Online Libraries", then "Chemical Sciences", and finally choose The Handbook of Environmental Chemistry.

You will find information about the

- Editorial Board
- Aims and Scope
- Instructions for Authors
- Sample Contribution

at springer.com using the search function.

Preface

Environmental Chemistry is a relatively young science. Interest in this subject, however, is growing very rapidly and, although no agreement has been reached as yet about the exact content and limits of this interdisciplinary discipline, there appears to be increasing interest in seeing environmental topics which are based on chemistry embodied in this subject. One of the first objectives of Environmental Chemistry must be the study of the environment and of natural chemical processes which occur in the environment. A major purpose of this series on Environmental Chemistry, therefore, is to present a reasonably uniform view of various aspects of the chemistry of the environment and chemical reactions occurring in the environment.

The industrial activities of man have given a new dimension to Environmental Chemistry. We have now synthesized and described over five million chemical compounds and chemical industry produces about hundred and fifty million tons of synthetic chemicals annually. We ship billions of tons of oil per year and through mining operations and other geophysical modifications, large quantities of inorganic and organic materials are released from their natural deposits. Cities and metropolitan areas of up to 15 million inhabitants produce large quantities of waste in relatively small and confined areas. Much of the chemical products and waste products of modern society are released into the environment either during production, storage, transport, use or ultimate disposal. These released materials participate in natural cycles and reactions and frequently lead to interference and disturbance of natural systems.

Environmental Chemistry is concerned with reactions in the environment. It is about distribution and equilibria between environmental compartments. It is about reactions, pathways, thermodynamics and kinetics. An important purpose of this Handbook, is to aid understanding of the basic distribution and chemical reaction processes which occur in the environment.

Laws regulating toxic substances in various countries are designed to assess and control risk of chemicals to man and his environment. Science can contribute in two areas to this assessment; firstly in the area of toxicology and secondly in the area of chemical exposure. The available concentration ("environmental exposure concentration") depends on the fate of chemical compounds in the environment and thus their distribution and reaction behaviour in the environment. One very important contribution of Environmental Chemistry to the above mentioned toxic substances laws is to develop laboratory test methods, or mathematical correlations and models that predict the environmental fate of new chemical compounds. The third purpose of this Handbook is to help in the basic understanding and development of such test methods and models.

The last explicit purpose of the Handbook is to present, in concise form, the most important properties relating to environmental chemistry and hazard assessment for the most important series of chemical compounds.

At the moment three volumes of the Handbook are planned. Volume 1 deals with the natural environment and the biogeochemical cycles therein, including some background information such as energetics and ecology. Volume 2 is concerned with reactions and processes in the environment and deals with physical factors such as transport and adsorption, and chemical, photochemical and biochemical reactions in the environment, as well as some aspects of pharmacokinetics and metabolism within organisms. Volume 3 deals with anthropogenic compounds, their chemical backgrounds, production methods and information about their use, their environmental behaviour, analytical methodology and some important aspects of their toxic effects. The material for volume 1, 2 and 3 was each more than could easily be fitted into a single volume, and for this reason, as well as for the purpose of rapid publication of available manuscripts, all three volumes were divided in the parts A and B. Part A of all three volumes is now being published and the second part of each of these volumes should appear about six months thereafter. Publisher and editor hope to keep materials of the volumes one to three up to date and to extend coverage in the subject areas by publishing further parts in the future. Plans also exist for volumes dealing with different subject matter such as analysis, chemical technology and toxicology, and readers are encouraged to offer suggestions and advice as to future editions of "The Handbook of Environmental Chemistry".

Most chapters in the Handbook are written to a fairly advanced level and should be of interest to the graduate student and practising scientist. I also hope that the subject matter treated will be of interest to people outside chemistry and to scientists in industry as well as government and regulatory bodies. It would be very satisfying for me to see the books used as a basis for developing graduate courses in Environmental Chemistry.

Due to the breadth of the subject matter, it was not easy to edit this Handbook. Specialists had to be found in quite different areas of science who were willing to contribute a chapter within the prescribed schedule. It is with great satisfaction that I thank all 52 authors from 8 countries for their understanding and for devoting their time to this effort. Special thanks are due to Dr. F. Boschke of Springer for his advice and discussions throughout all stages of preparation of the Handbook. Mrs. A. Heinrich of Springer has significantly contributed to the technical development of the book through her conscientious and efficient work. Finally I like to thank my family, students and colleagues for being so patient with me during several critical phases of preparation for the Handbook, and to some colleagues and the secretaries for technical help.

I consider it a privilege to see my chosen subject grow. My interest in Environmental Chemistry dates back to my early college days in Vienna. I received significant impulses during my postdoctoral period at the University of California and my interest slowly developed during my time with the National Research Council of Canada, before I could devote my full time of Environmental Chemistry, here in Amsterdam. I hope this Handbook may help deepen the interest of other scientists in this subject.

Amsterdam, May 1980 *O. Hutzinger*

Twenty-one years have now passed since the appearance of the first volumes of the Handbook. Although the basic concept has remained the same changes and adjustments were necessary.

Some years ago publishers and editors agreed to expand the Handbook by two new open-end volume series: Air Pollution and Water Pollution. These broad topics could not be fitted easily into the headings of the first three volumes. All five volume series are integrated through the choice of topics and by a system of cross referencing.

The outline of the Handbook is thus as follows:

1. The Natural Environment and the Biochemical Cycles,
2. Reaction and Processes,
3. Anthropogenic Compounds,
4. Air Pollution,
5. Water Pollution.

Rapid developments in Environmental Chemistry and the increasing breadth of the subject matter covered made it necessary to establish volume-editors. Each subject is now supervised by specialists in their respective fields.

A recent development is the accessibility of all new volumes of the Handbook from 1990 onwards, available via the Springer Homepage springeronline.com or springerlink.com.

During the last 5 to 10 years there was a growing tendency to include subject matters of societal relevance into a broad view of Environmental Chemistry. Topics include LCA (Life Cycle Analysis), Environmental Management, Sustainable Development and others. Whilst these topics are of great importance for the development and acceptance of Environmental Chemistry Publishers and Editors have decided to keep the Handbook essentially a source of information on "hard sciences".

With books in press and in preparation we have now well over 40 volumes available. Authors, volume-editors and editor-in-chief are rewarded by the broad acceptance of the "Handbook" in the scientific community.

Bayreuth, July 2001 *Otto Hutzinger*

Contents

Foreword

Why a special issue on water pollution in the Rhine River? Worldwide there are a great number of rivers comparable to the Rhine in terms of length and volume of water.

We present here a comprehensive book beginning with hydrological facts -development through the centuries - political demands and early warning strategies, accompanied by detailed contributions on modern analysis and the environmental fate of a wide spectrum of micropollutants.

The Rhine (German: Rhein, French: Rhin, Dutch: Rijn,) is one of the longest and most important rivers in Europe. Its length (1320 km or $\sim$ 820 miles) and flow (at Basel an average of 1100 m^3 and at Lobith an average of 2300 m^3 of water flows down the river every second), makes it the third biggest in Europe. With regard to the size of the catchment area – the Rhine's basin covers an area of 185 000 km^2 – the Rhine is Europe's fourth biggest river.

As a vital navigable waterway, trade and goods are carried far inland. The many castles and prehistoric fortifications along the Rhine testify not only to its importance as a waterway but also as a frontier.

Over the last few centuries, the Rhine has changed from being a natural, mostly undisturbed river into an anthropogenically influenced area. The Rhine brought growth into Europe through industrialization. An increase in trade and industry was followed by the pollution of the river, the sediments, and the fish. The changes in hydrology, the growing industry, and the peak of pollution in the 1970s resulted in political demands: a purification process was started, which has led to a relatively clean river in a highly industrialized region.

The importance of the Rhine as a supply for drinking water has made it the best analyzed river in the world from its source in the Alps to its estuary in the North Sea. This has brought about critical awareness of the consumers of drinking water and legislative bodies.

In order to avert dangers and to detect and investigate the causes of incidental pollution, such as spills and industrial or shipping accidents, (semi-) continuous measurements of water quality are performed by the water authorities and drinking water companies. Both operate systems in which various chemical screening techniques are used in combination with biological early warning systems.

A result of this investigation is that most of the relevant papers concerning pollution by micropollutants first published in the literature were about the Rhine and its tributaries.

Gradually the loadings of inputs have shifted from more persistent chemicals, such as HCH, PCBs and PAHs, to more polar persistent compounds – the so-called emerging contaminants, such as surfactants and pharmaceuticals. In this book, you will also find methods and data concerning pesticides, synthetic aromatic sulfonates, aminopolycarboxylate chelating agents, MTBE, and the group of substances that pollute the Rhine as by-products and degradation products of chemical syntheses. Most persistent organic compound concentrations have decreased, although a few, including brominated flame retardants, are increasing. A new EU directive brings natural and artificial radioactivity under discussion.

The successful management of the Rhine as a common project of industry, water suppliers, and legislative bodies shows a practicable way to solve the problems of a highly industrial and highly populated area. The results and experiences of the processes undertaken could be adapted to similar problems in other countries. Large cities in developing countries with growing industrialization should be able to avoid problems that have already been solved for the Rhine. The content of this book is suitable for analysts, industries, utilities, and politicians who wish to gain knowledge from concise but comprehensive information in all fields concerning the Rhine.

Idstein, January 2006 Thomas P. Knepper

Part I
General Aspects

Hdb Env Chem Vol. 5, Part L (2006): 3–45
DOI 10.1007/698_5_034

Published online: 23 November 2005

Hydrology

Walter Jülich[1] (✉) · Klaus Lindner[2]

[1]Riouwstraat 136, 2585 HH Den Haag, The Netherlands
walter.juelich@mac.com

[2]Tannenweg 15, 53340 Meckenheim, Germany
kblindner@gmx.de

Abstract The River Rhine is one of the major rivers of Europe – the only river that connects the Alps with the North Sea. It flows 1320 km from two small headstreams in the Alps of east-central Switzerland north and west to the North Sea, passing six countries (Austria, Liechtenstein, Switzerland, France, Germany, and the Netherlands) and receiving tributaries from three more countries (Italy, Belgium, and Luxembourg). Since the Treaty of Vienna in 1815 and the Convention of Mannheim in 1868, the Rhine is an international waterway, navigable from the North Sea as far as Rheinfelden on the Swiss-German border. Its catchment area, including the delta area, exceeds 185 000 km^2.

Keywords Drainage · Floods · Hydrology · Lakes · Modifications of hydrological flow · River Rhine · Tributaries · Water discharge regime

The course of the River Rhine linking north and south and the excellent transportation possibilities by boat made the river the mediator of trade and culture at an early stage, first between the Roman Empire and Gaul and later between Italy, Switzerland, Germany, and the Netherlands. Much of this important role is thanks to the well-balanced discharge of the Rhine, which makes the river navigable throughout the year. This is influenced by special hydrological conditions, the winter regime in the alpine areas and the summer regime in the lowlands.

1 The River Rhine

With a length of about 1320 km (approximately 820 miles), the Rhine River (German *Rhein*, French *Rhin*, Dutch *Rijn*) is one of the longest rivers in Europe. In terms of catchment size, the Rhine is Europe's fourth biggest river, after the Volga, the Danube, and the Vistula (Fig. 1). In terms of length and flow, however, it is the third biggest. The Rhine's basin covers an area of 185 000 km^2, some 25 000 km^2 of it in the Netherlands (Fig. 2). At Basel an average of 1100 m^3, and at Lobith, an average of 2300 m^3 of water flows down the river every second (Figs. 3 and 4).

1.1 Headstreams

The Rhine rises in two headstreams high in the Swiss Alps in the canton of Graubünden (east-central Switzerland). The "Vorderrhein" (Anterior

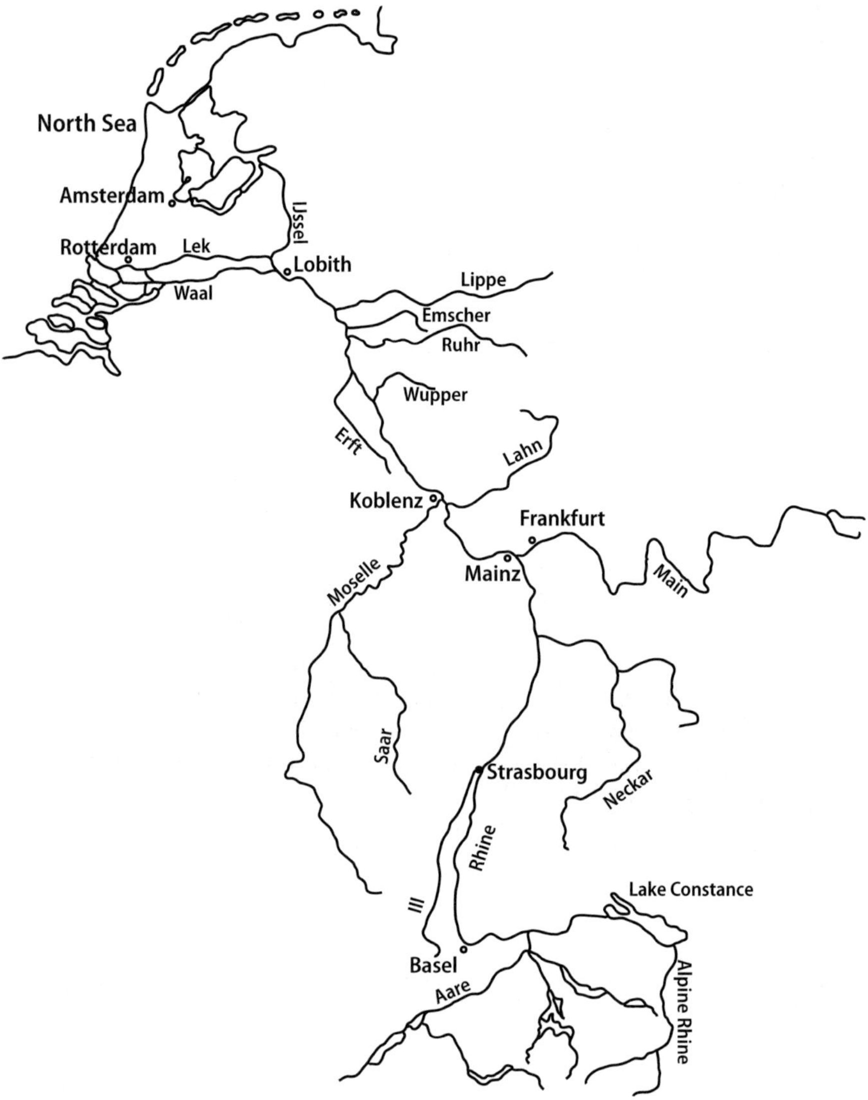

Fig. 1 Rhine catchment area with the principal tributaries

Rhine) springs from Lake Toma at 2344 m, near the Oberalp Pass in the Central Alps, and then flows eastward passing Disentis (Fig. 5). Near Reichenau (still in Graubünden) it is joined by the Hinterrhein from the south. The "Hinterrhein" (Posterior Rhine) is formed west of San Bernardino Pass, near the Swiss-Italian border, where it receives water from the Paradies glacier near the Rheinquellhorn and is joined by the Albula River below Thusis (Fig. 6).

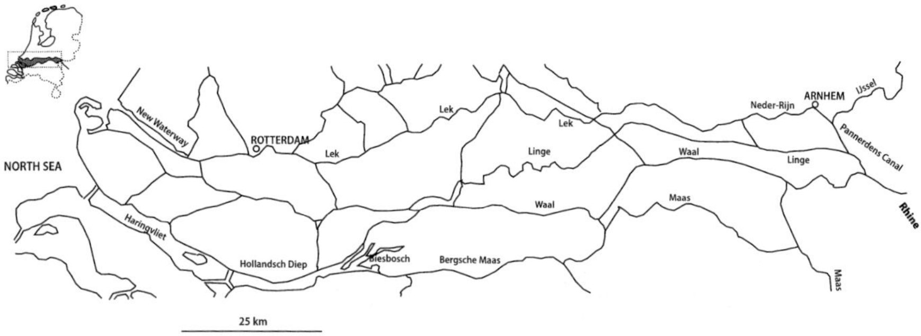

Fig. 2 Branches of the Rhine in the delta region

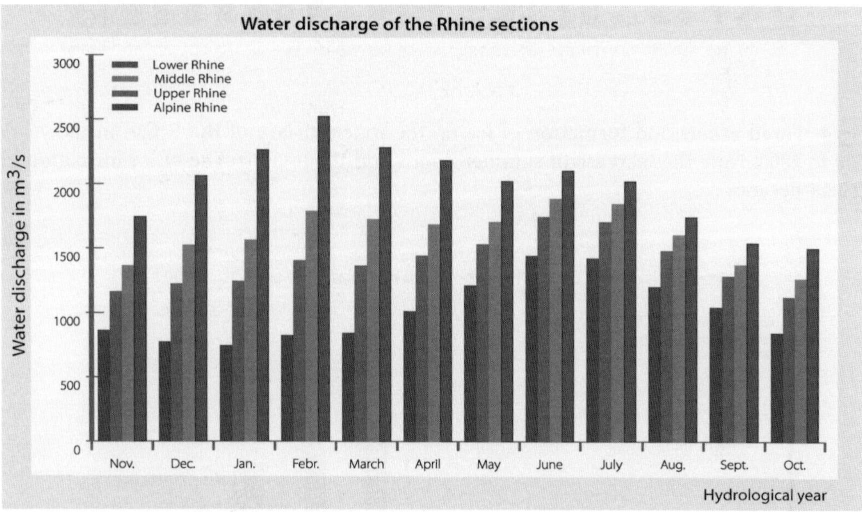

Fig. 3 Water discharge of the Rhine sections (Alpine Rhine, Upper Rhine, Middle Rhine, Lower Rhine). Maximum discharge lies during summer months (Alpine Rhine) or in spring (Lower Rhine)

1.2 Alpine Rhine

That part of the Rhine between the confluence of the Anterior and Posterior Rhine and its entry into Lake Constance is called the Alpine Rhine ("Alpenrhein"); it is 102 km long. The Alpine Rhine is an energetic mountain river. From Reichenau to Sargans it flows entirely within Swiss territory; for the next 30 km it forms the border between Switzerland and Liechtenstein; and the last part separates Switzerland from Austria. The lower part was forming a delta as the current slackened before the entrance to Lake Constance. In

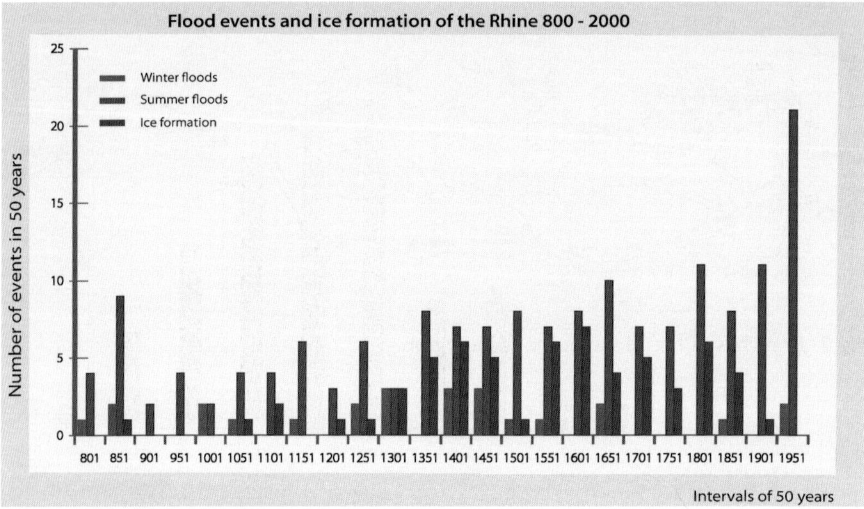

Fig. 4 Flood events and formation of ice on the water surface of the Rhine in the years 800 to 2000. Note the increase of summer floods and the total decline of ice formation in recent decades

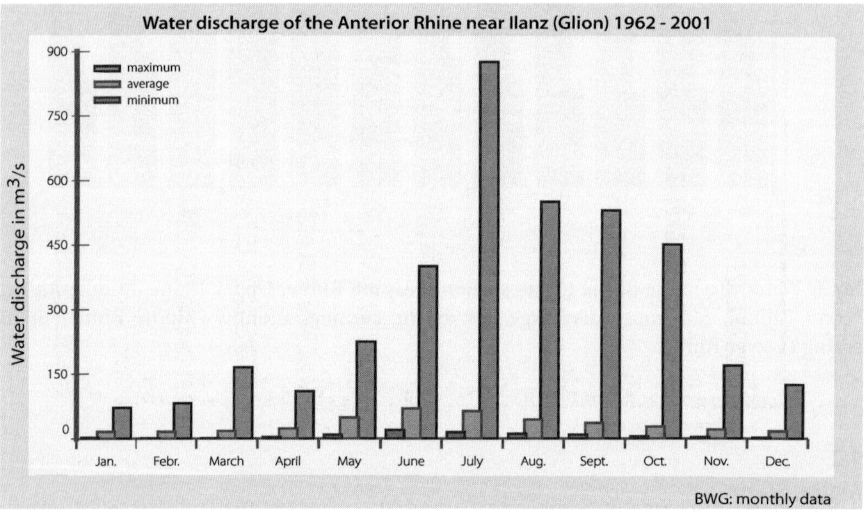

Fig. 5 Water discharge of the Anterior Rhine 1962–2001. Maximum discharge during summer months

this flat-floored section the Rhine has been straightened into a canal and the banks reinforced to prevent flooding. That part of the river transports lots of sediments (Fig. 7).

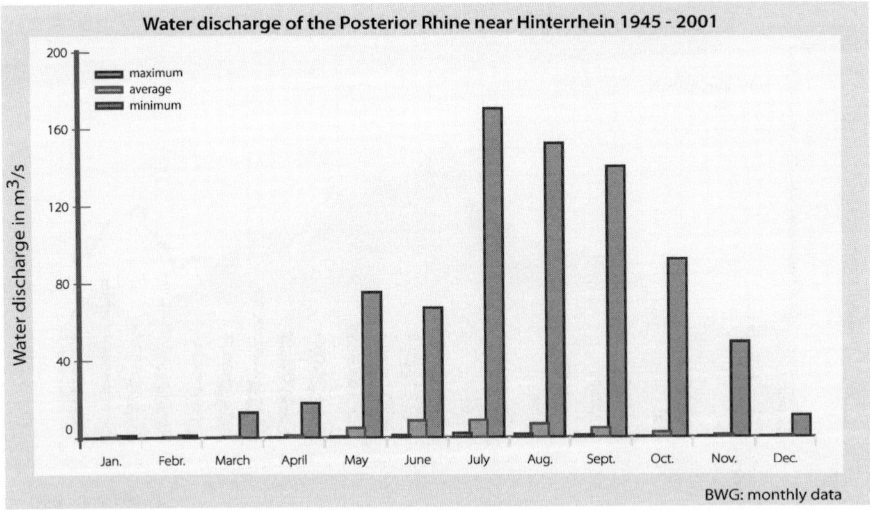

Fig. 6 Water discharge of the Posterior Rhine 1962–2001. Maximum discharge during summer months

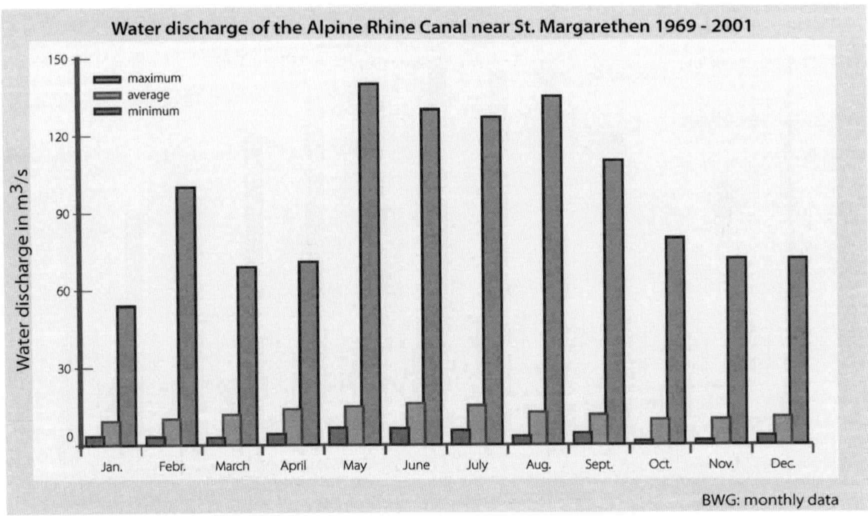

Fig. 7 Water discharge of the Alpine Rhine Canal (Rheintalischer Binnenkanal) 1969–2001, close to Lake Constance. Maximum discharge during summer months

1.3 Lake Constance

The Alpine Rhine enters Lake Constance ("Bodensee") in the southeast near Bregenz (Austria). Germany shares Lake Constance, its largest lake, with

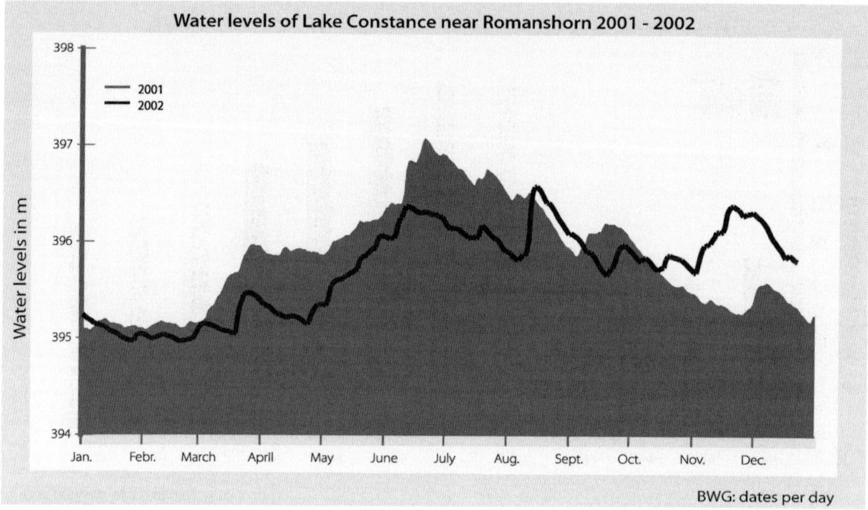

Fig. 8 Water level of Lake Constance near Romanshorn 2001–2002. Maximum water levels in summer and autumn

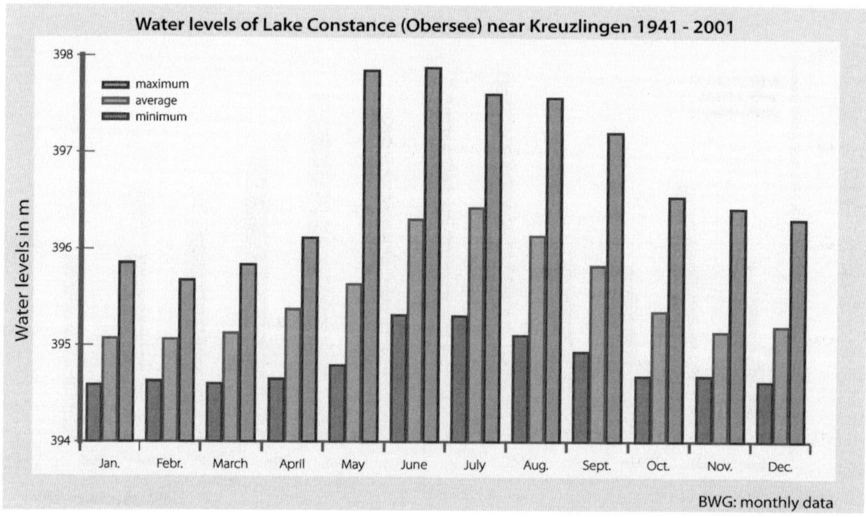

Fig. 9 Water levels of Lake Constance in Upper Sea (Obersee) 1941–2001. Maximum water levels in summer

Switzerland and Austria. Lake Constance occupies an old glacier basin at an elevation of 396 m. It has an area of 536 m^2 and is 63 km long and up to 14 km wide; average depth is 90 m, maximum depth 254 m, average surface height 395 m above sea level. The lake borders Switzerland, Germany, and Austria and has 273 km of shoreline (Fig. 8).

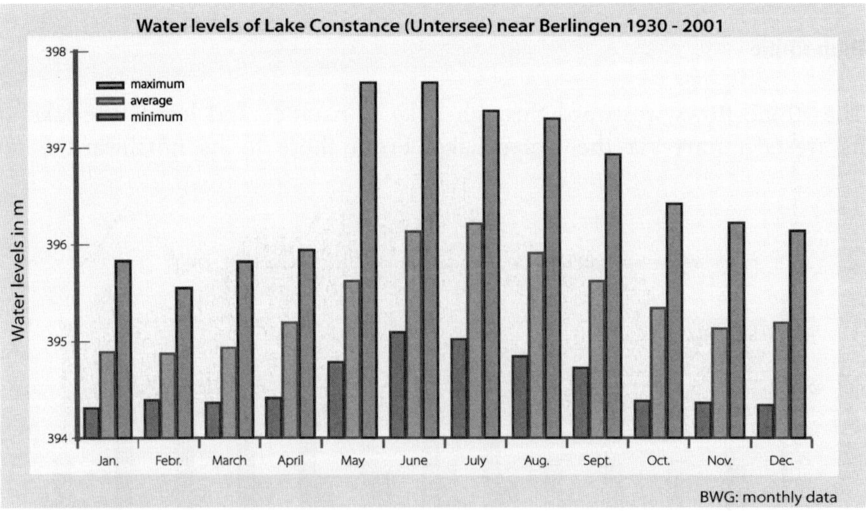

Fig. 10 Water levels of Lake Constance in Lower Sea (Untersee) 1930–2001. Maximum water levels in summer

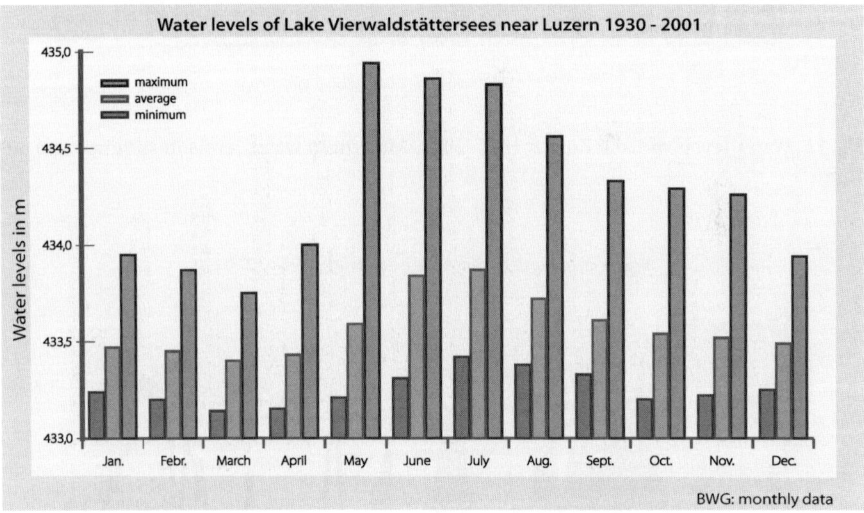

Fig. 11 Water levels of Lake Lucerne 1930–2001. Maximum water levels in summer

In the west, the lake is divided by the Bodan mountain ridge into two parts: the Unter Lake ("Untersee") (south) and the Überlinger Lake ("Überlinger See") (north). The lake's main body southeast of Constance ("Konstanz") is called the Ober Lake ("Obersee") (Figs. 9 and 10).

1.4
High Rhine

The Rhine flows westward through Lake Constance and leaves the lake at its western part via the Unter Lake. From there to its northward bend

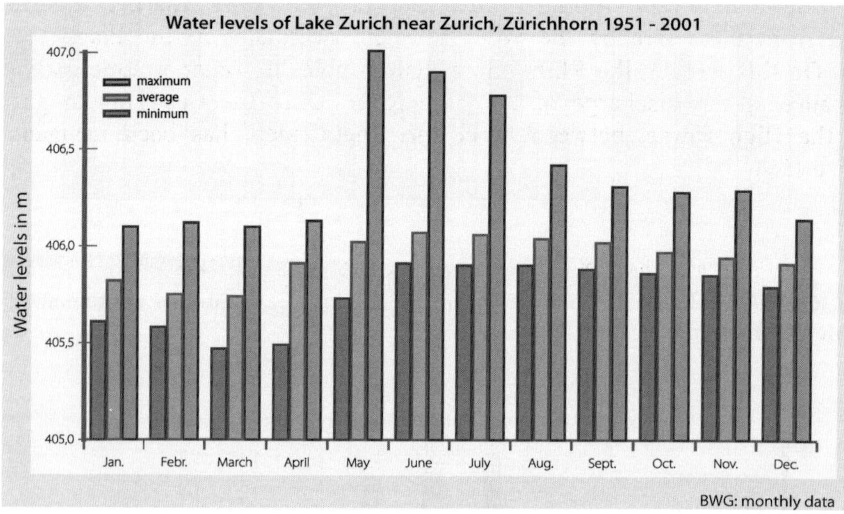

Fig. 12 Water levels of Lake Zurich 1951–2001. Maximum water levels in summer and autumn

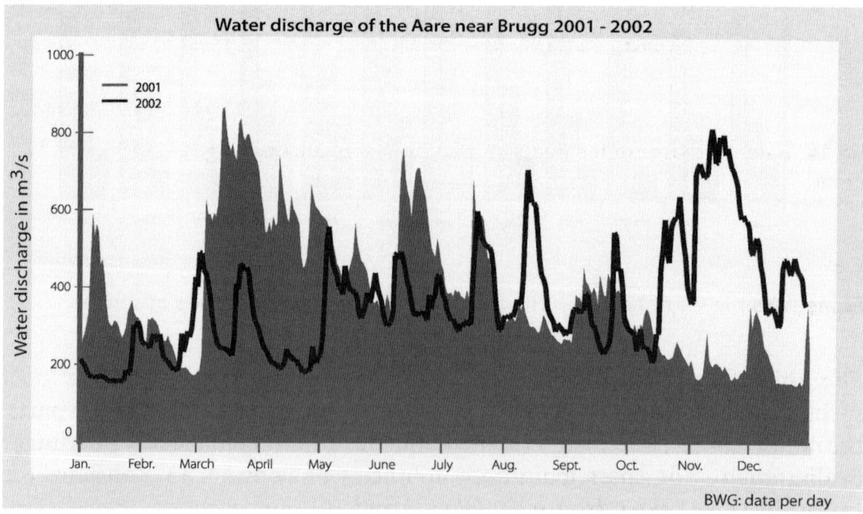

Fig. 13 Water discharge of the Aare 2001–2002. Maximum discharge in spring and autumn

at Basel, the river is called the High Rhine ("Hochrhein") and forms the Swiss-German frontier, except for the area below Stein am Rhein. The Rhine Falls at Schaffhausen are entirely within Switzerland. Downstream the Rhine flows swiftly between the Alpine foreland and the Black Forest region, its course interrupted by small rapids, where – as at Laufenburg (Switzerland) and Säckingen and Schwörstadt (Germany) – barrages (dams) have been built. In this part of the catchment, the Rhine is joined from the south by several Alpine tributaries, the Thur, Töss, Glatt, and Aare, and from the north by the Wutach, which originates in the Black Forest. On this stretch, the Rhine more than doubles its water volume mainly because of the discharge of the Aare River. Only the westernmost part of the High Rhine, between Basel and Rheinfelden, has been navigable since 1934.

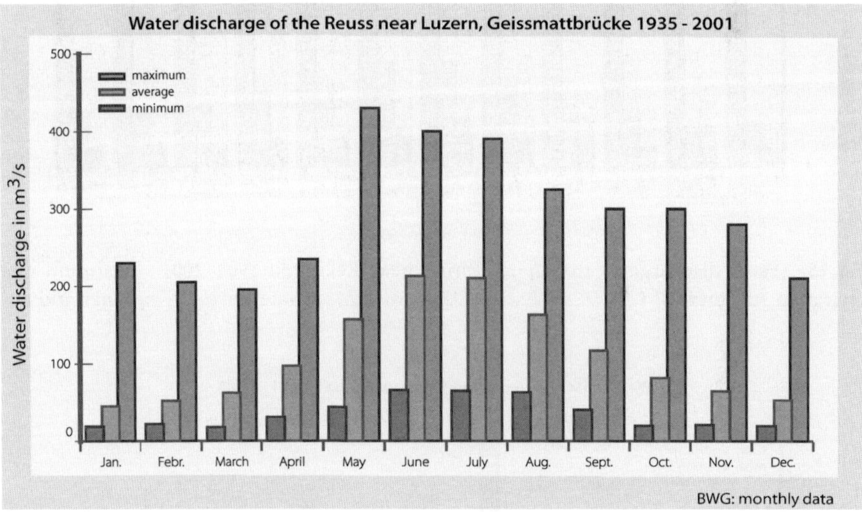

Fig. 14 Water discharge of the Reuss 1935–2001. Maximum discharge in summer and autumn

1.5 Rhine Falls

The Rhine falls ("Rheinfall") is, particularly with high water flow, the most spectacular waterfall in central Europe, situated on the upper Rhine River just below Schaffhausen in northern Switzerland. The total fall of the cataract, including the rapids upstream, is about 30 m and the width 150 m. There are two main falls divided by a pillar-like rock formation in the middle, that on the right bank dropping about 15 m, and that on the left, 20 m.

1.6 Upper Rhine

Downstream of Basel, the Rhine is called Upper Rhine ("Oberrhein") and turns northward to flow through a broad, flat-floored valley, some 30 km

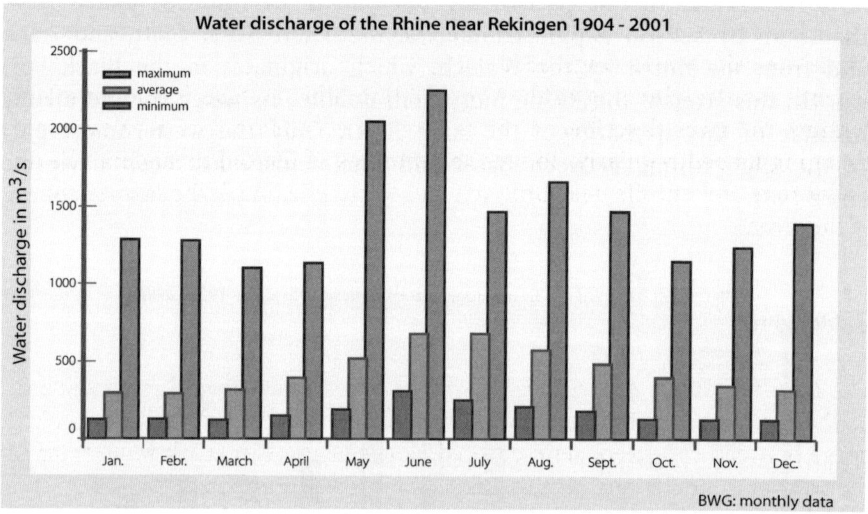

Fig. 15 Water discharge of the Upper Rhine near Rekingen 1904–2001. Maximum discharge in summer

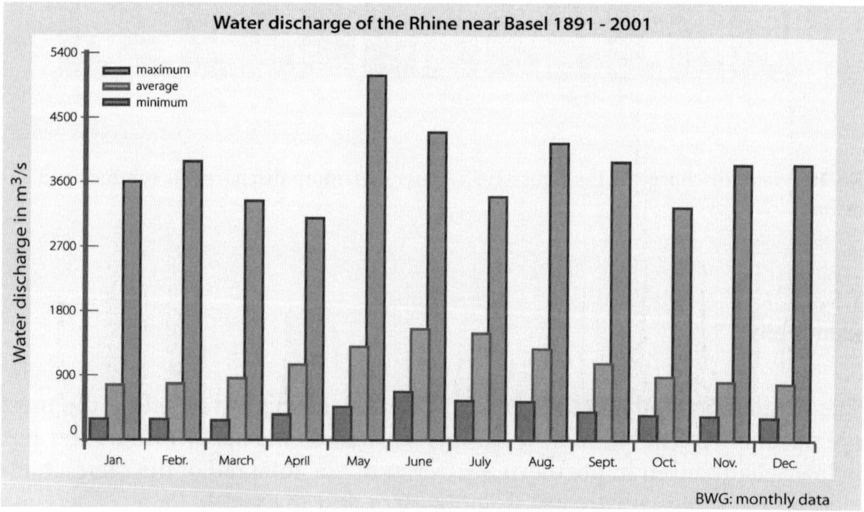

Fig. 16 Water discharge of the Upper Rhine near Basel 1891–2001. Maximum discharge in summer

wide, held between the ancient massifs of the Vosges Mountains to the west and the Black Forest ("Schwarzwald"), the Haardt Mountains ("Haardtberge") and the Oden Forest ("Odenwald") to the east. The main tributary from Alsace (France) is the Ill, which joins the Rhine at Strasbourg. Various shorter tributaries, such as the Dreisam and the Kinzig, drain from the Black Forest (Germany). Downstream, the regulated Neckar, after crossing the Oden uplands, enters the Rhine at Mannheim. Further downstream, the Main leaves the plain of lower Franconian Switzerland and flows into the Rhine opposite Mainz (Figs. 15 and 16).

Until the straightening of the Upper Rhine, which began in the early 19th century, the river described a series of great meanders over its floodplain, changing its bed regularly. Today the remnants of the old meanders, the old backwaters and cutoffs near Breisach and Karlsruhe, mark the former course of the river.

1.7 Middle Rhine

The Middle Rhine ("Mittelrhein") is the most spectacular and romantic part of the river. In this 145 km stretch the Rhine has cut a deep and winding gorge between the steep, slate-covered slopes of the Hunsrück Mountains to the west and the Taunus Mountains to the east. At Coblenz ("Koblenz"), the Moselle River, which has its source in France, joins the Rhine (Figs. 17 and 18). On the right bank, the Lahn tributary enters. Downstream the hills re-

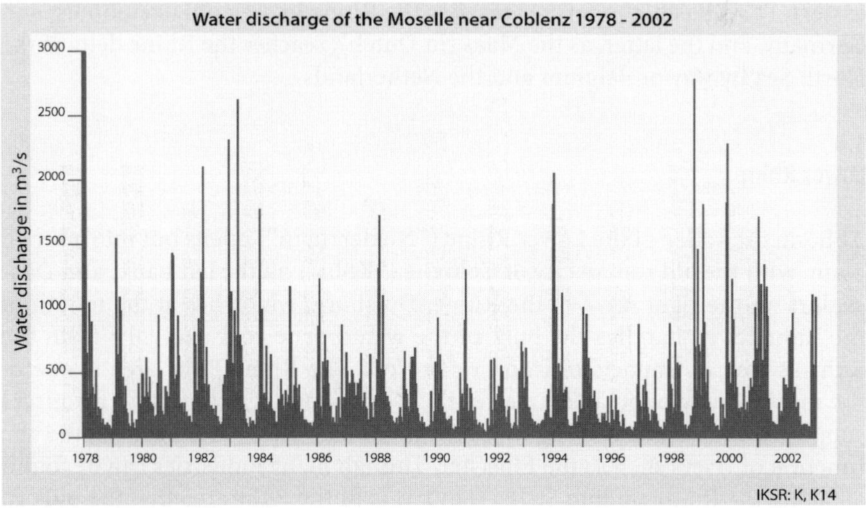

Fig. 17 Water discharge of the Moselle near Coblenz 1978–2002. Distinct high and low water periods, frequent discharge peaks

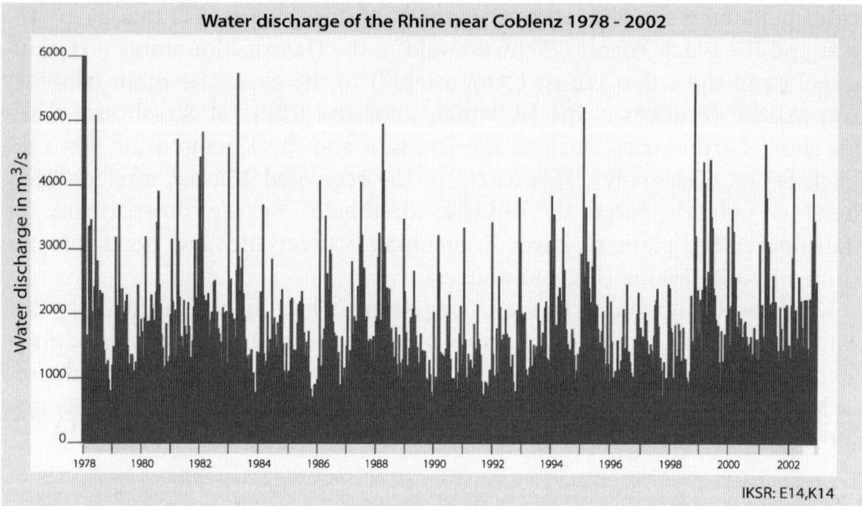

Fig. 18 Water discharge of the Middle Rhine near Coblenz 1978–2002. Frequent discharge peaks

cede, the foothills of the volcanic Eifel region lying to the west and those of the Wester Forest ("Westerwald") to the east. Between Bingen and Bonn, the Rhine flows through the steep-sided Rhine Gorge, a formation created by erosion, which happened at about the same rate as an uplift in the region, leaving the river at about its original level, and the surrounding lands raised.

Two tributaries, the Moselle, partly canalized, and the Meuse drain the eastern Paris Basin; the former reaches the Rhine by way of Luxembourg and Germany, and the latter, as the Maas (in Dutch), reaches the Rhine delta at the North Sea by way of Belgium and the Netherlands.

1.8 Lower Rhine

At Bonn the valley of the Lower Rhine ("Niederrhein") opens out into a broad plain, with the old roman city of Cologne ("Köln") on the left bank, and Düsseldorf on the right bank of the Rhine. Duisburg, which lies at the mouth of the Ruhr River, handles the bulk of the waterborne coal and coke from the Ruhr as well as imports of iron ore and oil. The Rhine-Ruhr area is one of the most heavily populated areas of the world, and its rich natural resources support a wide variety of highly developed industries – and have altered the function of rivers such as the Emscher. Though many industries can be found all along the Rhine up into Switzerland, it is in the Ruhr area that the bulk of them are concentrated, with all of its streams – chief among them the Ruhr itself – draining into the Rhine (Figs. 19 and 20).

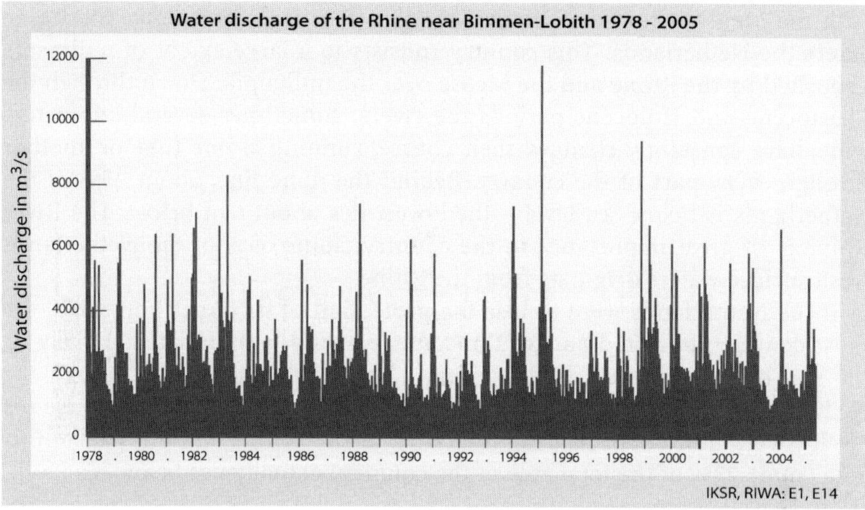

Fig. 19 Water discharge of the Lower Rhine near Bimmen/Lobith 1978–2005. Frequent discharge peaks, extreme peaks in 1995 and 1999

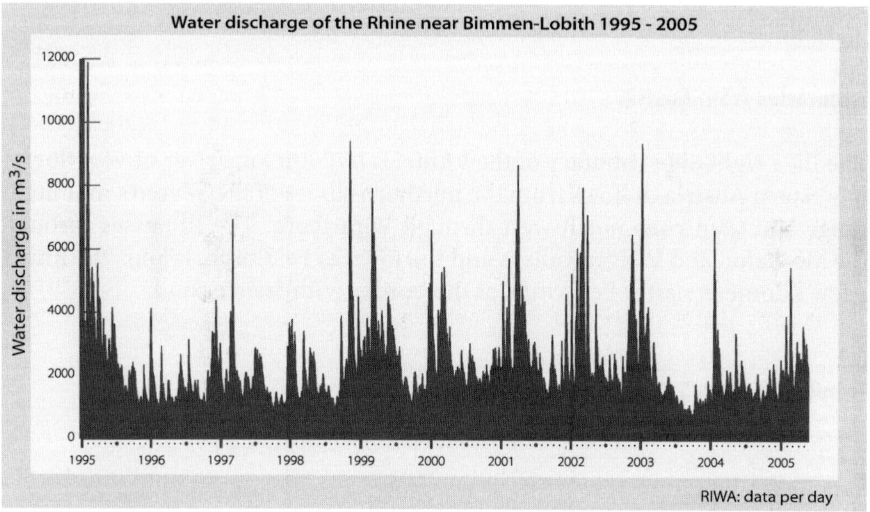

Fig. 20 Water discharge of the Lower Rhine near Bimmen/Lobith 1995–2005. Frequent discharge peaks, extreme peaks in 1995 and 1999–2003; noticeable high discharge between 1999 and 2003

1.9 Delta Rhine

The last section of the Rhine, the Delta Rhine, lies west of the German frontier town of Emmerich in the low regions of the Netherlands, where together with the Meuse it forms an extensive delta (Fig. 2). Near Lobith, the Rhine enters the Netherlands. This country consists to a large extent of sediments deposited by the Rhine and the Meuse over the millennia. Down through the Pleistocene and Holocene periods the rivers, sometimes several kilometers wide, have constantly changed their course, running at one time or another through every part of the country. Behind the dune line, about 25% of the Netherlands is below sea level – the lowest lies about 6 m below. The River Rhine is of great importance to the country: some 65% of the Netherlands fresh surface waters originate from the Rhine.

If the Netherlands were to lose the protection of its dunes and dikes, the most densely populated part of the country would be inundated (largely by the sea, but also in part by the rivers). This highly developed part of the Netherlands, which generally does not lie higher than about 1–2 m above sea level, covers more than half the total area of the country. About half of this area (some 27% of the total area of the country) actually lies below sea level.

2 Tributaries of the Rhine

2.1 Tributaries from Austria

The Ill, a right-side tributary of the Rhine, is a 72 km long river of Vorarlberg, in western Austria. It flows from the northern slopes of the Silvretta mountain range and then runs north-west through Vorarlberg. The Ill passes through the Montafon and Walgau valleys and the town of Feldkirch. It joins the Rhine a few kilometers after Feldkirch, at the border with Switzerland.

2.2 Tributaries from Switzerland

2.2.1 Aare River

This tributary of the Rhine, also spelled Aar, is with 295 km the longest river entirely within Switzerland. It descends 1565 m from its source to the junction with the Rhine and drains an area of 17 620 km^2. The river rises in the Aare Glacier of the Bernese Alps in the canton of Bern, below the Finsteraarhorn and west of the Grimsel Pass, in the south-central part of Switzerland. It runs east to the Grimsel Hospice, and then northwest through the Hasli valley, forming on the way the waterfall of the Handegg (46 m), and pierces the limestone barrier of the Kirchet by a grand gorge, before reaching Meiringen. After turning west, it expands into the glacial Lake Brienz, where it becomes navigable. The river is canalized at Interlaken above its entry into Lake Thun, from where it flows northwest in a deeply entrenched valley and passes the city of Bern. It turns west to Lake Wohlen and then flows north to Aarberg, where it is diverted westward by the Hagneck Canal into Lake Biel. Continuing north-eastward, the river runs parallel to the foot of the Jura Mountains. Below Brugg, the Aare receives first the Reuss (right), and very shortly afterwards the Limmat (right). It now turns north, and soon becomes itself an affluent of the Rhine, which it surpasses in volume when they unite at Coblenz (Switzerland), opposite Waldshut (Germany) (Figs. 13 and 14).

Hagneck Canal

The Hagneck Canal is one of the big Swiss engineering achievements of the 19th century. The River Aare which had turned the countryside north of Bern into swamps with its frequent flooding, was diverted by the Hagneck Canal into Lake Biel, from the upper end of which it issues through the Nidau Canal and then runs east to Büren. The lake swallows the huge amounts of gravel and melted snow that the river brings from the Alps and the former swamps have become valuable agricultural plains.

2.2.2 Thur River

The Thur is a 131 km long river in north-eastern Switzerland. Its source is near the mountain Säntis in the south-east of the canton of St. Gallen. In this canton it flows through the Toggenburg region, continues to flow through the canton of Thurgau and finally for the last 19 km through the canton of Zurich. It flows into the Rhine on the border with Germany, south of Schaffhausen.

2.3 Tributaries from France

2.3.1 Ill River

The Ill is a river of Alsace, in north-eastern France. It is a left-side tributary of the Rhine. It originates in the Vosges mountains and then runs northward through Alsace, flowing parallel to the Rhine, passing through Mulhouse, Colmar, and Sélestat, before merging with the Rhine at Strasbourg. Flowing through the city of Strasbourg, the river forms part of the 17th century fortifications and passes through a series of locks and channels in the old town, where its waters were once used to power mills and tanneries.

2.3.2 Moselle River (German: Mosel)

This west-bank tributary of the Rhine, flowing for 545 km across northeastern France, Luxembourg, and western Germany, has its origin on the forested slopes of the Vosges massif. The river gave its name to two French departments: Moselle and Meurthe-et-Moselle. The river meanders past Épinal, Toul, Frouard, Metz, and Thionville before leaving France to form the frontier between Germany and Luxembourg for a short distance. The river enters Germany and flows past Trier to its confluence with the Rhine at Coblenz (Fig. 17). Chief tributaries are the Madon, Orne, and Sauer (French: Sûre) on the west and the Meurthe, Seille, and Saar (French: Sarre) on the east. Downstream of Metz, the Moselle has been navigable since the 19th century, connecting at Toul and Frouard with the Rhine-Marne Canal. From Metz to Thionville the river has been navigable since 1932, but downstream Thionville it was not navigable until 1964 when the Moselle Canal from Metz to Coblenz was inaugurated.

2.4 Tributaries from Germany

2.4.1 Murg River

The Murg is a right-side tributary of the Rhine, located in Baden-Wuerttemberg, Germany. It rises in the Black Forest, flows through Baiersbronn, Forbach, Gaggenau, and Gernsbach before reaching the Rhine at Rastatt. Its length is 96 km; its drainage area has a size of 637 km^2.

2.4.2 Neckar River

The 367 km long right-bank tributary of the Rhine in southwestern Germany rises in the Black Forest near Villingen-Schwenningen am Neckar, near the headwaters of the Danube River. The Neckar flows north and northeast, along the northwestern edge of the Swabian Jura ("Schwäbische Alb"), passes Tübingen, changes at Plochingen to a northwesterly course and flows through Stuttgart. The river continues northward, its valleys becoming broader and deeper. The river flows by Heilbronn to Eberbach, where it takes a westerly course, cutting through a steep valley in the Oden Forest hills. It sweeps beneath the Königstuhl peak (566 m), passes Heidelberg, and enters the Rhine at Mannheim.

2.4.3 Main River

This important right-bank tributary of the Rhine in Germany flows through the federal states of Bavaria and Hesse. Its catchment basin competes with the Danube for water; as a result, much of its boundaries are identical with the European watershed. The river is formed, near Kulmbach, by the confluence of its two headstreams: the White Main ("Weisser Main") rises in the Fichtel Mountains ("Fichtelgebirge") and has a length of 41 km; the Red Main ("Roter Main") rises on the eastern slope of the Franconian Jura ("Fränkische Alb") and has a length of 50 km. The Main River flows in southwesterly direction around the northern end of the Franconian Jura to Bamberg, where the Regnitz River enters it from the south. Further downstream, the Main flows south and east among vine-clad hills to Würzburg and then north between the forest-covered Spessart and Oden Forest ranges to Gemünden, where the Frankish Saale River ("Fränkische Saale") enters from the north. Turning abruptly to the south, it receives the Tauber River from the south at Wertheim and continues west and then north to Aschaffenburg. From there, a more or less westward course takes the river through Frankfurt am Main to its junction with the Rhine River upstream of Mainz after a course of 524 km. Extensive works have made the Main navigable for river shipping from its mouth at the Rhine via the Rhine-Main-Danube Canal and the highly regulated Altmühl River, all the way to the Danube.

2.4.4 Nahe River

The Nahe is a tributary to the Rhine in Rhineland-Palatinate. The Nahe separates the northern part of the Palatinate from the Hunsrück. It rises in the area of Nohfelden (Saarland), flowing through Rhineland-Palatinate and joining the Rhine River in Bingen. Its length is 116 km.

2.4.5 Lahn River

The source of this right-bank tributary of the Rhine River lies on the Jagd Berg (676 m), a summit of the Rothaar Hills ("Rothaargebirge") in western Germany. The river (245 km long), first flows eastward and then southward to Giessen, before turning southwestward and reaches the Rhine at Lahnstein. Navigation by small barges is possible on the partly canalized river. The lower part of its valley divides the Taunus Mountains from the Wester Forest Mountains.

2.4.6
Ahr River

The Ahr is a tributary of the Rhine beginning at a height of approximately 520 m above sea level in Blankenheim (Eifel) in North Rhine-Westphalia. After few kilometers it reaches Rhineland-Palatinate and merges at about 50 m above sea level with the Rhine. Its length is approximately 86 km.

2.4.7
Sieg River

The Sieg is a tributary of the Rhine in North Rhine-Westphalia and Rhineland-Palatinate. It has a length of 152 km. The source is located in the Rothaar Hills. From here the river runs southwestwards to the city of Siegen and the hills of Siegerland. Further west the Sieg valley forms the boundary of the northern Bergisches Land and the southern Wester Forest. After passing the cities of Hennef and Siegburg it joins the Rhine just a few kilometers north of the center of Bonn.

2.4.8
Erft River

The Erft is a river in North Rhine-Westphalia. Its origin is near Nettersheim, it flows through the foothills of the Eifel and joins the Rhine in Neuss. The river is now 103 km long, significantly shorter than it was originally. Due to the open-pit mining of lignite (brown coal) in the Hambacher Loch the flow of the river had to be changed.

2.4.9
Wupper River

The Wupper is a tributary to the Rhine in North Rhine-Westphalia. It emerges near Meinerzhagen, traverses the town of Wuppertal on its course of about 112 km and enters the Rhine south of Düsseldorf. The river is not navigable. Early industrialization profited from waterpower provided by the many rivulets to the Wupper.

2.4.10
Ruhr River

The Ruhr is a large river in North Rhine-Westphalia starting near the town of Winterberg in Sauerland and ending in the Rhine at the city of Duisburg, passing through the largest European inland harbor. The river forms the southern edge of the Ruhr area, flowing through Hagen, Witten, Essen,

Mülheim, and Duisburg. (The River Ruhr is not to be confused with the River Rur, in western North Rhine-Westphalia, Belgium, and Netherlands).

2.4.11 Lippe River

The Lippe is a river in North Rhine-Westphalia. The source is located at the edge of the Teutoburg Forest ("Teutoburger Wald") in Bad Lippspringe close to the city of Paderborn. It runs at first westward and then along the northern edge of the Ruhr area. The river is 255 km in length and finally enters the Rhine at Wesel. Due to industrial effluents and straightenings of the river's course the Lippe was formerly in an ecologically disastrous condition.

3 Branches of the Delta Rhine

Shortly after passing the Dutch-German border, the Rhine divides repeatedly on its way through the low lands of the Netherlands and forms an extensive delta – thus forming by far the most important river system in the Netherlands.

It is interesting to note that, shortly after the Dutch-German border, the situation becomes rather complicated, as the name "Rhine" no longer coincides with the main flow of water. Most of the Rhine water flows further west through the Waal and then via the Nieuwe Waterweg and, merging with the Meuse, through Hollands Diep and Haringvliet estuaries into the North Sea. The IJssel branch carries its portion of the water north into the IJsselmeer while the Nederrijn flows west parallel to the Waal.

The name "Rijn" (Rhine) is used inside the Netherlands only for smaller streams further to the north which together once formed the main River Rhine in Roman times. Though they retained the name Rijn, these streams do not carry water from the Rhine anymore, but are used for draining the surrounding land and polders. From Wijk bij Duurstede, the old north branch of the Rhine is called "Kromme Rijn" (Crooked Rhine) and past Utrecht, first "Leidse Rijn" (Leiden Rhine) and then "Oude Rijn" (Old Rhine). The latter flows west past Leiden into a sluice at Katwijk, where its waters can be discharged into the North Sea.

3.1 Branches of the Rhine in the Delta

At Pannerden, 5 km downstream of Lobith, the river splits into the artificial Pannerden Canal and the Waal. The Pannerden Canal divides south of Arnhem to form the Nederrijn and the IJssel. The weir at Driel divides the river water between the Nederrijn and the IJssel, insuring that, when water levels are low, a sufficient proportion flows into the IJssel. Beyond Wijk bij Duurstede, the Nederrijn changes its name and becomes the Lek. It flows further west to rejoin the main flow into the New Waterway Canal ("Nieuwe Waterweg").

In about 10 B.C. the Roman governor Drusus built a dam at the point where the Rhine divides to form the Waal and the Nederrijn. Drusus's Dam was designed to reduce the amount of water flowing into the Waal and increases the amount entering the Nederrijn, thereby giving the Batavi, who were Rome's allies, a more defensible border with the Germanic tribes.

Canalized along almost its entire length, the Nederrijn is wider than the IJssel, but not as wide as the Waal; it is called Lek further downstream.

3.1.1 IJssel River

The IJssel is a narrow branch of the Rhine in the Dutch provinces of Gelderland and Overijssel. It is one of the three major distributary branches into which the Rhine divides itself shortly after crossing the German-Dutch border, the other two being the Nether Rhine and Waal Rivers. The IJssel leaves the Lower Rhine (Nederrijn) near the city of Arnhem and flows northeastward for about 120 km to enter Lake IJssel ("IJsselmeer"), until 1932 known as the Zuiderzee and situated between the Northeast (Noordoost) and East Flevoland polders.

The River IJssel is navigable, but because of its slow current it deposits much silt. In places its flood plain has yet to be excavated and is several kilometers across. Water levels in the lower reaches of the IJssel are determined partly by the water level in Lake IJssel.

The IJssel once was the lower part of the Old IJssel ("Oude IJssel", German "Issel"), a small river that rises in Germany and is now a 70 km long tributary of the IJssel, merging at Doesburg. The connection between the Rhine and the IJssel was probably artificial in origin, constructed as a canal by the Roman general Nero Claudius Drusus as a defense against Germanic tribes.

3.1.2 Lake IJssel

The Lake IJssel "IJsselmeer" drains into the Waddenzee via a sluice in the Afsluitdijk (the sea dike closing off the Lake IJssel). The Zuiderzee was originally an estuary of the Rhine. By natural action it then became a shallow inland sea, biting deep into the land, and eventually it was hollowed into an almost circular shape by the action of winds and tides. In 1920 work was begun on the Zuiderzee project, of which the Lake IJssel Dam (Afsluitdijk), begun in 1927, was a part. This 340 km long dam, or dike, running north-eastward to connect the provinces of Noord-Holland and Friesland, was completed in 1932 to finally seal off the Zuiderzee from the Waddenzee and the North Sea. In the Lake IJssel, formed from the southern part of the Zuiderzee, four large polders with a total area of about 1700 km^2 were constructed around a freshwater basin fed by the IJssel and other rivers and linked with the sea by sluices and locks in the barrier dam.

3.1.3 Waal River

In terms of flow, the Waal, seemingly a direct continuation of the Rhine coming from Germany, is naturally the most important branch of the Rhine. The Waal is a broad, free-flowing river; it regularly inundates its flood plain, within which there has been large-scale gravel and sand extraction. The river has been linked directly to the Meuse since Roman times. When the Waal was running high, the connecting canal at Heerewaarden diverted water into the Meuse. However, the Meuse was not able to contain the extra water and repeatedly broke its banks. Furthermore, the reduced flow in the Waal encouraged sedimentation, causing higher water levels in the lower reaches. Below Gorinchem, the two rivers merged, but the Merwede lacked the capacity to cope with all the water.

3.1.4 Meuse River

The Meuse (Dutch "Maas") is not a distributary of the Rhine, although it is intimately connected with that river. The river rises in the Langres Plateau (France) about 200 km northeast of Dijon, some 409 m above sea level, and flows north past Sedan (the head of navigation) and Charleville-Mézières into Belgium. It is joined by the Sambre River at Namur. From Namur the Meuse winds eastward skirting the Ardennes, passes Liège, and turns north. Then it forms part of the Belgian-Dutch border. Inside the Netherlands it continues north along Venlo, then turns westward. Near the coast, it merges with the Rhine into an extensive delta, eventually flowing into the North Sea via the New Waterway Canal and the Hollands Diep.

The Meuse is 935 km long. Its basin is roughly a sixth of the area covered by the Rhine basin. Flow in the Meuse averages 230 $m^3 s^{-1}$ near the Dutch-Belgian border. Within the basin, there are no glaciers or snow accumulations to feed the river. Because the river is fed by rainfall all year round, flow is much higher in the winter than in the summer, when evaporation levels are highest. Weirs more or less permanently control water levels in the Dutch Meuse, in order to make the river navigable. In addition, canalization has rendered part of the French Meuse and the entire length of the Belgian Meuse navigable.

The Dutch Meuse is 250 km long. Between Eijsden and Maastricht (9 km) and between Borgharen and Stevensweert (47 km), the Meuse forms the border between the Netherlands and Belgium. This part of the river is known as the Border Meuse. From Maastricht to Maasbracht, the Meuse meanders over shallow gravel banks; this stretch is uncanalized, fast flowing and virtually unnavigable. Therefore, shipping has to use the parallel Juliana Canal. Upstream of Mook, the river flows along a valley whose sides slope gently upwards, forming natural dikes. There are no polders that need to be protected against flooding. At Heerewaarden, the Meuse and the Waal are separated only by a barely 2 km wide stretch of land. In the past, when the Waal discharge was high, water spilled from the Waal into the Meuse. The last stretch of the river, the Getijde Maas (Tidal Meuse) downstream of Lith, is not canalized at all, and the water is freely flowing. Both, Rhine and Meuse are linked to the sea by the New Waterway Canal, consequently both rivers exhibits tidal influences.

3.1.5 Delta Project

With the completion of the huge Delta Project in 1986 – constructed to prevent flooding in Zeeland, the south-western coastal area of the Netherlands – the main branches of the Rhine were closed off; sluices and lateral channels now allow river water to discharge into the North Sea. Since 1872, however, the New Waterway Canal, constructed to improve access from the North Sea to Rotterdam, has been the main navigation link between the Rhine and the sea.

3.1.6 New Waterway Canal ("Nieuwe Waterweg")

The New Waterway Canal is the only direct link between the river system and the sea; all the other river openings have been closed by weirs and discharge sluices. Through this canal, the distance from the city of Rotterdam to the sea is 30 km, and along this canal was built Europoort, one of the world's largest ports.

3.2 The Haringvliet Sluices and the Biesbosch

Haring River ("Haringvliet"). This freshwater channel in southwestern Netherlands is a distributary of the Hollandsch Canal and ultimately has its origin in the Lower Rhine (Neder Rijn) River. The Haring River, which discharges into the North Sea, flows for about 32 km between the joined islands of Voorne and Putten and the island of Beijerland to the north and the joined islands of Goeree and Overflakkee to the south. On February 1953, its low shores were devastated during the disastrous tidal surge floods. As part of the Delta Plan for land reclamation and both Rhine and tidal flood protection, a dam with numerous sluices was completed at the Haring mouth in 1971. Since completion of the Haring River Dam, the formerly saline Haring River has become a fresh water body. In addition, the difference between high water level and low water level has gone from 2 m to a few decimeters. Shipping remains possible through a large lock built as part of the Haring Dam.

In their lower reaches close to the Atlantic Ocean, the branches of the Rhine (Lek, Waal, Merwede) and Meuse are increasingly subjected to influences of the sea. Flow and water levels are affected by the tides, with occasional extreme high water levels during storm tides. At high tide, a wedge of salt water enters the New Waterway Canal and, if the flow in the river is low, gradually makes its way upstream.

3.3 Regulation of the Water Flow of the Rhine in the Netherlands

The flood plains on the Rhine branches cover some 28 000 ha of Dutch soil. In many places, low summer embankments, protecting the flood plain from summer flooding, bound the river's main channel. In addition to the flood plains, the river system includes pools, clay ponds, and old cut-off arms of the river.

When the river is low (i.e., the flow is less than 1300 $m^3\ s^{-1}$), three weirs are closed on the Nederrijn and the Lek. By these methods the proportion of Rhine water entering the Waal and the IJssel is increased, while the amount flowing into the Nederrijn is reduced. This has beneficial effects on the Waal as an important shipping route and ensures that sufficient fresh water reaches Lake IJssel.

When water flow rises to between 1300 and 2400 $m^3\ s^{-1}$, the weirs at Driel are gradually opened and the amount of water flowing through the Nederrijn increases.

When the Rhine is running high flows (more than 2400 $m^3\ s^{-1}$), the weirs at Driel are opened fully and the distribution of water between the river branches is not actively controlled by man. During this period, about two-thirds of the flow is discharged through the Waal, while two ninths finds its way into the Nederrijn and one ninth into the IJssel.

Haring River sluices and the Biesbosch: The sluices can be used simply for discharge purposes, or opened only when the river runs high (normally in the winter). Alternatively, the sluices could be open at all times except when the river flow is very low, or when a flood tide is imminent.

Natural processes in the tidal river could be restored to a certain degree by partial opening of sluices of the Haring River. A certain degree of tidal influence would make the environment more attractive to various plants and animals, while also allowing better access for migratory fish.

3.4 Characteristics of the Water Flow of the Rhine Branches in the Delta

The physical differences between the three Rhine branches Waal, Nederrijn and IJssel, and the Meuse (which is intimately connected to the Rhine in the lower parts of the delta) affect the way ecotopes are distributed over the flood plain. To start with, the four rivers differ in terms of water flow and sediment burden. Considerable differences exist between the Rhine and the Meuse. First, far more water flows down the Rhine: an average of 2300 $m^3 s^{-1}$ of water enters the Netherlands at Lobith. Flow in the Meuse is not only much lower (only 230 $m^3 s^{-1}$ on average at Eijsden) but exhibits much greater seasonal variance. The Meuse's maximum flow is 150 times greater than its minimum flow, whereas the difference between the minimum and the maximum in the Rhine is merely a factor of twenty.

In the canalized Meuse and Nederrijn, there is very little water flow for a large part of the year. By contrast, the Waal and the IJssel are free-flowing rivers, carrying large quantities of sandy sediment.

The IJssel has developed meanders; its discharge was a lot higher than it is today and it was strong enough to change the course of the river quite frequently. During the 15th century, silt accumulation became an increasing problem in the IJssel, interfering with the commercial traffic on which the Hanseatic towns along the river depended. Today, however, flow in the IJssel is much lower than it once was. With its artificial ripraps on the banks and reduced dynamism, the river is barely able to build up natural levees any longer. Differences in flood plain height are also significant.

The Waal flows through flat land and has a relatively low flood plain, which would be even lower if the excess sedimentary clay and the summer embankments were removed.

The Meuse, on the other hand, has cut a deep channel through its own sediment. On the relatively high Meuse terraces, which rarely inundate, the natural vegetation is therefore more wooded than that along the Waal.

4 Important Lakes in the Rhine Catchment Area

4.1 Alpine Lakes

Eleven significant European lakes fringe the Alps. During the uplift of the mountain chain of the Alps, most of the valleys were formed that harbor Alpine lakes. During the Ice Age of the geologically recent Pleistocene, less than 2 million years ago, numerous glaciers flowed through these valleys, of which only few glaciers are left. These glaciers deepened and excavated the ground, leaving moraines (deposits of waste material) while shrinking at the end of the glacial period. Filling of the excavations, which were often dammed by the moraines, formed alpine lakes. Those lakes that originated in mountain valleys are generally long, narrow and very deep.

The lakes are divided into a northern and a southern group by the Alpine watershed. The southern group, which lies in an Alpine environment, is made up of Lake Geneva and the Insubrian lakes (Maggiore, Lugano, Como, and Garda). Parts of the northern lakes (Lakes Neuchâtel, Lucerne, Zurich, Constance, Chiem, Attersee) are situated in the foothill zone of the Alps or even beyond that zone.

The water composition of the Alpine lakes is quite uniform. About 100 years ago almost all the Alpine lakes were poor in plant nutrients, especially in phosphates. During the 20th century the water wastes from households and cities polluted many lakes. The phosphorus content increased, causing phytoplankton to multiply (eutrophication). The extreme growth of phytoplankton under these conditions made the water turbid and intensified oxygen consumption in the deep layers of the lake, owing to the increased decomposition of dead algae.

4.1.1 Lake Zurich ("Zürichsee")

This Swiss lake, situated southeast of the city of Zurich, lies at an altitude of 406 m, has an extreme length of 29 km, is up to 143 m deep and covers an area of about 90 km^2. The Linth River flows into it and emerges as the Limmat.

The Linth and Rhine glaciers flowing north-westward excavated the basin that the lake occupies today. At the location of what is now Zurich, a terminal moraine created the lake by damming the flow of water. Another moraine between Pfäffikon and Rapperswil separates the navigable lower portion of the lake (Untersee) from the shallow upper part (Obersee) (Fig. 12).

4.1.2 Lake Lucerne ("Vierwaldstätter See")

The lake, named after the city of Lucerne ("Luzern") at its western end, is surrounded by steep limestone mountains (e.g. Rigi to the north, and Pilatus to the west). It lies at an elevation of 434 m, occupies an area of 114 km^2 and is about 39 km long, with a maximum depth of 214 m. The Reuss River enters the lake at the south-eastern part near Flüelen and leaves it at Lucerne(Fig. 11). Other influents are from northeast the Muota and from the south the Engelberger Aa and the Sarner Aa. Lake Lucerne is composed of four topographically distinct main basins, the former are glaciated valleys connected by narrow and tortuous channels.

Lake Constance and Lake IJssel are dealt with elsewhere in this chapter.

5 Important Canals Connected to the Rhine

5.1 Amsterdam-Rhine Canal

The Amsterdam-Rhine canal ("Amsterdam-Rijnkanaal") connects the port of Amsterdam with the Lek River (via Utrecht) and the Waal River (at Tiel) in the Netherlands. Inaugurated in 1952, it has a total length of 72 km. It is one of the most heavily used canals in Western Europe.

5.2 Rhine-Main-Danube Canal, also called Europe Canal ("Rhein-Main-Donau-Kanal")

This commercial waterway in southern Germany (Bavaria), completed in 1992, has a length of 171 km, and runs from Bamberg on the Main River (a tributary of the Rhine) to Kelheim on the Danube River, permitting navigation between the North Sea and the Black Sea. It thus creates a 3500 km waterway connecting many Eastern European countries with the Rhine River and the North Sea. The canal, one of the largest civil engineering projects ever undertaken, has a total of 16 locks and reaches a height of more than 406 m over the Swabian Alps ("Schwäbische Alb"), south of Nürnberg. The idea for such a canal dates back to 793, when Charlemagne wished to open a route through the center of Europe for his battle fleet. In 1921 the German government and the state of Bavaria formed a company to build the much larger Rhine-Main-Danube Canal, but the construction took place much later, between 1960 and 1992.

6 Water Regime

Because the Rhine is fed from different sources at different times of year, the amount of water flowing down the river is relatively stable, making it navigable all year round.

6.1 Seasonal Discharges

The *Alpine Rhine*, with its steep gradient, pronounced winter minimum, and high water in spring from snowmelt, has a characteristic alpine regime. Variations in discharge are balanced by Lake Constance, which is not only fed by the Rhine but also by several minor upland tributaries.

In the *High Rhine*, upstream of Basel, the amount of water flowing down the river is determined not only by the large volumes of rainwater which drain into it, but also by melt water from snow and some 150 glaciers. As a result, flow in the Rhine's upper courses peaks in the early summer. Discharge is increased particularly by the confluence with the Aare, which on an average carries more water than the Alpine Rhine itself.

In the *Upper and Middle Rhine*, downstream of Basel, however, the tributaries from the uplands in Germany and France, with their spring maximums at higher and winter maximums at lower elevations, increasingly moderate the unbalanced water flow. Thus, at Cologne the average deviations from mean flow are less pronounced, and the regime is favorable to navigation. Winters in the navigable regions of the river, moreover, are generally mild, and the Rhine freezes only in exceptional winters (formerly more often than nowadays).

Lower Rhine: At Lobith, average flow is highest in winter, when evaporation levels are low and rainwater accounts for the bulk of the river's volume. During the winter, only around 30% of the water in the Rhine comes from the Alps.

During the summer months, more than 70% of the flow at Lobith originates from the Alps, while the amount of water originating from other parts of the basin is relatively small. This is because much of the winter precipitation falls as snow, which lies unmelted until late spring. Only then, when the thaw sets in, does this precipitation find its way into the Rhine (Figs. 3 and 4).

6.2
Situation in the Netherlands

Among the countries of the Rhine catchment, the Netherlands is particularly dependent on Rhine water. The Netherlands has a temperate maritime climate, with precipitation occurring throughout the year. Average annual rainfall totals 750 mm: in terms of flow, the equivalent of 950 $m^3 s^{-1}$. This figure is approximately 30% of the total amount of water entering the Netherlands on average. No less than 65% of the Netherlands' fresh surface waters come from the Rhine and a further 8% from the Meuse. The remainder comes from smaller cross-border rivers. The Rhine is therefore the most important source of drinking water.

6.3
Climate Change

Climate change has consequences for the water cycle and thus for the amount of water in rivers. If temperatures in the Alps rise by one or two degrees, the snow line will retreat dramatically in some areas, and melt water flows will fall sharply. The changes of the climate will also affect the water regime of the important River Rhine. Some discussion is still going on concerning the exact causes of the climate change, whether it can be attributed solely to natural changes or by how much man is the cause.

The consequences of gradual changes in the earth's climate could ultimately be so far-reaching that politicians are already taking them into account. It is believed that climate changes, soil subsidence, sea level rise, and so forth are slow but sure processes, which cannot be directly influenced by man. On the basis of measured data sets and climate models, the Intergovernmental Panel on Climate Change (IPCC) has concluded that global rises in temperature and sea level are likely to continue in the years ahead. For the Netherlands it has been calculated that average temperatures will be between 0.5 and 2 °C degrees higher in 2050 than they are today, and a rise of between 2 and 5% in annual precipitation is expected.

According to these calculations, flow at Lobith on the lower reaches of the Rhine is forecast to drop by 10–20% in the summer. In the winter, however, with increased rainfall and the greater likelihood that the snow in the Alps will melt soon after falling, the river will more frequently run high.

Climate change is expected to cause more frequent and prolonged periods of low water in the late summer and autumn. During winter, on the other hand, the chances of extreme flood will be greater because precipitation increase is expected to be most marked in the winter. Such variations in river flow could have implications for transport along the Rhine route between Rotterdam, Germany, and Switzerland. Without intervention, shipping between Basel and Rotterdam will have to contend with low river levels more often. In rather dry summers, it is more likely that shippers will not be able to fully load their vessels; it might even be possible that there will be times when the rivers become unnavigable for large vessels.

Thus, the Rhine will become much more of a river with a pluvial regime. In the late summer and autumn, low water will be more likely, if only because the higher temperatures will cause more evaporation. Although the overall annual volume of water flowing down the rivers will remain much the same, extremely high and low flows will become more common.

7 Geomorphology of the Rhine Basin

7.1 The Alps

The Alps emerged during the Alpine orogeny, an event that began about 70 million years ago at the end of the Mesozoic era.

At the end of the Paleozoic era, about 245 million years ago, eroded Hercynian mountains, similar to the present Massif Central in France, stood at the present-day location of the Alps. A large landmass (Tyrrhenia), formed of crystalline rocks, occupying the western Mediterranean basin, was slowly leveled by the forces of erosion during the Mesozoic (245–66 million years ago). The eroded materials were carried southward by rivers and deposited at the bottom of a vast ocean (Tethys Sea), forming horizontal layers of limestone, clay, shale, and sandstone.

During the middle Tertiary period (about 44 million years ago), pressures from the south folded the deposited layers of the Tethys-Sea around and against the crystalline bedrock, first forming the Pyrenees and then the Alps, up to heights approaching the present-day Himalayas. These tectonic movements lasted until 9 million years ago. At the beginning of the Quaternary period, about 1.6 million years ago, Tyrrhenia sank. Throughout the Quaternary period, erosive forces steadily diminished the newly folded and upthrust mountains, forming the general outlines of the present-day landscape.

During the Quaternary, the landscape was further modeled by Alpine glaciation and by expanding ice tongues, which filled in the valleys and overflowed onto the plains, forming U-shaped valleys, and elongated lakes of great depth such as Lake Annecy in France, Lake Constance (bordering Switzerland, Germany, and Austria), and the lakes of the Salzkammergut in Austria. During glaciations, enormous quantities of sands and gravels were deposited by the melting glaciers. When the ice left the main valleys, down cutting by the new rivers occurred both in the lateral and transverse valleys. The morainic deposits carried huge water reservoirs and the more fertile soils encouraged settlements.

Vigorous glacial erosion continues in modern times. Many hundreds of square kilometers of Alpine glaciers are still found in the Alps. The summer runoffs from these ice masses is instrumental in filling the lakes and improving the water discharge of rivers.

7.2 Hercynian Orogenic Belt (also called Variscan Orogenic Belt)

The collision between Africa and a North American–North European continent formed a series of mountain ranges, the Hercynian orogenic belt. The Hercynian extends in Europe for more than 3000 km from Portugal, Ireland, and England in the west through Spain, France (Brittany, Massif Central, Vosges, and Corsica), and Germany (Black Forest, Harz) to the Czech Republic in the Bohemian Massif. The southern part of the belt was extensively deformed by the collision tectonics that gave rise to the Alps and Pyrenees.

7.2.1 Black Forest ("Schwarzwald")

This is a mountainous region in Baden-Wuerttemberg (southwestern Germany), source of the Danube and Neckar Rivers and of several smaller tributaries of the Rhine. It occupies an area of 6000 km^2 and extends toward the northeast for about 160 km from Säckingen on the Upper Rhine River (at the Swiss border) to Durlach (east of Karlsruhe). Structurally and topographically it forms the counterpart of the Vosges, which lie west of the Rhine Valley. The Black Forest drops abruptly to the Rhine plain but slopes more gently toward the Neckar and Nagold valleys to the east.

It is mainly a granite highland with rounded summits (its highest peak is the Feldberg, 1493 m), divided into two parts by the deep Kinzig Valley. The northern part of the Black Forest is characterized by forested sandstone, whereas the southern part exhibits a narrow band of lower and more fertile limestone.

7.2.2 Vosges

This massif extends west of the Upper Rhine Valley in eastern France. Of ancient rocks, the dome-shaped mountains spread westward for more than 64 km towards the Moselle Valley and northward for more than 114 km parallel to the Rhine. The chains of granite in the south and of red sandstone in the north fall abruptly to the Rhine Valley, whereas to the west the forested slopes descend more gently. In the south the high Vosges summits, known as "ballons", culminate in the "Ballon de Guebwiller" (1424 m). The higher mountains are covered with snow for several months of the year.

7.3 Rhenish Slate Mountains ("Rheinisches Schiefergebirge")

These mountainous highlands lying mainly in north-western Germany, extend westward as the Ardennes through south-eastern Belgium and northern Luxembourg, with an overlap into eastern France beyond the Meuse River. The plateau is dissected by the gorge of the Rhine River and the deep meanders of its tributaries, particularly the Moselle, but also other smaller rivers. Although slates and sandstones are the main rocks, there are volcanic outcrops and considerable areas of limestone where the relief is smoother, the soils more fertile, and the land more densely settled.

8 Drainage of Adjacent Countries

8.1 Austria

About 96% of Austrian territory drains to the Danube River system. The main watershed between the Black Sea and the North Sea runs across northern

Austria, while to the west the watershed between the Danube and the river systems emptying into the Atlantic and the Mediterranean match with the western political boundary of Austria. In the south the Julian and Carnic Alps together with the main Alpine range form the watershed of the region draining into the Po River of northern Italy.

Austria is a land of lakes, many of which occupy basins formed during Pleistocene glacial erosion. The largest lakes, lying partly in the territory of neighboring countries, are Lake Constance in the west and the Neusiedler Lake ("Neusiedlersee") to the east.

8.2 Switzerland

Switzerland is the source of many major rivers, springing from what is called the hydrographical center of Europe. The two most important are the Rhône, which flows into the Mediterranean, and the Rhine, which discharges into the North Sea. Switzerland is composed of three topographically distinct regions: the Jura Mountains in the northwest, the Alps to the south, and the Mittelland or central plateau between the two mountain ranges.

The Jura occupies about 12% of the country. It was formed under the extended impact of the general Alpine folding, which created the folded Jura next to the Mittelland and the tabular plateau Jura at the northern edge. Characteristics are Jurassic limestone and marl with rich fossil content. The limestone has been eroded in typical karst fashion, forming sinkholes, caves, and allowing extensive underground drainage.

The Alps were built of large complexes of extremely varied, massed overthrusts that were shaped by glaciation. The northern and southern Swiss Alps are separated by the trough formed by the Rhône and upper Rhine valleys, the narrowest portion being the Urseren Valley. The sources of the Rhône and the Reuss (Furkareuss), an important tributary of the Rhine, are only 2 km apart, and those of the Rhône and the Rhine about 20 km apart.

In the central Alpine region of Switzerland, its role as the European watershed is striking: from the St. Gotthardt massive, the Rhône River flows west, the Rhine River east, the Ticino River south to the Po, and the Reuss River north to the Aare (one of the most important tributaries of the Rhine). Further southeast, another major Alpine source point is located in the upper Engadine valley at the Piz Lunghin, from which streams flow toward the Adriatic Sea and from which the headwaters of the Inn River flow toward the Danube and ultimately into the Black Sea.

Between the Jura and the main Alpine ranges lies the hilly Mittelland, covering some 23% of the country and enclosed by two mountain ranges and the two largest lakes of Europe, Lake Geneva ("Genfer See") in the west and Lake Constance in the east. The Mittelland is the agricultural heartland of the country and has become highly urbanized.

8.3
France

The river systems of France are divided into an eastern and a western group by a line running from the southern end of the Vosges down the eastern and south-eastern edge of the Massif Central to the Noire Mountains ("Montagne Noire"), the south-western promontory of the massif. Along the divide originate most of the rivers of the larger, western part of the country, including the Seine and the Loire. Other major rivers include the Garonne, originating in the Pyrenees, and the Rhône and the Rhine, originating in the Alps.

8.3.1
The Rhine System

The Rhine forms the eastern boundary of France for some 190 km. The water regime of this section of the river is dominated by the melting of snow and ice from Alpine headstreams, giving it a pronounced late spring-summer peak and often generally low water in autumn. Southern Alsace is drained by the Ill, which joins the Rhine at Strasbourg. To improve navigation and to produce electric power, France has considerably modified the Rhine valley by the construction on the French side of the lateral Grand Canal d'Alsace. Two tributaries of the Rhine drain the eastern Paris Basin: the Moselle, partly canalized, and the Meuse ("Maas"). The former reaches the Rhine by way of Luxembourg and Germany, and the latter reaches the Rhine delta at the North Sea by way of Belgium and the Netherlands.

8.3.2
Lakes

The French hydrographical system includes a number of natural lakes of different origin. Some occupy depressions carved out by glaciations at the western periphery of the Alps. Others occur on the surfaces of ancient massifs and include the lakes of the Vosges. Some lakes are caused by structural faults and are lodged in narrow valleys, as are the Jura lakes.

8.4
Germany

With the exception of the Danube, which rises in the Black Forest and flows eastward, most German rivers follow the general north-north-westward inclination of the land, finally discharging into the North Sea. The most important of the rivers flowing through Germany is the Rhine. River flow relates mainly to climate: in all but Alpine Germany, maximum river flow occurs in winter when evaporation is low, though in the lowlands the peak rainfall is in summer.

The Rhine belongs to two types of river regimes. Rising in the Alps, it profits first from the torrential Alpine regime, which causes streams to be swollen by snowmelt in late spring and summer, and with very low water flow in winter. Then, by means of its tributaries to the east (Neckar and Main) and west (Moselle) the Rhine receives the drainage of the Central German Uplands and the eastern part of France, which contribute to a maximum flow during the winter. As a result, the river has a remarkably powerful and even flow, a physical endowment that caused it to become the busiest waterway in Europe.

River courses in the northern lowlands are influenced by remnants of the ice age – rivers partly follow the glacial valleys, which had been carved outside the fringes of the retreating ice sheets before breaking through the next moraine ridge to the north. These trenches greatly facilitated the cutting of canals linking the Rhine River with the city of Berlin and with the Elbe and Oder Rivers.

Apart from Lake Constance, Germany has relatively few lakes. The greatest concentration comprises the shallow lakes of the postglacial lowland of the northeast; others are concentrated at the extreme south-eastern corner of Upper Bavaria.

9
Modifications to the Hydrological Flow

Modern rivers are no longer free to meander, to push back their banks or alter their courses. They are either canalized, enclosed by dikes or regulated by groynes.

Without man's interference, rivers could be extremely dynamic. Flood plains would be alternately inundated and drained, sediment would be deposited at times and later moved on again, floating ice would erode the high-water bed. The locations of riverbeds would change, agriculture near the rivers would be difficult to maintain, human settlements sometimes endangered, and the nuisance of mosquitoes more frequent.

For centuries, man has been seeking to control the rivers. With the growth of cities and expansion of industries, the river landscape has inevitably become an artificial landscape, with considerable differences between various river branches. Although the rivers today are far from natural, they are still and probably cannot be completely controlled by man, which is demonstrated by unexpected floods and inundations.

The Rhine has created a number of problems in the past, particularly in the regions of the Upper Rhine, the Lower Rhine and the Delta Rhine. To address all these problems, extensive river improvement schemes began around 1850. Main channels were systematically fixed and narrowed, navigation channels were dredged, sandbanks were removed, groynes were created to fix the riverbanks, and the rivers straightened at necessary points.

All these achievements had strong impacts on the hydrological characteristics of the Rhine. Inevitably, human interventions produce responses from the river. River managers consequently have to ensure that the actions undertaken achieve the intended goal, but have no undesirable side effects.

9.1 Modifications near Bingen

The meandering section of the Upper Rhine as well as the narrow valley near Bingen presented serious handicaps to navigation in former times. At Bingen, two navigation channels were blasted out in 1830–1832, and later improvements doubled the width and deepened the riverbed, thus eliminating the need for a pilot. This did not greatly alter the hydrological conditions of the Rhine.

9.2 Tulla and the Upper Rhine

The regulation of the river course seemed to be particularly urgent in the Upper Rhine basin. There, in 1800, Johann Gottfried Tulla started the standardization (fixation of the main riverbed) of the Upper Rhine. Maps depicting the area at Breisach before and after Tulla's correction around 1828 and after the last canalization in 1963 show dramatic achievements. The Rhine no longer meandered, inundations were brought back to a minimum, the surrounding land could more easily be used for agriculture, and navigation was considerably improved. As a result of canalizations less space was left for the River Rhine and the length of the river in this area was reduced by about 80 km.

For the upper Rhine basin, these measures were beneficial to man, although the ecosystem changed considerably. However, the impact was by no means limited to the part of the river that was modified. The canalization and diking of the Upper Rhine between Basel and Maxau involved the loss of 60% of the original flood plain area. Once the canalization and diking work was complete, flood waves propagated more quickly along this stretch of river. Such waves also became steeper, since canalization reduced water storage along the river. Discharge peaks along the Rhine increased sometimes over 20% owing to the lack of such a dampening effect.

9.3 Great Alsace Canal

By the Treaty of Versailles, France had obtained the right to divert Rhine water below Basel into a canal to improve navigation of the Rhine (formerly hindered by a rapid current) and thus increase traffic upstream from Strasbourg – and to procure hydroelectric power through locks and weirs; the canal was to rejoin the Rhine at Strasbourg. This Great Alsace Canal ("Grand Canal d'Alsace") was completed with the building of a dam at Kembs in 1932 and greatly improved navigation. After World War II, construction was resumed, but in a treaty (1956) France consented to terminate the canal at Neu Breisach, in return for West Germany's agreement to the canalization of the Moselle.

The canal utilizes the greatest part of Rhine water, leaving only a small amount to feed the older Rhine bed, with a dramatic decline of the ecology of the flood plain.

9.4 Dikes along the River

In other parts of the river, particularly in the Netherlands, dikes were erected. The oldest dikes on the Dutch river system were built in the 10th century. Raised more or less at right angles to the river, these early flood defenses were placed upstream of village polders.

After the creation of the first water authorities in the 12th and 13th centuries, the process of enclosing the rivers within dikes began on a large scale, and with it the formation of polders. By 1450, the great rivers had been more or less completely diked.

9.5 Waal and Meuse

To prevent the Rhine from forming new meanders and inundating surrounding areas, bifurcations of the Rivers Rhine, Waal, and IJssel and the cut-off canals were dug in the 18th century. Especially important was the Pannerden Canal near the Dutch-German border that was dug in 1707 to improve water distribution between the river branches.

The normalization of the River Waal was undertaken in the 18th century. The width of the main channel was reduced from more than 500 m to 260 m, islands and sandbanks were removed, and the banks were protected from erosion by regular arrays of groynes.

To prevent the Waal from spilling water into the Meuse during high floods, a lock was installed in 1856, which separated the Waal and the Meuse at Heerewaarden. To improve discharge from the Waal, the Nieuwe Merwede was cut at Gorinchem in 1876, which enabled the Waal to discharge via the Haring River. The Bergse Maas was built in 1904, which created a shorter route via the Amer and the Hollands Diep to the Haring River. Between 1918 and 1929, the Meuse was canalized from Grave to Maasbracht. In 1935, the Juliana Canal was opened, thus bypassing an almost unnavigable 47 km stretch of the Grensmaas. During the 1930s, the meander cut-off shortened the course of the Meuse downstream of Grave by nearly 30%.

9.5.1
The Waal Project

Modernization of the Waal as Western Europe's busiest inland waterway is on its way. The aim is to create a navigation channel wide enough and deep enough for two six-unit multiple barge sets and one twin-vessel set to pass one another at any point. For this purpose, the channel will have to be enlarged from the Pannerden bifurcation to Zaltbommel, so that, even with low water, the river is at least 170 m wide and 2.8 m deep. Downstream of Zaltbommel, the channel already meets these new specifications.

9.6
Lake IJssel

Given the lake's importance as a central fresh water basin, maintaining an adequate supply of water to the Lake IJssel via the IJssel River will have a high priority. If climate change results in hotter summers, resulting in summer river flows falling, then fresh water may become scarcer in the future. Under such circumstances, demands for water are likely to rise and adequate measures have to be taken.

9.7
Shipping

The Rhine is one of the main transport arteries linking the world port of Rotterdam to the Ruhr and central Europe as far as Basel. The water level in the Rhine is almost always high enough to be navigable for freight vessels.

The River Waal transports nearly half the freight shipped to and from Germany: the busiest inland waterway in Western Europe. Located at the mouths of two of Europe's rivers, the Netherlands is a natural access point to a large part of the European mainland.

Variations in water level have a major impact on the navigability of the river. When water levels are very high, shipping is not allowed on the river in Germany, in case the vessels' bow waves and wakes should damage buildings and other structures. When the river is running very low, shippers have to limit their loads to avoid grounding.

The volume of goods transported per vessel will increase considerably in the years ahead. Promotion of water freight transport is desirable on the one hand for environmental reasons, on the other hand as a relief of the pressure on the crowded road network. More, faster and, above all, bigger ships are therefore expected. To make these developments possible, the rivers' navigation channels will need to be modified to facilitate quick and safe passage as well as the use of larger vessels.

9.8 Change of Flow Characteristics

Large volumes of water flowing down a river generate enormous forces, which dislodge and entrain huge quantities of sand and gravel from the riverbed. Measures designed to improve discharge have led to even stronger currents than formerly. As a result, the Rhine's main channel has been eroded considerably over the last 100 years. The faster the river flows – increasingly so in heavily canalized or shortened rivers – the greater the force on the material forming the riverbed. Along fast-flowing stretches of riverbed, therefore, only the largest pebbles and grains will be left undisturbed.

On its way towards the sea, a river gradually slows down and loses its energy to move the sediments. So first the coarse particles and later the finer grains settle to the bottom or cease to move; as a result, the riverbed becomes covered with finer sediment further downstream. Approximately a third of the sediment transported by the Rhine reaches the North Sea. The rest is deposited in the river's lower reaches, where the current is not strong enough to carry the particles any further.

After periods of serious flooding, the river can leave up to 10 cm of sediment along its banks. A few millimeters per flood cycle is deposited on the rest of the flood plain. At various places, a layer of sedimentary clay more than a meter thick has been built up over the centuries.

How much sediment can be transported depends on the hydrology and the morphology of the river. It has become obvious in recent years that changes in the riverbed upstream can have far-reaching consequences downstream.

9.9 Tidal River Area

The area containing the former estuarine stretches of the Lower Rhine and the Meuse is known as the tidal river area. A direct connection to the North Sea exists via the New Waterway Canal, but the Haring River Dam closes off the southern outlet. Because of the construction of that dam, the southern part of the tidal river area has become a fresh water basin, and thus a major source of water for agriculture and drinking water supply.

The lock in the Haring River Dam regulates water levels in the tidal river area. When flow in the Rhine at Lobith is less than 1700 $m^3\ s^{-1}$, the locks are closed. As flow increases, they are gradually opened. In addition, water is distributed in such a way as to ensure that flow in the New Waterway Canal does not fall below 1500 $m^3\ s^{-1}$. Otherwise, the incursion of salt water would reach undesirable levels. During storms, the locks are closed.

9.10 A Vicious Circle

By embanking the rivers within dikes, man has created a vicious circle. Dikes limit the space occupied by a river and thus alter dramatically its discharge capacity; they also restrict the area over which silt can be deposited. With the rise of the ground levels between the dikes higher water levels were inevitable, which forced the population to repeatedly increase the height of the dikes. It is therefore imperative that this vicious circle of higher flood plains between the dikes resulting in ever-higher dikes is broken.

In the coming decades, large-scale initiatives to increase river capacity will be required in the countries of the Rhine catchment. Efforts to establish a sustainable flood protection policy are based on the assumption that to encourage nature development and to improve navigation routes, measures are most likely to be successful if they tie in with natural processes as far as possible. According to plan, the space available to the rivers will be increased so that greater volumes of water can move downstream without further dike reinforcement being required.

Now that water quality is improving, the river ecology is recovering as the plants and animals naturally associated with the rivers are returning. Greater emphasis is now placed upon developing the Rhine as a natural habitat. Although the established ecological objectives are to remain in force, water levels must not be allowed to fall to a level that would disadvantage shipping.

Truly integrated river management has to extend beyond national borders. Like ecological recovery and pollution control, effective flood protection depends on policies implemented across the international river basins. The countries along the Rhine and Meuse therefore need to continue their concerted actions.

10 Floods

Until the last century, at certain places of the Upper Rhine, the Lower Rhine, and the Delta Rhine, the river's main channels were wide and shallow, with islands and sandbanks. As a result, the meandering river often inundated the surrounding area and the Rhine was sometimes difficult to navigate.

Many parts of the Rhine catchment have been severely flooded in the past. Down through the centuries, people living near the rivers have had to live with the possibility of flood disaster. Dikes have repeatedly failed, not only at times of floods but also because of insufficient maintenance.

Efficient flood defenses are therefore absolutely vital.

Throughout the centuries, the river dikes have repeatedly been made higher and stronger and meters of sediment have accumulated on the flood plains. Today, the prospect of raising the dikes much further is no longer seen as the only solution. Alternative strategies are required to counter the risk of flooding in the long term, such as increasing the absorptive capability of the river basins and the storage capacity of the rivers and adjoining water bodies. Thus, emphasis is placed on activities such as removing obstacles from the flood plains, and creating retention basins. The river needs to be given more space in order to increase the amount of water it can cope with.

Rivers are part of the natural water cycle. The amount of water in a river primarily depends on the amount of precipitation, in form of rain and snow, in the river basin. Dependent on weather conditions, snow may lie on the ground before melting and draining into the river. And not all rain falling on the catchments area enters the river: some is absorbed and subsequently transpirated by vegetation. That precipitation which is not retained in the soil either evaporates or drains into watercourses (Fig. 4).

Changes in land use affect the volume of precipitation that can be absorbed and evaporated by the vegetation. The ability of the soil to absorb water can also be affected by land changes. Where woodland is cleared for agriculture, water drains more easily from the soil into the rivers. As a result, less precipitation finds its way into the groundwater reservoir. Urbanization has a similar but more marked effect. Precipitation falling on sealed surfaces drains immediately into the sewer system and reaches the watercourses more quickly.

10.1 Floods of 1993 and 1995

Are the floods of 1993 and 1995 signs of climate change? In these years the rivers rose to extreme heights. At the end of 1993, discharge in the Meuse hit a previously unrecorded water flow of 3120 $m^3\,s^{-1}$; in the province of Limburg (the Netherlands), the water level was 6 m higher than normal. Similar events were seen in 1995, and this time the Rhine was also very high. The discharge at Lobith reached approximately 12 000 $m^3\,s^{-1}$ – the second highest recorded figure ever. In the province of Gelderland, some 200 000 people were evacuated from areas close to the rivers. The main reasons for these floods were the warm, wet winters of 1993–1994 and 1994–1995. Persistent depressions over the North Sea caused extremely heavy rainfall in the Rhine and Meuse basins. Simultaneous peaks in discharge from the Moselle, the Sieg, the Ruhr, and the Lippe caused a wave of extreme flood in the area of Cologne and in the Netherlands (Figs. 5 and 6).

Although it is tempting to attribute the high floods of the years 1993–1994 and 1994–1995 to climate changes, flood events are historically not uncommon. Records show that similar events took place in 1824–1825,

1844–1845, 1918–1920, 1925–1926, and 1982–1983. Hence, it is not possible to say whether we are already witnessing the greenhouse effect in action.

10.2 Need for Flood Protection in the Netherlands

Of all the countries along the River Rhine, the Netherlands are particularly interested in reducing the likelihood of flood events. In this low country, a great part of which is lying below sea level, flood protection is vital. Without the coastal dunes and river dikes, two-thirds of the country would be flooded at times of high water.

When the first settlers came to the Netherlands around 5000 years ago, they found a marshy river delta. They raised terps or mounds to keep dry and constructed primitive river dikes. Over the centuries since, water management has been perfected, with the development of dikes, overflows, drainage pools, pumping stations, and storm flood barriers. Now, the Rhine branches are diked along their entire length. Behind the dikes is mostly low-lying polder. If the dykes were to fail during events of high floods, large parts of the provinces of Gelderland, Utrecht, South Holland, and North Brabant would soon be covered by meters of water.

In response to the floods seen in 1993 and 1995, the Great Rivers Delta Plan was set up to improve the safety of areas adjacent to the rivers. Action plans setting out flood defense measures were developed for the branches of the Rhine, the Meuse, and their basins. It has been decided to give the rivers more space – and not to increase the height of the dikes even further.

10.3 Reducing Peak Discharges

Flood waves can be reduced in a variety of ways: all involve increasing the absorptive or storage capacity of the river basin and river system. Proposed is an increase of the infiltration capacity of the soil, a reduction of the drainage rate, and promoting better land use. Such measures are effective mainly on small tributaries. Concerning larger rivers, it is possible to increase the amount of water that can be stored in the river system by creating or reestablishing retention basins or overflow areas.

When dealing with an international river such as the Rhine, flood protection initiatives are less effective if confined to one country only. Accordingly, international action plans have been elaborated for the Rhine and the Meuse basins to minimize the damage caused by future high water levels. Under these plans, peak flows are to be reduced by increasing the absorptive capabilities of the river basins, as well as the storage and drainage capacities of the rivers.

10.4
Water Retention along the Rhine

Along the Upper Rhine in Germany, retention areas – large polders beside the river, which inundate only at times of flood – are being created. These will be able to store up to 270×10^6 m^3 of water and are also planned in the state of North Rhine-Westphalia. About 100×10^6 m^3 of capacity is already available, and was first utilized during the high-water episode of 1988. By diverting water into these retention areas, it was possible to reduce the water level by 23 cm at Maxau and by 5 cm at Cologne, which indeed prevented the city center from being flooded. To create the retention areas, the state government will be clearing about 4000 ha of riverside land in the years to come, enough to accommodate 170×10^6 m^3 of water.

10.5
Ice Jams

Down through the centuries, people living near the rivers have had to live not only with the possibility of flood disaster. Drifting ice and resulting ice jams could sometimes create even greater problems. Drifting ice would damage the tops of the dikes and often led to the formation of ice jams, damaging bridges or raising the water levels and thus increasing the risk of dikes overflowing.

In the past, many dike failures were caused by ice jams. However, thermal pollution and the normalization activities undertaken since 1850 have reduced the risk of ice-floe formation considerably. Cold winters still occur, but the number of days with ice on the rivers has dropped dramatically.

References

1. ARW (2003) Jahresbericht. Köln
2. AWBR (2004) Jahresbericht. Stuttgart
3. Grabs W (ed) (1997) Impact of climate change on hydrological regimes and water resources management in the Rhine basin. CHR, Lelystad
4. IGKB (2004) Der Bodensee
5. IKSR (2002) Zahlentafeln. Koblenz
6. Jülich W (2003) Die Wasserqualität des Rheins und seiner Nebengewässer. IAWR, Nieuwegein
7. KHR (1993) Der Rhein unter der Einwirkung des Menschen – Ausbau, Schiffahrt, Wasserwirtschaft. CHR, Lelystad
8. RIWA (2003) Jaarverslag Rijn. Nieuwegein
9. VenW (1984) De waterhuishouding van Nederland. Staatsuitgeverij, Den Haag
10. VenW (1999) New deal with an old enemy. Water management in the Netherlands. Den Haag
11. Witte W, Krahe P, Liebscher H-J (1995) Rekonstruktion der Witterungsverhältnisse im Mittelrheingebiet von 1000 n. Chr. bis heute anhand historischer hydrologischer Ereignisse. CHR, Lelystad

Hdb Env Chem Vol. 5, Part L (2006): 47–87
DOI 10.1007/698_5_036

Published online: 25 October 2005

The Recovered Rhine and Its History

Rolf-Dieter Wilken

Rhine-Main Water Research, Institute AT Mainz University,
R&D Competence Center of Hessenwasser, Kurfürstenstr. 6, 65203 Wiesbaden, Germany
wilken@uni-mainz.de

Abstract The Rhine has suffered tremendous changes: from a natural river to a heavily contaminated one, then through a purification process to a rather clean river in a highly industrialized region. The river brought wealth into Europe as a motor of global industrialization. An increase in trade and industry was followed by pollution of the river, the sediments, and the fish. Outlined is the history of the Rhine from the Congress of Vienna in 1815 to the present, the year 2005. The changes in hydrology, the growing industry, and the peak of pollution in the 1970s are described. Then follows the success story of the Rhine, whereby the International Commission of the Protection of the Rhine, triggered by the Sandoz accident, converted it to a nearly uncontaminated stream. Hopefully another decline in the quality of river water, sediments, flora, and fauna can be prevented.

Abbreviations

AOX	Adsorbable organic halogens
ARW	Arbeitsgemeinschaft Rhein-Wasserwerke – Association of Waterworks in the Rhine basin
AWBR	Arbeitsgemeinschaft Wasserwerke Bodensee–Association of Waterworks at Lake Constance and Rhine
BOD	Biological oxygen demand
COD	Chemical oxygen demand
DKR	German Commission for the Protection of the Rhine
DOC	Dissolved organic carbon
DVGW	German Association of Gas and Water
EU	European Union
IAWR	International Association of Waterworks in the river Rhine basin
ICPR/CIPR/IKSR	International Commission for the Protection of the Rhine
IPPC	Integrated pollution and prevention control
LOD	Limit of detection
NGO	Non-government organizations
P^3	Polar persistent pollutants
RAP	Rhine Action Program
RIWA	Association of river water companies
TOC	Total organic carbon
WFD	European Water Framework Directive

1 Introduction

The most important and famous river in Europe has suffered a tremendous contamination and re-creation. “Father Rhine”, as the Germans call him, has the duties that fathers have: he gives the family food and drink, cleans the surroundings of his children and the children themselves, brings money for all who live in his neighborhood, and is also a witness to the fighting of those who are entrusted to him.

If a father is not properly treated, he can get ill. When the rooms of his children are dirty and in bad shape, he is perhaps mourning and angry, which leads to anger and high tides threatening all in his environment. When his children want to change him he is very wise: he let them, but has his own power to show who is the most powerful.

But, unlike a human father, Father Rhine cannot die. Even if the children are not able to appreciate his grandiosity, and torment him with their trash, he can re-create with pride. On the other hand, a change in the minds of his children can contribute to a good relationship and to the good health of the father.

So the story has to begin with the recovery of the father, and his good health in the new millennium.

Now, in the year 2005, the river Rhine is as unpolluted as in 1800. At the beginning of the 20th century the industrial revolution brought wealth into Europe as a motor for global industrialization. In those times the river was used as a trading artery, showing both sides of the coin: an increase in trade and a decrease in nature around it. So we have to hold in mind that rivers can be used for contradictory tasks, as outlined in Table 1.

Table 1 What rivers and their waters can be used for

Shipping and trade
Recreation
Fishing
Producing hydro-energy
Water can be used
• For farmland irrigation
• For drinking water purposes
• For industrial activities
– As cooling water
– Process water
– For cleaning
As a border to other countries

2
Development of the River Rhine

2.1
The Rhine as an International Trade Highway

In 1815 the Congress of Vienna placed the Rhine under an "international regime" designed to accelerate the free flow of trade. So the Rhine started to be an offspring of the French and industrial revolutions. Conceived by Napoleon and designed by engineers, the river acquired its canal-like profile during the 19th century. Three events in rapid succession marked its birth. In 1816 the Prince of Orange steamed upstream from Rotterdam to Cologne, inaugurating the age of coal and iron. Then, in 1817, the Baden engineer Johann Gottfried Tulla began the most ambitious rectification work ever undertaken on a European river. Celebrated as the "Tamer of the Wild Rhine," Tulla is best known for the simple maxim that guided his work: "No stream or river, the Rhine included, needs more than one bed; as a rule, multiple branches are redundant." Cooperation, coal, and concrete: together they started a riparian revolution that has determined Rhine affairs ever since [1].

None of the Vienna Congress delegates had any feeling as to the real significance of what they had just created. All they meant to do was foster trade among the riparian states after 25 years of war and bloodshed. To this end, they established the Central Commission for Rhine Navigation (the Rhine Commission) and gave it the task of eliminating the river's commercial border points – human ones such as the innumerable toll booths, and natural ones such as the Bingen reef and Lorelei cliffs – which had hindered river traffic for centuries. "The Rhine can count more tolls than miles," went a popular rhyme of the time, "and knight and priestling block its path."

Placing the river in the foster care of the Rhine Commission proved a mixed blessing. On the positive side, the new river regime stimulated economic growth and free trade on its banks. The Rhine is today one of the world's greatest commercial water ways in volume of traffic second only to the Mississippi. It transports millions of tons of coal, steel, chemicals, pharmaceuticals, textiles, and other goods each year, many of which are produced directly on its banks. The Rhine Commission, now headquartered in Strasbourg, can justifiably lay claim to being the oldest continuous interstate institution in Europe and the first step in the long march of diplomacy that culminated in the Common Market and European Union (EU).

On the negative side, the multinational engineers who took possession of the river in 1815 were strict disciplinarians, whose idea of a well-behaved river was not a river at all: it was a canal, utterly and completely suited to the needs of transport. They did not view themselves as custodians of the Rhine's fish stocks and alluvial forests, although salmon and timber were the mainstays of river commerce at the time. Nor did they see themselves as protectors of the Rhine's broad floodplain, although it was an integral part of the river's drainage system and home to a rich variety of flora and fauna. The birth of the new Rhine thus spelled doom for the old one. First, engineers severed the river's arms and legs from its trunk as dictated by the Tulla maxim. Then industries and cities introduced slow-acting poisons into its water system. The result was a truncated river deprived of its biological diversity.

The international river idea has survived remarkably well for nearly 200 years, even if it was (until 1945) mauled occasionally between the jagged teeth of France and Germany. That General George Patton publicly urinated in the Rhine to display his contempt for Nazi Germany meant nothing to the river's health. That millions of Europeans, before and after Patton, regularly flushed their raw sewage down the river meant everything. Round-the-clock Allied aerial bombing during World War II had a relatively short-term impact on the river's life, for the Rhine bridges and harbors were quickly repaired at war's end. But continual assaults from Rhine-based coal, steel, and chemical industries left the river's ecology crippled and disfigured over a long time period.

Humans live on rivers for much the same reason that other organisms do. Rivers provide a ready supply of nourishment and a convenient mode of transport. And any river with human inhabitants is much more than just a physical and biological entity: it is also the site of political, economic, and cultural activity.

2.2 The River in the Hands of Men

The Rhine would look different today from the Rhine of former years even if it had never become an "international river", or no steamers had entered its waters, or no engineers had tinkered with its hydrology, or indeed if all its human inhabitants had collectively moved away. But it would be hard to imagine that under these circumstances the Rhine would now be virtually denuded of its islands and forests, or that salmon would have vanished from its waters. Today's Rhine, in other words, is not just a geological entity – a construct of plate tectonics, volcanic activity, climatic variation, soil erosion, and other natural processes. It is above all a human artifact, a techno-river, a thoroughly anthropomorphized stream in Europe; and also a jointly administered Euro-river promoting transnational trade and commerce.

Ultimately, the new Rhine was constructed not just by Eurocrats and engineers but also by the millions of people who lived and traveled on its banks every day – by laborers who preferred the tugboat to the tow-line, by boaters who preferred a smooth channel to a rocky one, and by residents and visitors who preferred the "water closet" to the outhouse – that is, by practically everyone who knew the river through work or leisure.

Europeans shook off the Congress of Vienna's yoke during the revolutions of 1848. But they left the Rhine Commission intact. To one degree or another, nearly everyone got caught up in the Rhine's faster current. The new Rhine was more productive than the old Rhine. The power of steam made the stench of industry tolerable – at least until the day arrived when nobody could see the steamboat for the smoke.

2.2.1 Changing the Hydrological System of the River

Tulla focused on flood control measures at the base of the Alps, where most of the river's worst floods occurred. Subsequent projects by later generations of engineers highlighted navigation, not flood control, especially once steam power began to be exploited to its fullest extent after 1850. Changes in objectives over time meant that some stretches of the river were revisited and reconstructed many times. The technocrats changed the river until it was strait-jacketed more than any large river before or since.

The Rhine's modern contours were designed and crafted by engineers, but the projects were largely tailored to the needs of the 19th century European economy, especially Germany's coal and chemical industries. An example is shown in Fig. 1.

In 1817, the year the Tulla Project began, the Rhine was enveloped in nearly 2300 km^2 of floodplain (more if the Alpine feeder streams are added). Tulla's Rhine Correction Project was one of the biggest technical accomplishments of the 19th century, and prevents flooding and flood damage in the Baden region. It was accomplished in 1879. These corrections involved the draining and cultivation of the flood plains, a fact that provided new space for industry and farming close to the bank.

2.2.2 Implementation of Industrialization

The processes of industrialization and canalization were mutually dependent: Inventions and new machines brought better possibilities to canalize, to pump, excavate and deepen, and to control the rivers, whereas the better navigability of the Rhine was an important condition for widespread industrialization.

Reasons for the canalization of the Rhine were:

- Connection between mining places of coal and ore deposits
- Supply of the big cities with food and goods
- Cultivation of the flood plains to provide more food

Germany's rapidly developing rail system played a key role in the development of the iron and steel industry by increasing the demand (Fig. 2).

It was during the 1860s when large iron and steel production in the Ruhr area took off. During these years the Rhine became Europe's major trans-

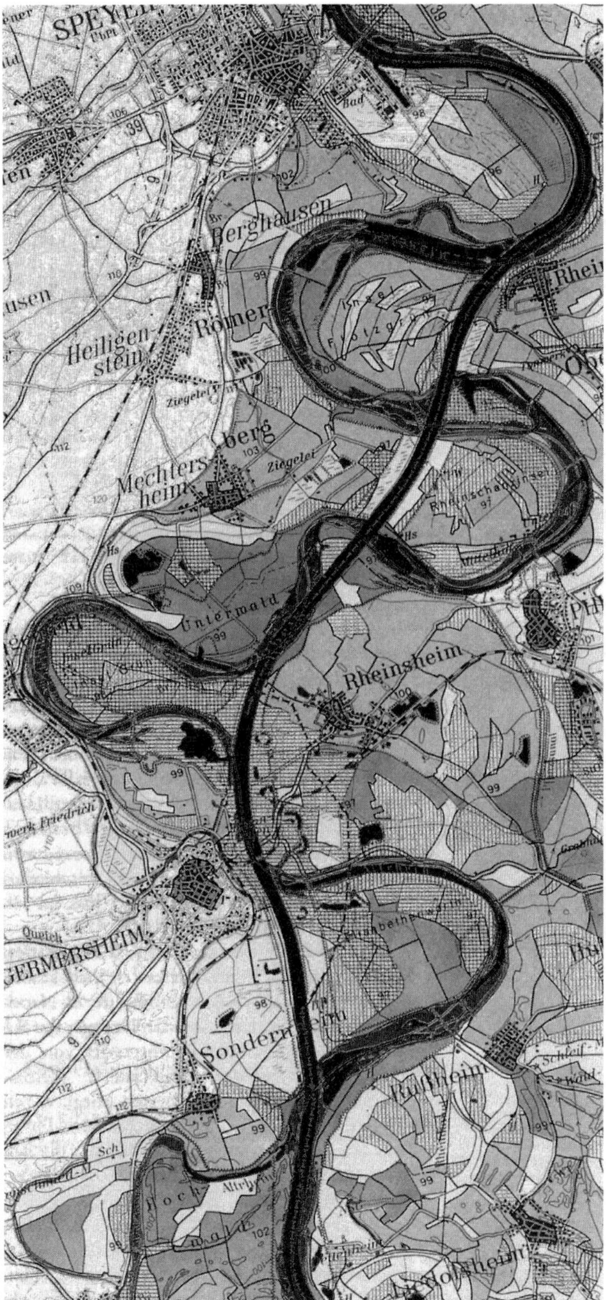

Fig. 1 Map of the Upper Rhine in the South of Speyer, showing the river after the corrections of 1817 by J.G. Tulla. The *straight line* is the Tulla correction, leaving old bends in the landscape [2]

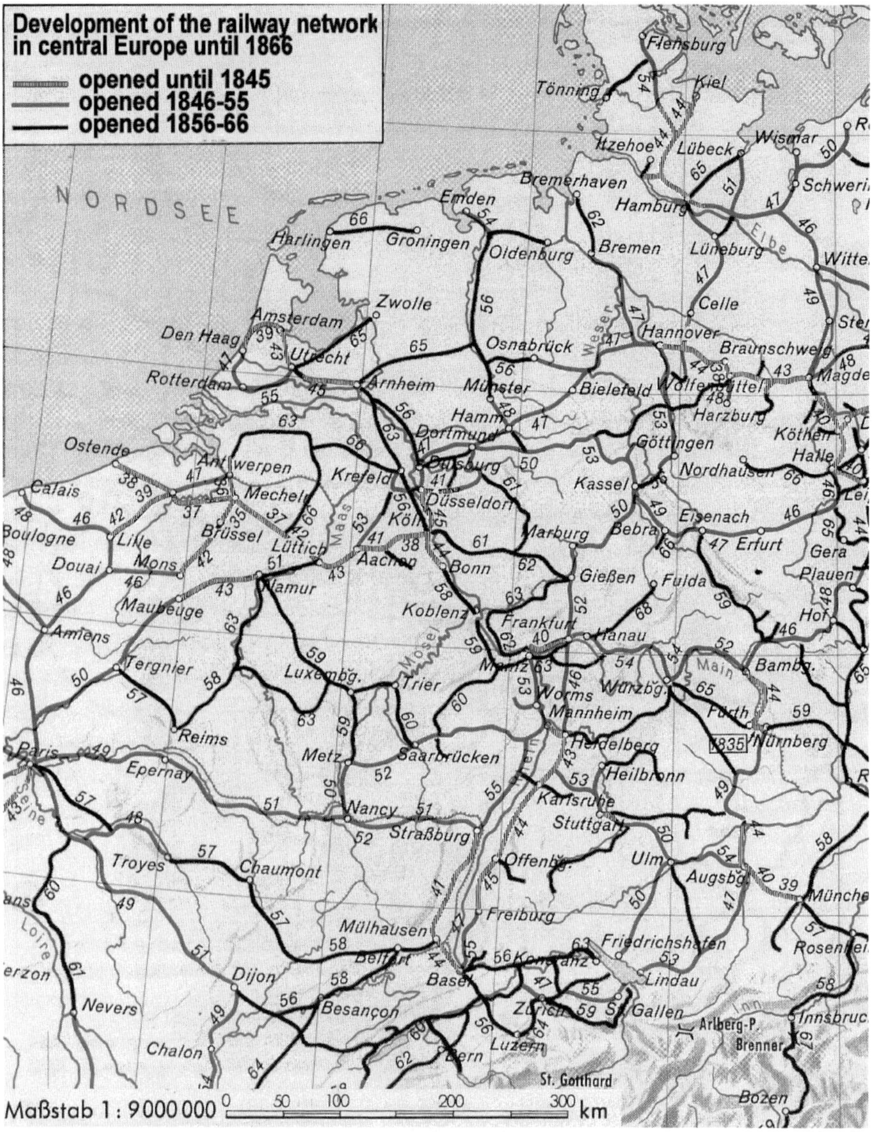

Fig. 2 Development of the railway network in Europe [3]. The railway network in Germany was developed until 1866 and is today, more or less, the same

port route, linking the Ruhr to its markets. Coal transport started about 1770. In 1815 there were 60 000 t, and in 1835 400 000 t of coal being transported on the Rhine, which was half of the mining during this time. Still, the importance and the spread of the inland ship navigation was bigger than of the paved road network. The independence of the Belgium Netherlands in 1830 brought additional capital in the Ruhr area. Germany's victory

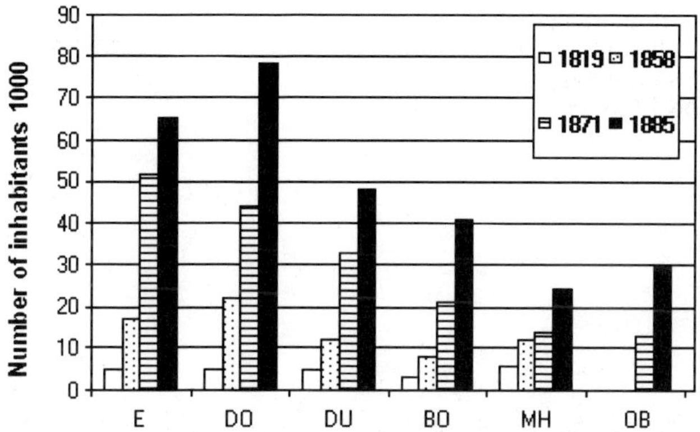

Fig. 3 Population development in different cities of the Ruhr area in the 19th century (*E* = City of Essen, *DO* = Dortmund, *DU* = Duisburg, *BO* = Bochum, *MH* = Mülheim, *OB* = Oberhausen) [4]

over France in the 1870/71 war gave the area access to the ore deposits in Lorraine.

The process of industrialization caused widespread urbanization of the Ruhr area. Oriented towards the stacks, huge urban agglomerations developed from 1850 to 1900. This involved population growth in the region due to natural growth and migration as, described in Fig. 3.

The first crisis of the coal industry followed the restoration after World War II, starting in 1960. The decline of the steel industry started about ten years later.

The steel industry brought the machine industry in its wake. Apart from the Ruhr area, machine industry cores on the Rhine developed in Cologne, Karlsruhe, Mannheim, Mainz, and Rotterdam. Another early industry was the chemical industry, producing particularly fertilizers and paints.

Today's industrial cores on the Rhine often have their origins in small family-structured manufactories. Some companies that were founded during these times are listed in Table 2.

Table 2 Small family-structured manufactories in the 19th century and their development

Foundation		Turnover in 2003 (million €)	
1811	Thyssen-Krupp, Düsseldorf	36.0	
1834	Südzucker, Mannheim	4.6	
1863	Bayer Leverkusen	28.6	(Revenues)
1865	BASF, Ludwigshafen	40.1	
1872	IWKA, Karlsruhe	2.3	(Revenues)

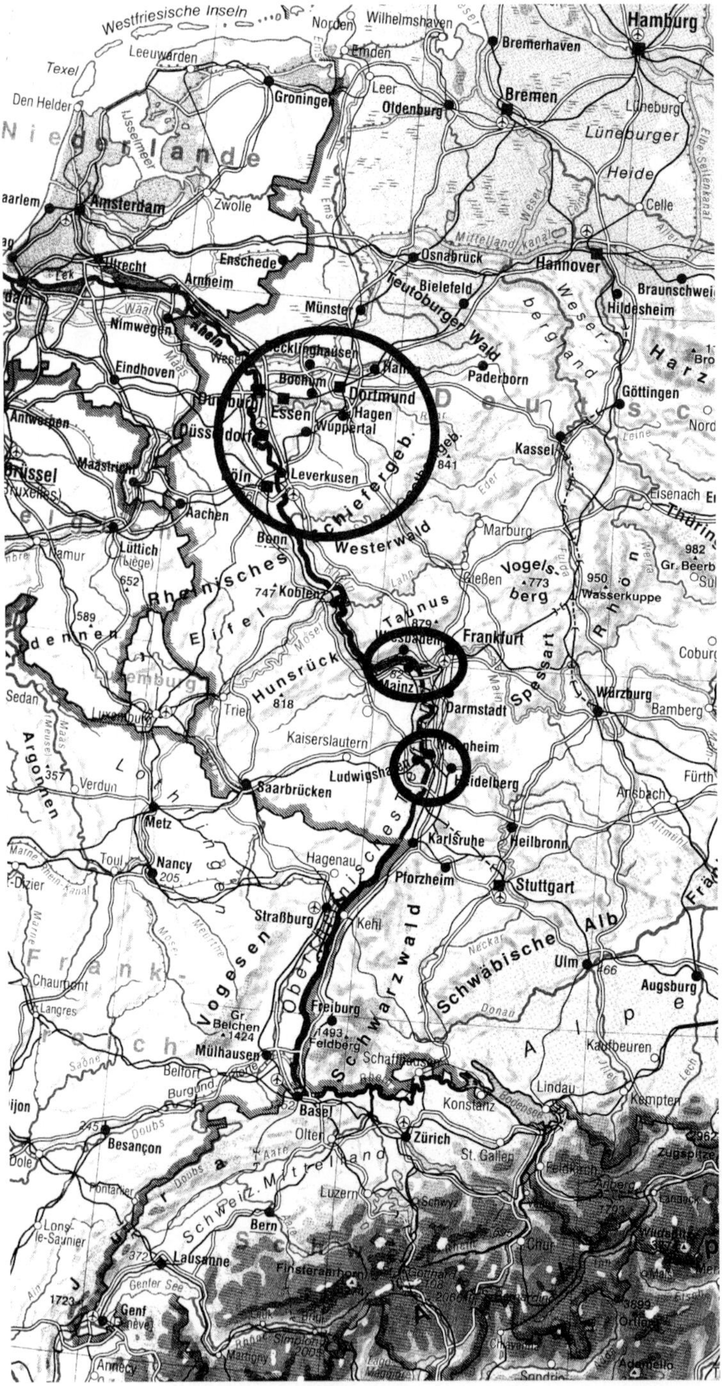

Fig. 4 Metropolitan areas on the Rhine, modified after [7]

In the mid-19th century and particularly after the German Reich was founded (in 1871), German industries reduced and finally matched the British lead. Electrical, chemical, and machine industries developed increasingly independently. Particularly in the first two sectors, inventions and new techniques put Germany in a leading position (e.g. with chemical fertilizers, paint, and pharmaceutical products) [5, 6]. Core regions on the whole river Rhine are still Basel, Strasbourg, Karlsruhe, Ludwigshafen, Mannheim, Mainz, and the whole lower Rhine from Bonn to Duisburg and Rotterdam.

Thus today's metropolitan areas on the Rhine (Fig. 4 are the:

- Rhine-Neckar
- Rhine-Main
- Rhine-Ruhr

which have their origins in the mid 19th century in the co-occurrence of good transport facilities and water supplies. Today the old industrial areas are in different phases of transformation from traditional industrial areas to high-tech regions. Karlsruhe, e.g., comes second with employees in high-tech companies in the whole EU.

As transport played a key role in the industrialization process of the Rhine-bordering countries there follows a short overview on the most important ports of the river.

Because of the increase of turnover of goods the port of Rotterdam was first extended in 1860. There was always a high demand for industrial estates because of the arriving raw material. As a result, Rotterdam is the biggest location for the petrol industry and since 1962 the biggest port in the world. In 2004, 352×10^6 t of goods were handled, half of which was crude oil. Chemical industry is seeking locations near Rotterdam because of the incoming raw materials.

With the beginning of the intensive coal and steel production in the 1860 the port of Duisburg expanded greatly. It was not only the transportation of industrial products but providing the fast-growing population with wheat that made the port so significant. Nowadays it is Europe's biggest river port. Ships up to 4500 t can go directly to Duisburg without turnover of their goods in Rotterdam or Hamburg.

2.3
Worry About the River's Health

The English Romantic, Samuel Taylor Coleridge, in a famous verse, saw nothing but trouble ahead [8]:

> In Köln, a town of monks and bones,
> And pavements fang'd with murderous stones
> And rags, and hags, and hideous wenches;
> I counted two and seventy stenches,

All well defined, and several stinks!
Ye Nymphs that reign o'er sewers and sinks,
The river Rhine, it is well known,
Doth wash your city of Cologne;
But tell me, Nymphs, what power divine
Shall henceforth wash the river Rhine?

In the years 1960–1980 water and sediment of the Rhine was so contaminated with pollutants – salts, cadmium, mercury, lead, nickel, antimony, chromium, zinc, phenols, organotins, and pesticides – that it could not be used neither for land reclamation nor farming nor drinking water purposes without costly treatment (see Fig. 5).

It is not simply that commercial fishing and crabbing have all but disappeared, or that spectacular catastrophes (such as the infamous Sandoz chemical spill of 1986) periodically brought havoc to the river's flora and

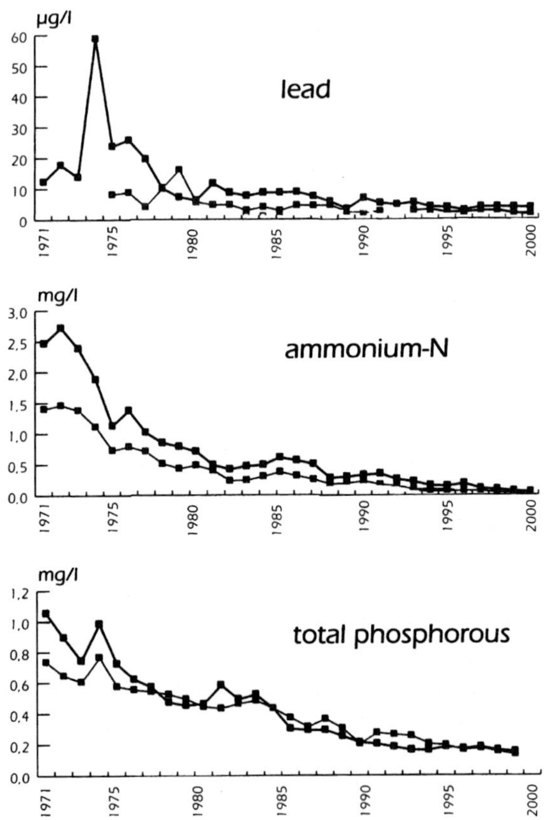

Fig. 5 Lead, ammonium, and total phosphorus in Rhine water over time at the monitoring stations Koblenz (*grey*) and Bimmen/Lobith (*black*) ([15], modified)

fauna. The Rhine of today is a fundamentally different riparian habitat from the river of 1815. The old riverbank vegetation was all but gone, as are most of the old-growth forests and all of the salmon runs.

Of the 47 fish species that swam in the Rhine, 200 years ago, only about half of them could still be found there by the 1970, and many of those survived only because of fish hatcheries. Salmon, shad, and sturgeon – the three most important commercial species – had all vanished, along with the fishermen who once earned their livelihood from them. Meanwhile, several new fish migrated to the Rhine, or were transported there – all of them less dependent on pure water, flood-plains, natural riverbanks, and meadowlands than their predecessors. The same is true of invertebrate macro-fauna, key indicators of water quality: in 1915, there were 80 known indigenous species, by 1956 only 42, and by 1971 only 27. Several native crab, mussel, and snail species became extinct, their ecological niches taken by species more resistant to salts and pollution. Dozens of bird species disappeared as well. "More faunal changes have occurred in the past one hundred fifty years," noted the biologist Ragnar Kinzelbach in 1984, "than in the previous ten thousand years". Only in the last few years, with the Rhine River Action Plan coming into force, has the situation shifted to the better.

It was not only the traffic and the stronger water currents, running down without detour from the Alps through Lake Constance to the North Sea, but also the water, which quality was determined often by factories and domestic effluents.

2.4
Efforts to Restore the River's Health: the International Commission for the Protection of the Rhine

Efforts to restore the river's health have been made in the last 40 years. Much of the rehabilitation work is being carried out by a new organization, the International Commission for the Protection of the Rhine against pollution (ICPR), which since 1963 has come to overshadow the original Rhine Commission. What drove the riparian states to reassess their Rhine regime was, above all, the recognition that the river could no longer handle the multitude of roles assigned it in the vast industrial-agricultural-urban-tourist nexus that had grown up on its banks. The river could not function as a conduit for industrial and agricultural wastes and still provide clean water to cities; it could not support endless urban sprawl and still be a favored destination for tourists; and it could not offer safety to anyone as long as it repeatedly flooded its artificial banks.

As the new organization's name suggests, pollution abatement rather than habitat restoration was uppermost in the minds of those who created it, at least initially. The progress to date – which can be seen in the form of cleaner water, biotic recovery, and floodplain restoration – has brought enormous benefits to humans and non-humans alike. There are, however, limits on what the new commission can accomplish now that millions of Europeans are wholly dependent on the river's goods, water, and energy. Full "restoration" is an impractical goal. All that the new commission can hope to achieve is a partial resuscitation – enough to allow the river of old to show its face more often here and there amid the harbors, hydro-dams, factories, and cities.

2.4.1
Natural Concentrations of Substances in the River Rhine

One of the biggest concerns was water and sediment quality regarding toxic substances. To assess the contamination of the water, sediment, and suspended matter of the Rhine we have to consider the natural regional concentrations or so called background concentrations. How difficult it is to identify such low background concentrations is discussed in this small section.

It is easy to determine the concentration of chloride in groundwater to be under 20 mg L^{-1}, as it is in rain, or for sulfate to be under 50 mg L^{-1}. The concentrations can exceed such values when groundwater is in contact with salt containing minerals such as gypsum or sodium chloride. Also, deep salty water can arise from the depth and cause elevated concentrations, which is the case in North Rhine Westphalia, and with the Mesozoic minerals in the Egge and Weser mountains.

The determination of background values, neither of simple parameters nor of complicated organic compounds, is not easy to obtain. It must also be noted that the analytical tools used often give results below the determination level, the limit of detection, LOD.

Geogenic elevated metal concentrations in groundwater can be found near ore and deposits and are also depending on pH and redox-condition 98% percentiles of all values of different areas along the Rhine are between 2 and 40 μg/L, what is a factor of 20 [9]. From these few examples it is obvious that background values cannot be fixed and contamination has a range not clearly defined in the lower concentrations.

2.4.2
Monitoring the River

Monitoring the Rhine river started with basic physical parameters such as temperature, pH, oxygen concentrations together with oxygen demands, then total element concentrations and heavy metal concentrations, followed later by different organic compounds. The chemical analytical tools developed rapidly and new substances such as organometallic-species were investigated. Today polar substances are of interest, as pharmaceutical residues or endocrine-disrupting substances.

Time of Highest Contamination in the River Rhine – the 1970s

Figure 6 can illustrate the ecological situation on the river during the peak contamination time. The values are taken from our own institute's measurements [10, 11]. The annual average values of ammonia, COD (chemical oxygen demand) and oxygen concentration, chloride concentration and smell, and water discharge are given.

Ammonia can only be measured in surface water when a lack of oxygen is observed, otherwise it is oxidized to gaseous N_2 and nitrate. This was the case in the years 1966–1977, i.e. for 11 years, with a maximum in 1972/1973. Paral-

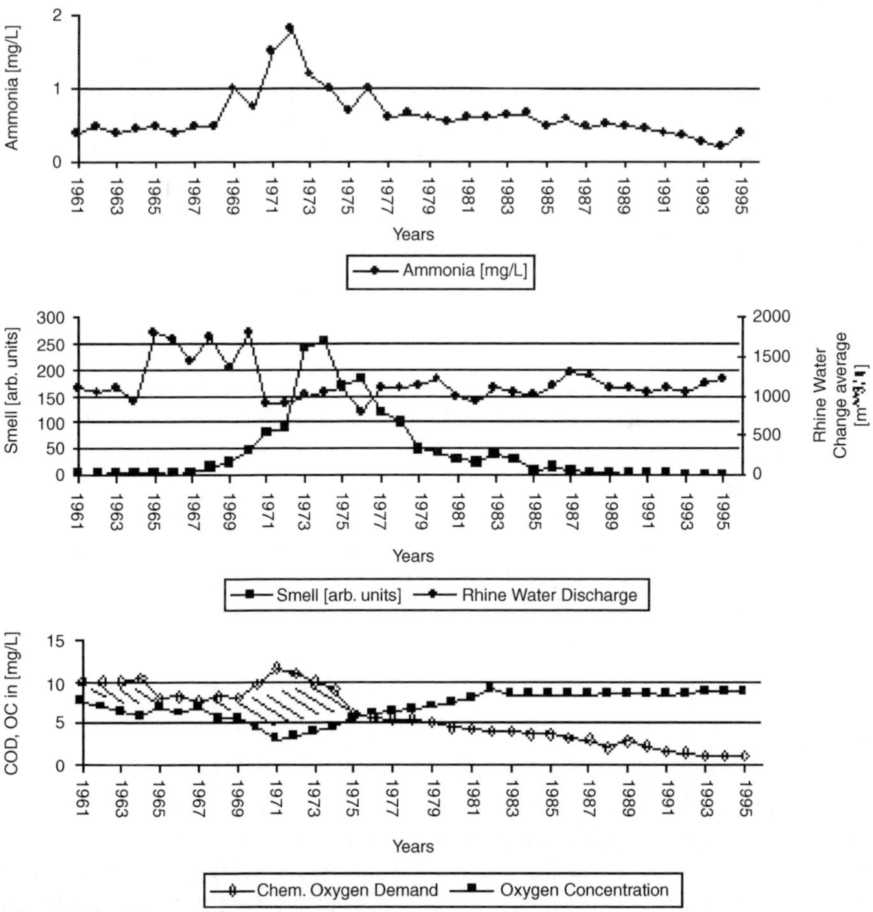

Fig. 6 Rhine at Wiesbaden, Germany, Rhine-km 506. Average annual values of ammonia, smell, and oxygen situation, with water discharge during 1961–1995. The *hachure* show zero oxygen concentrations in the river, where oxygen demand exceeds oxygen concentration [10, 11]

lel to this the oxygen concentration in the water was lower than the chemical oxygen demand. The chemical oxygen demand is understood as the demand for oxygen in a water sample to decompose all substances by oxidizing chemicals such as $K_2Cr_2O_7$.

Most noticeable for everybody walking near the rivers banks during this time was the bad smell of the river water, quantified by dilution of river water by non-smelling clean water, shown in the diagram as the most increasing and several years later decreasing line. The odor threshold value is defined as that amount of non-smelling water, compared to the amount of sample, that must be added to get a non-smelling impression.

The development of the odor threshold value in the years shown was dramatic. From 1966 on the values increased exponentially. Each year they doubled compared to the year before. The maximum was reached in 1974 with a odor threshold value of over 250, that is, River Rhine water had to be diluted by 250 times more non-smelling water than the sample water. The situation changed in 1975, when the big combined waste water treatment plant of BASF and the city of Ludwigshafen came into operation.

In detail, the oxygen situation gives an important impression of the ecological situation in the river through the years. This is shown in Fig. 7, where the biological oxygen demand is depicted. Contrary to COD, the BOD_5 is measured not by the consumption of certain oxidizing chemicals, but by the oxygen demand of bacteria decomposing organic substances in X, in this case 5 days.

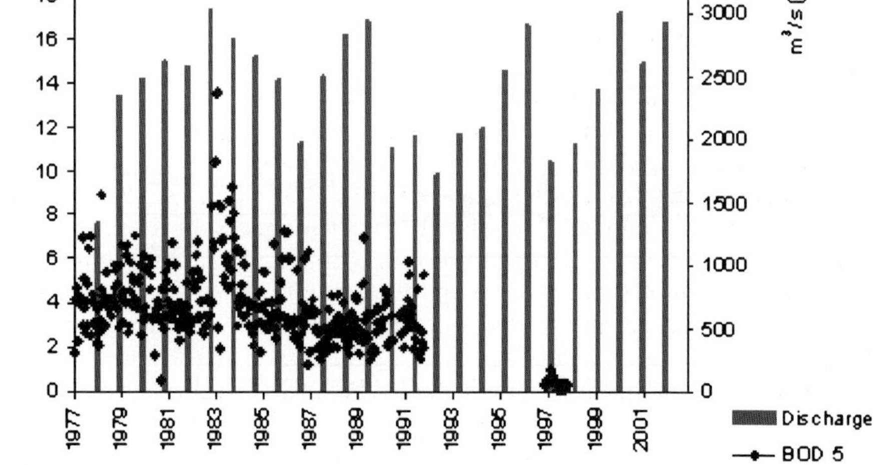

Fig. 7 Biological oxygen demand (BOD) at Kleve/Bimmen monitoring station together with water discharge [13, 14]

Nitrogen and Its Compounds

In 1985, more than 500 000 t of nitrogen, mainly as nitrate, flowed down the Rhine. In 2000, nitrogen loads still amounted to 360 000 t, one third of which originated from sewers and two thirds from diffuse discharges.

In the North Sea, nitrogen in combination with phosphorous acts as fertilizer for algae, which may regionally reproduce in masses and, when dying, will consume the oxygen content and may cause suffocation of fish.

The target fixed with a view to North Sea protection, i.e. to cut down inputs by 50%, has not been achieved. Due to measures taken in industry and to the construction of denitrification zones in municipal waste water treatment plants, it has been possible to reduce point source inputs.

However, diffuse nitrogen inputs as a result of leaching from agricultural fertilized soil via drainage and the ground water almost remained unchanged. Reduced fertilizing will only show its effects after several years, as soil and ground water passage are slow.

Heavy Metals

Regarding heavy metals, there are a lot of measuring stations along the Rhine. The values are measured in a defined program at distinct time intervals. Average values cannot be calculated by the mean of all. It must be taken into account that in different stations different programs are running, which differ in amount of numbers and results. Some stations are measuring often, others seldom. A homogeneous and representative data set is prepared in weighted results of a number of stations.

Nickel is a critical element because the threshold for nickel in drinking water is now 20 $\mu g\,L^{-1}$ after the EU Drinking Water Directive of 1998 [12]. Therefore nickel concentrations are chosen for comparison. It is obvious that the concentrations of nickel in the river water do not exceed the ground water concentrations of 10 $\mu g\,L^{-1}$ mentioned above.

Heavy metal concentrations over time are shown in Figs. 8–13. These data are extracted from DKR [13] and ICPR [14] with Rhine water or suspended material samples taken from the Bimmen/Lobith station (Rhine-km 865) on the border of Germany and the Netherlands. The discharge of the Rhine is given because substances can be differently diluted by the amount of river water. The concentrations given as points are single values from different single measurements. This has the advantage of showing the variety of values measured over the year, and not one average of all values. They all start in 1990, after the peak of contamination in the Rhine. This has been chosen for two reasons: the values measured earlier do not comply with the current analytical protocols for comparison of different compounds in the same time range. The author is aware of many other possibilities to describe monitored data over time, including different sampling strategies.

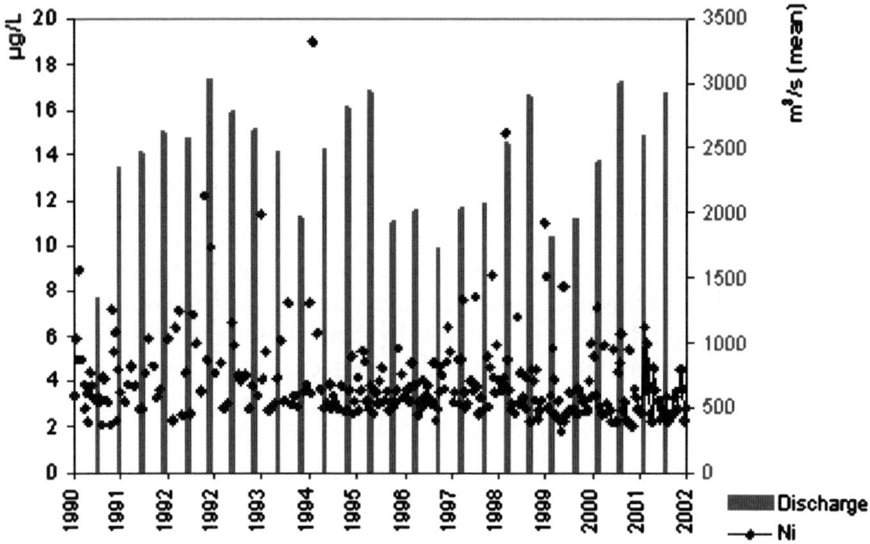

Fig. 8 Nickel in water, with water discharge of the Rhine (Bimmen/Lobith) 1990–2002 [13, 14]

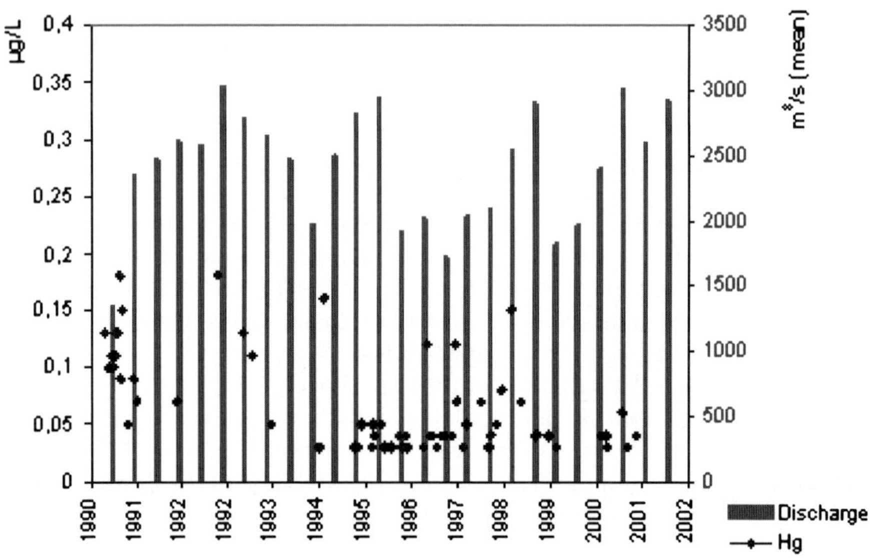

Fig. 9 Mercury in water with water discharge of the Rhine (Bimmen/Lobith) 1990–2002 [13, 14]

The nickel data seem to have a basic line of about 3 $\mu g\,L^{-1}$, with outliers of higher contamination.

The concentrations of the toxic element mercury in water also show a basic line in the last ten years with a concentration of 0.03 $\mu g\,L^{-1}$, and outliers

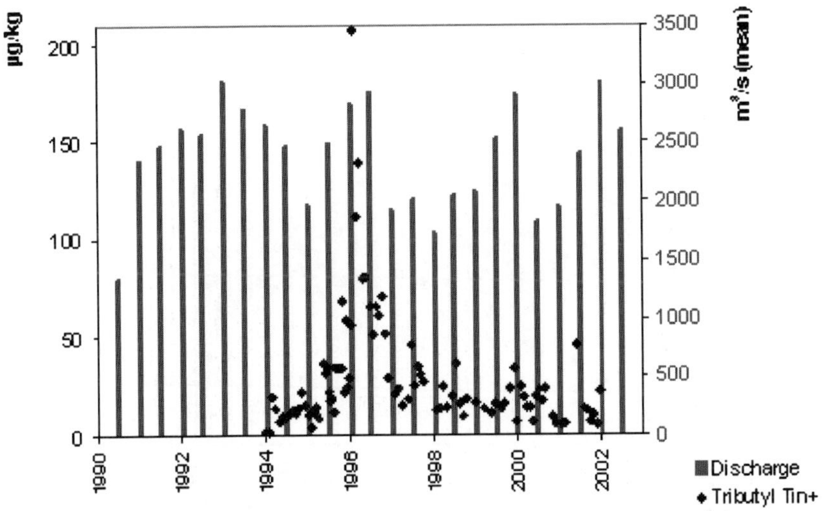

Fig. 10 Tributyl-tin$^+$ concentrations in suspended solids with water discharge of the Rhine (Bimmen/Lobith) [13, 14]

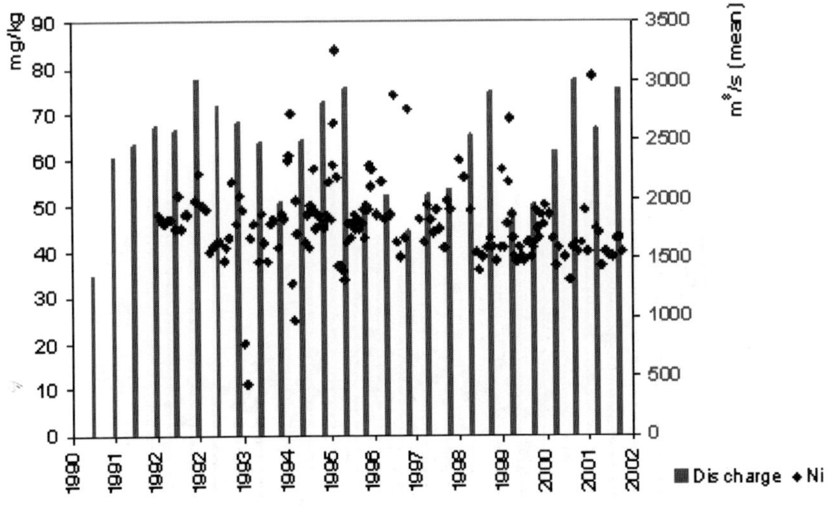

Fig. 11 Concentrations of nickel in the suspended load of the Rhine at Kleve/Bimmen 1990–2002 [13, 14]

sometimes six times higher. More interesting are the data of mercury in suspended particulate matter, as this element bonds itself mainly to particles. This is shown in Fig. 9.

One of the most toxic substances in water, for oysters in the ng L^{-1} range, is a component of antifouling paints for ships. The tributyltin$^+$ compound was both produced and used on the Rhine. The monitoring of this substance

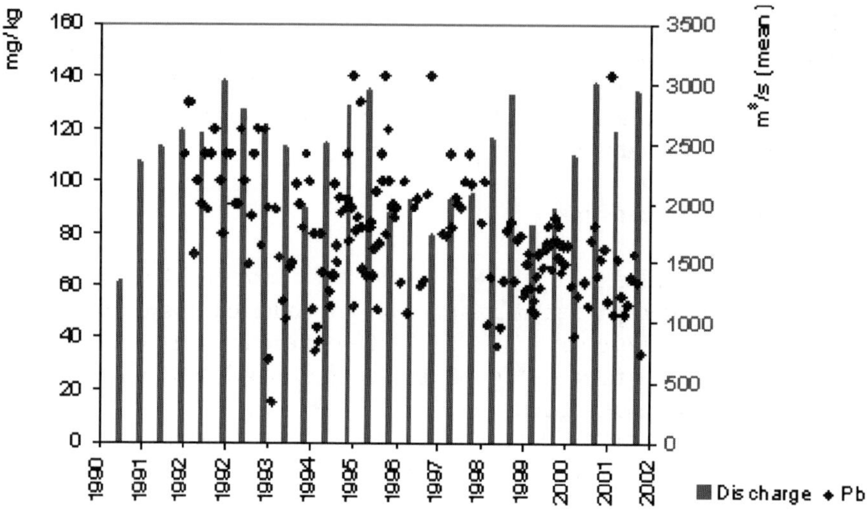

Fig. 12 Concentrations of lead in the suspended load of the Rhine at Kleve/Bimmen, 1990–2002 [13, 14]

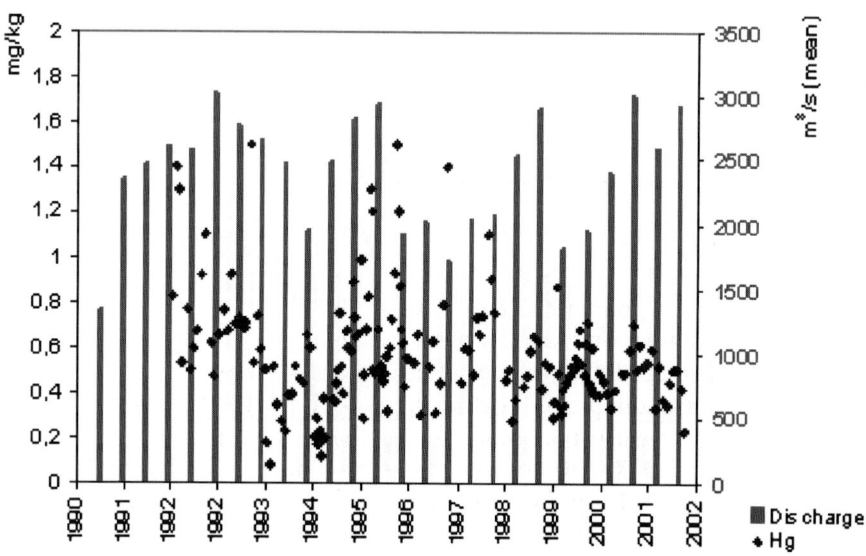

Fig. 13 Concentrations of mercury in the suspended load of the Rhine at Kleve/Bimmen, 1990–2002 [13, 14]

started in the early 1990s and shows that it has also decreased through better waste water treatment and regulations on use for the ships hull. Astonishing is a peak with highest concentrations in 1995. Whether this is due to a tributyltin$^+$ factory near Worms or from shipyards along the Rhine is not clear.

Not all data of all heavy metals are shown in the presentations. But a general conclusion may be allowed: it seems today that the concentrations of many metals in groundwater (in the river basin area of the Rhine) are comparable to the concentrations in Rhine river water, so in this case we can ascertain that a contamination of river water by most heavy metals is history.

Contaminants in Suspensions

Heavy metals or unpolar organic substances (not described here) are often bound to suspended particulate matter and are collected in sediments. The concentrations of those contaminants show no clear trend to lower values. This can be demonstrated in the figures showing nickel, lead, and mercury concentrations in suspended matter.

Even though all heavy metal contents in Rhine water have dropped in their base line, the contents of some heavy metals in suspended matter of the Rhine is still too high.

The contents of lead and mercury are now near the ICPR target values, but cadmium, copper, and zinc have failed to achieve the target. These heavy metals go back to numerous, in particular diffuse sources, e.g. roofs, water supply lines, inorganic fertilizer, abrasion of car tyres, mining activities, and flue gas. Cadmium accumulates in organisms, e.g. in algae and mushrooms. Fish belongs also to this kind of "pollutant sink hole", particularly fish which have a high proportion of fatty tissue, such as eels which adsorb heavy metals and chlorine organic contaminants. Therefore, fish are continually examined with respect to bioaccumulation of these contaminants.

Organic Pollutants

Regarding organic pollutants, there can be 100 000 of them in use. Finally they reach the sewer and the surface or ground water, when they are not degraded or adsorbed. Only few of them are checked for their environmental toxicity. As an example of the development of organic pollutants the sum parameter AOX (adsorbable organic halogens) is shown in the Fig. 14. The concentrations of AOX in water samples taken at Mainz (Rhine-km 498) show a significant decline. The reduction is due to different measures, mainly introduced by the chemical industries.

The Situation of Organic Compounds

The main organic substances in river water nowadays are the humic substances, regarded as decay products of plants and decomposed animal tissues. Total organic carbon, TOC, is the value measured for the estimation of total organic freight, whereas DOC, the dissolved organic carbon, is accessible as a nutrient. The DOC development in the river is shown in Fig. 15.

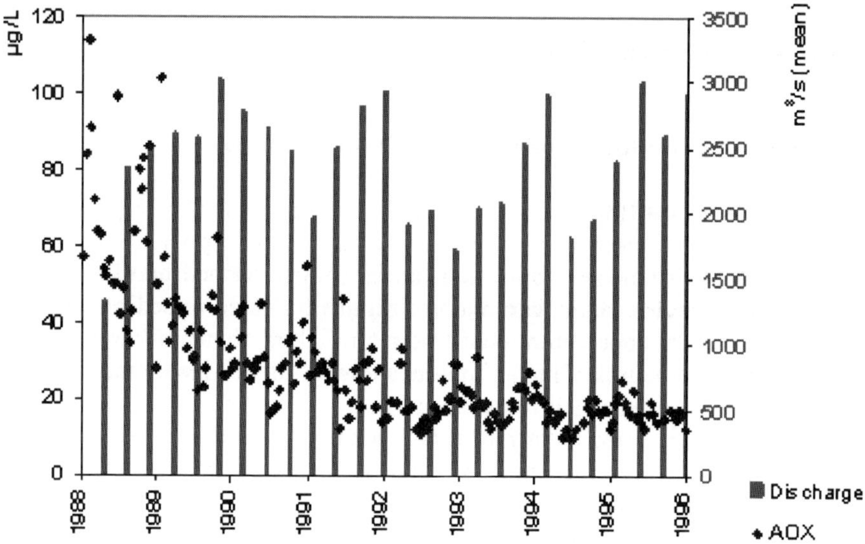

Fig. 14 Adsorbable organic halogenides (AOX) with discharge at Mainz monitoring station [13, 14]

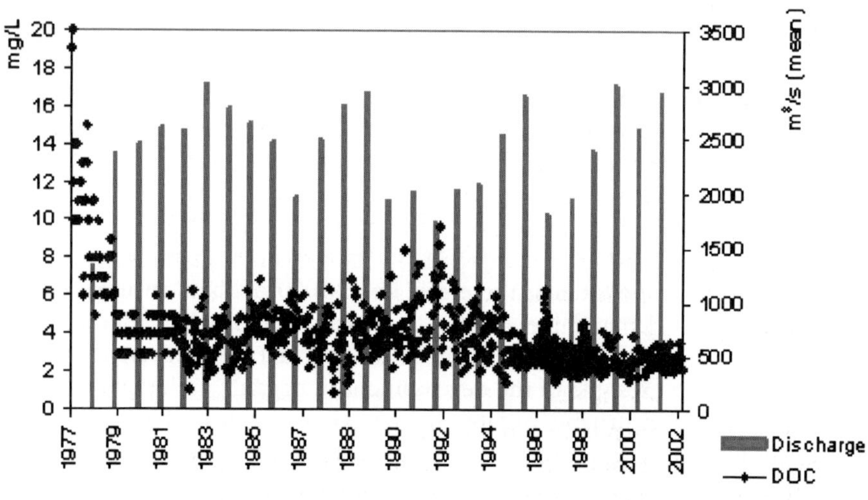

Fig. 15 Dissolved organic carbon DOC in the Rhine at Kleve/Bimmen, 1977–2002 [13, 14]

The contents of many chemical substances used as herbicides, fungicides, and insecticides have dropped in the Rhine. Some substances, however, have not achieved the target values: the most important are four of the PCBs and cadmium, zinc, and the pesticide diuron.

Upcoming in the last years are pharmaceuticals, pharmaceutical residues and endocrine disruptors.

2.5 Measures Taken to Reduce Rhine Contamination

2.5.1 The Sandoz-Schweizerhalle Accident and Reducing Risks

Since the 1950s, the Rhine had been playing the sad role of being "Europe's greatest sewer". The reverse of the economic boom and of population growth became evident. From 1975 on, when the first waste water treatment plants were put into service, there were the first signs of a slight improvement. But still, due to regular inputs and to additional accidents the river continued to cope with large amounts of hazardous substances. Then, a major accident occurred. On November 1st in 1986 a warehouse of the chemical production site Sandoz at Schweizerhalle near Basel burnt down. Some 20 t of highly toxic pesticides entered the Rhine together with the fire extinction water and caused the death of fish and micro-fauna as far as Koblenz.

Direct water intake from the Rhine serving drinking water purposes was stopped right down to the Netherlands. The public was deeply shocked. The media criticized that "along the Rhine, an accident is part of everyday life". The riparian states of the Rhine were forced to act. The Rhine Ministers and the Representative of the EU "... unanimously assessed that the warehouse fire in Schweizerhalle had considerably damaged the Rhine ecosystem, which is disturbed over a longer period of time. They are deeply concerned and share the concern of the public. The implementation of appropriate measures is aimed at avoiding such accidents in future, or at least at avoiding their consequences for the Rhine by applying the best technical means".

Shortly after the accident the Ministers in charge of the Rhine met, first in Zurich, then in Rotterdam and charged the ICPR to draft a plan which would ultimately eliminate the image of the Rhine as a sewer. In 1987, the Rhine Ministers approved the Rhine Action Program (RAP) in Strasbourg. It was designated to achieve the following targets by the year 2000:

- Fauna species which had vanished from the Rhine, such as the salmon, should again return to live in the Rhine
- Drinking water production from Rhine water should continue
- Pollutant contents of river sediments should be reduced
- Measures should be designed to accelerate the reduction of permanent pollution from direct and diffuse discharges
- Reduce the risk of accidents
- Improve hydrological, biological and morphological conditions.

2.5.2 Farming: Diffuse Sources Input

After closing or rehabilitation of point sources, diffuse sources came into sight. This is mainly farming and diffuse inputs from farmland such as pesticides or nutrients as phosphorus or nitrogen compounds. An overview of diffuse sources is given in Fig. 16.

As an example, the nutrient phosphorous is mainly transported into the water bodies due to soil erosion and drainage. Drainage and ground water are the most important diffuse pathways for the input of the nutrient nitrogen into the Rhine. Even in 2000, and in all Rhine bordering countries, most of these nitrogen inputs mainly had their origin in agriculture. Nitrogen sieves from heavily fertilized fields into the ground water and slowly moves in the direction of surface water bodies. Even though, in recent years, distinctly less fertilizer has been applied to many agriculturally used surfaces, this does not yet have any effect upon the Rhine. Nitrates are considerably retarded before

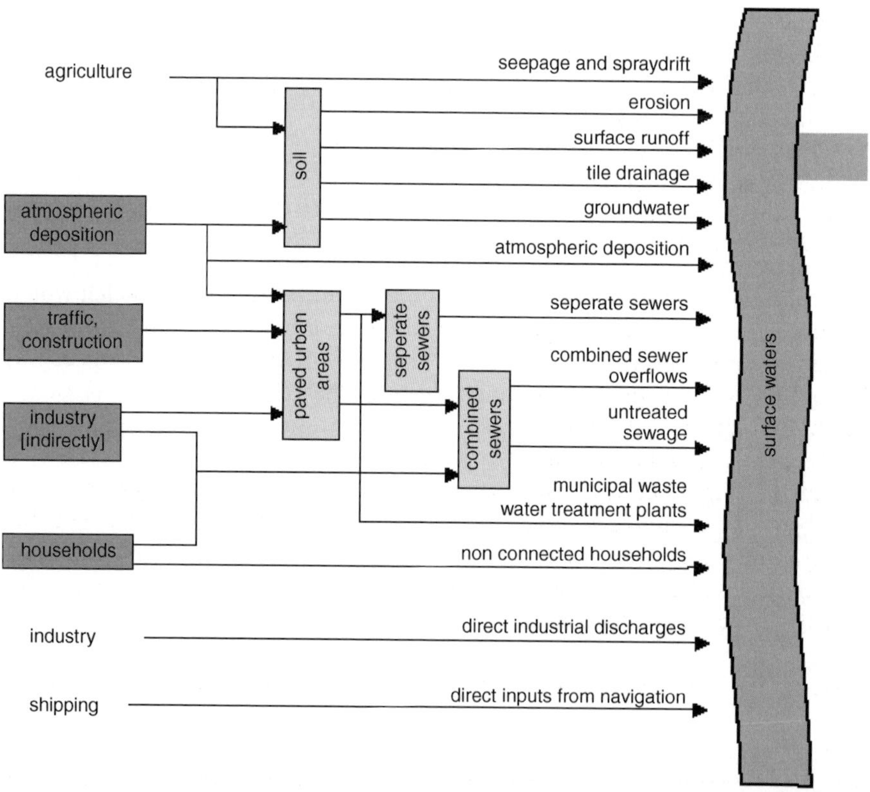

Fig. 16 Sources of diffuse input in a river, after [15]

they reach the river through the groundwater aquifer. The behavior of nitrogen compounds in surface runoff is different, reaching the river during storm water events in big quantities.

So, it is obvious that such diffuse sources are not really under control and improvements difficult to obtain.

2.5.3
The Rhine Action Program

The Rhine Action Program (RAP) was split into three phases:

- *Phase 1* (1987–89). The ICPR drafted a list of "priority substances", took stock of the source and amount of inputs and submitted proposals for their reduction. It demanded the implementation of the "state of the art" in industrial production and municipal waste water treatment plants. Properties of priority substances have been defined, e.g., that they harm organisms in water, accumulate in river sediment, accumulate in fish, and endanger drinking water production.
- *Phase 2* (until 1995). The discharges of priority substances were to be reduced by 50%, for some heavy metals even by 70%. The starting point chosen was the amount discharged in 1985. Additionally, in this phase, the ICPR was invited to draft a concept for the reduction of diffuse inputs from large surfaces.
- *Phase 3* (until 2000). Additional measures were to be implemented after an intermediate stocktaking. These measures even included the interdiction of dangerous substances in order to achieve the aims of improved Rhine water quality.

The RAP target of improved water quality was achieved (see Table 3).

In particular the discharges of toxic substances by municipalities and industry fell distinctly. Inputs of most priority substances were reduced by 70–100% or were no longer detectable in 2000. Today about 95% of the population in the Rhine catchment are connected to municipal waste water treatment plants. In 1985, no more than 85% had been connected. It is the little things that still cause problems: there are still a few substances of which too great amounts flow down the Rhine and into the North Sea. Recently, others have become the focus of attention, such as pharmaceutics and pharmaceutical residues, and certain substances with hormonal effects, the so-called endocrine disruptors.

Costs of 50 Billion € for the Rhine Action Program: Positive Effects on the Rhine

During the past 25 years the construction of waste water treatment plants at a cost of more than 50 billion € has had very positive effects on the Rhine. Since distinctly less pollutants flow through the industrial and municipal sewers and

Table 3 ICPR results for priority chemical substances reaching target values

Clearly been achieved	Almost been achieved	Not been achieved	Not detectable as under limit of detection
Non-volatile hydrocarbons			
2-Chloroaniline	AOX	Hexachlorobenzene (HCB)	4-Chloroaniline
3-Chloroaniline	Benzo(a)pyren	PCB (7 substances)	
1-Chloro-2-nitrobenzene			
1-Chloro-3-nitrobenzene			
1-Chloro-4-nitrobenzene			
2-Chlorotoluene			
4-Chlorotoluene			
3,4-Dichloroaniline			
Hexachlorobutadien			
1,2,3-Trichlorobenzene			
1,2,4-Trichlorobenzene			
1,3,5-Trichlorobenzene			
Volatile hydrocarbons			
Benzene			1,4-Dichlorobenzene
1,2-Dichloroethane			2,4-Dichlorophenoxyacetic acid
Tetrachloroethylene (PER)			Trichloromethane (Chloroform)
Tetrachloromethane			
1,1,1-Trichloroethane			
Trichloroethylene			

Table 3 (continued)

Clearly been achieved	Almost been achieved	Not been achieved	Not detectable as under limit of detection
Pesticides			
Adrin	Atrazin	Diuron	Azinphos-methyl
Azinphos-ethyl	Bentazon	Fenitrothion	Dichlorvos
DDT-group	Isoproturon	Lindane (γ-HCH)	
Dibutyltin cation	Tributyltin cation		Endosulfan
Dieldrin			Fenthion
Endrin			Mecoprop-P
α-HCH			Parathion-ethyl
β-HCH			Parathion-methyl
δ-HCH			Trifluralin
Isodrin			
Malathion			
Pentachlorophenol (pcp)			
Simazine			
Tetrabutyltin			
Triphenyltin cation (tpt)			
Metals			
	Arsenic	Cadmium	
	Lead	Copper	
	Chromium	Zinc	
	Nickel		
	Mercury		
Nutrients			
	Ammonium-nitrogen		
	Total phosphorus		

into our brooks and rivers, diffuse pollution increasingly becomes the focus of attention. Through numerous pathways large amounts of substances of diffuse origin reach our waters, through the air and through rain, acting as rinsing water. Nutrients and plant protection agents are washed out of the agricultural soil or are washed away by eroded soils in sloped positions. Foul water is washed off the streets and squares. Combined sewers equally receiving rain water overflow after heavy rainfall, as they are not conceived for large amounts of water.

The successful Rhine program is being continued. The ICPR and the Rhine-bordering countries have a new vision of more room for the Rhine. They plan to open the old alluvial areas to the river and to thus combine nature protection and flood prevention. "Rhine 2020", a program aimed at the sustainable development of the Rhine is meant to implement these targets. At the same time, it serves the EU water framework directive and its main objective, to achieve a "good chemical and ecological state" of the European water bodies.

Rhine 2020 – Program on the Sustainable Development of the Rhine

The Rhine 2020 program continues the extremely successful Rhine Action Program of 1987–2000 and fixes the general objectives of Rhine protection policy as well as the required measures for its implementation during the next 20 years. Focal points are the continued implementation of the Ecological Master Plan for the Rhine (1991), the improvement of flood prevention by implementing the Action Plan on Floods (1998), and the indispensable further improvement of water quality and ground water protection.

The targets of the Rhine 2020 program are:

- Sustainable development of the Rhine ecosystem
- Secure the use of Rhine water for drinking water production
- Improve sediment quality in order to be able to dispose dredged material without causing any harm
- Comprehensive flood prevention and protection taking into account ecological requirements
- Depollution of the North Sea

On December 22nd 2000 the European Water Framework Directive (WFD) came into force. In central fields of action the implementation of the Water Framework Directive will contribute to materializing the Rhine 2020 program.

Switzerland, which is not an EU-member but a bordering country, organizes its water policy comparable to that of the EU. Therefore, within their own policy, the Swiss support the implementation of the WFD by the EU Member States [15].

Both the Rhine 2020 program and the EU Water Framework Directive are important tools for the future development of the Rhine river.

3
Technological Alternatives and Innovations of Waste Water Treatment

The purpose of engineered treatment involved in waste water is to ensure water quality for human consumption downstream of a river. It is possible to define several commonly used treatment levels as follows:

- Primary treatment for the removal of suspended solids
- Secondary treatment for biological oxidation of carbon and nitrogen pollution, in some cases combined with phosphorus removal
- Micropollutant oxidation
- Removal of heavy metals and dissolved salts by physico-chemical processes
- Disinfection for elimination of pathogenic microorganisms and parasites

The treatment level related to waste water discharge into a transboundary river is determined by rules, regulations, and standards in Europe, especially by the European Water Framework Directive. Within this context, the same objectives can be achieved using extensive or intensive technologies depending on the analysis of different economic criteria and regional conditions (e.g. investment cost, operation and maintenance costs, floor occupation, and infrastructure).

With this treatment nutrients and contaminants are removed, degraded, or scavenged in the sludge, which has to be treated separately. Whereas the incoming concentrations of heavy metals and pesticides have decreased since the International Commission for the Protection of the Rhine (IRPC) set new rules, new emerging substances are under suspect and may be harmful in drinking water, especially the polar persistent pollutants (P^3), well soluble in water and hardly decomposed by bacteria. They must not even be persistent to reach the drinking water if their decomposition goes hand in hand with new inputs of such compounds from the effluent of the next sewage plant so to reach a kind of steady state in the river. Pharmaceutical residues and endocrine disruptors belong to this group. The concentrations found are in the ng/L range or even lower, and are therefore not contributing to any health effect in drinking water, but they have to be discussed because the relevant doses and impacts are still subject of modern water research.

4
Drinking Water Safety

Not all contaminants can reach the drinking water supply (Table 4). They can be classified by their abilities to be removed or degraded by bacteria in the river, soil passage, or by water treatment in waterworks. If they pass these barriers, they can reach the drinking water. New substances or substances which have not been approved can be qualified in a test filter system, which consists of a water loop filled with river water, where the substances have been added. The solution is pumped through an adsorptive material coated with natural bacteria from the river water and the behavior is analyzed.

Table 4 Abilities of waterborne substances to reach drinking water

1. Biologically degradable substances
   - Removable by bank filtration and soil passage
2. Relevant for waterworks
   - Biologically not removable or difficult to remove
   - Not removable in soil passage
3. Relevant for drinking water
   - Not removable totally by degradation or adsorption

For the river ecosystem this classification for drinking water safety has to be handled differently. In any case, if the substance is degradable, without toxic degradation products, this can be performed in a sewage plant so that only mineralized products reach the receiving waters.

Most chlorinated products are not degradable and are named persistent. So these substances have been of great concern in the river Rhine's pollution history. The concentrations of such compounds has gone down due to banning or replacement by less persistent chemicals.

5 Assessment of Quality under the European Water Framework Directive

5.1 Introduction to the New EU Water Framework Directive

The increasing demand by citizens and environmental organizations for cleaner rivers and lakes, groundwater, and coastal beaches is evident. This demand by citizens is one of the main reasons why the EU Commission has made water protection one of the priorities of its work.

On 23rd October 2000, the "Directive 2000/60/EC of the European Parliament and of the Council establishing a framework for the Community action in the field of water policy", or the EU Water Framework Directive (WFD) for short, was finally adopted by the member states of the European Community [16]. The purpose of the directive is formulated in Article 1:

Article 1 of the European Water Framework Directive

The purpose of this Directive is to establish a framework for the protection of inland surface waters, transitional waters, coastal waters and groundwater which:

(a) Prevents further deterioration and protects and enhances the status of aquatic ecosystems and, with regard to their water needs, terrestrial ecosystems and wetlands directly depending on the aquatic ecosystems
(b) Promotes sustainable water use based on a long-term protection of available water resources
(c) Aims at enhanced protection and improvement of the aquatic environment, inter alia, through specific measures for the progressive reduction of discharges, emissions and losses of priority substances and the cessation or phasing-out of discharges, emissions and losses of the priority hazardous substances
(d) Ensures the progressive reduction of pollution of groundwater and prevents its further pollution
(e) Contributes to mitigating the effects of floods and droughts and thereby contributes to the provision of the sufficient supply of good quality surface water and groundwater as needed for sustainable, balanced and equitable water use:

  - Significant reduction in pollution of groundwater
  - Protection of territorial and marine waters
  - Achieving the objectives of relevant international agreements, including those which aim to prevent and eliminate pollution of the marine environment, ...

The following section provides an overview on the development, present state and future of European Water Policy, especially for the Rhine River.

5.2
An Early Beginning

Early European water legislation began in 1975 with standards for those of our rivers and lakes used for drinking water abstraction, and culminated in 1980 in setting binding quality targets for our drinking water. It also included quality objective legislation on fish waters, shellfish waters, bathing waters, and ground waters. Its main emission control element was the Dangerous Substances Directive.

In 1988 the Frankfurt ministerial seminar on water reviewed the existing legislation and identified a number of improvements that could be made and gaps that could be filled. This resulted in the second phase of water legislation, the first results of this were, in 1991, the adoption of:

- The Urban Waste Water Treatment Directive, providing for secondary (biological) waste water treatment, and even more stringent treatment where necessary
- The Nitrates Directive, addressing water pollution by nitrates from agriculture
- Other legislative results of these developments were Commission proposals for action on:
  - A new Drinking Water Directive, reviewing the quality standards and, where necessary, tightening them (adopted November 1998)
  - A Directive for Integrated Pollution and Prevention Control (IPPC), adopted in 1996, addressing pollution from large industrial installations

5.3 Getting Europe's Waters Cleaner, Getting the Citizen Involved: the European Water Policy

Pressure for a fundamental rethinking of EU Community water policy came to a head in mid-1995: The Commission, which had already been considering the need for a more global approach to water policy, accepted requests from the European Parliament's environment committee and from the Council of environment ministers.

The outcome of this consultation process was a widespread consensus that, while considerable progress had been made in tackling individual issues, the current water policy was fragmented, in terms both of objectives and of means. All parties agreed on the need for a single piece of framework legislation to resolve these problems. In response to this, the Commission presented a Proposal for a Water Framework Directive with the following key aims:

- Expanding the scope of water protection to all waters, surface waters and groundwater
- Achieving "good status" for all waters by a set deadline
- Water management based on river basins
- A "combined approach" of emission limit values and quality standards
- Getting the prices right
- Getting the citizen involved more closely
- Streamlining legislation

The outline below shows how these elements are made operational within the Directive.

5.4 A Single System of Water Management: River Basin Management

The best model for a single system of water management is management by river basin – the natural geographical and hydrological unit – instead of according to administrative or political boundaries. Initiatives taken by the states concerned for the Maas, Schelde, and Rhine river basins have served as positive examples of this approach, with their cooperation and joint objective-setting across Member State borders, or in the case of the Rhine even beyond the EU territory. For each river basin district – some of which traverse national frontiers – a "river basin management plan" will need to be established and updated every six years, and this will provide the context for the coordination requirements identified above.

5.5 Coordination of Objectives – Good Status for All Waters by a Set Deadline

There are a number of objectives in respect of which the quality of water is protected. The key ones at European level are general protection of the aquatic ecology, specific protection of unique and valuable habitats, protection of drinking water resources, and protection of bathing water. All these objectives must be integrated for each river basin. It is clear that the last three – special habitats, drinking water areas,and bathing water – apply only to specific bodies of water (those supporting special wetlands, those identified for drinking water abstraction, and those generally used as bathing areas). In contrast, ecological protection should apply to all waters; the central requirement of the treaty is that the environment be protected to a high level in its entirety.

5.5.1 Surface Water

Ecological Protection

For this reason, a general requirement for ecological protection, and a general minimum chemical standard, was introduced to cover all surface waters. These are the two elements "good ecological status" and "good chemical status". Good ecological status is defined in Annex V of the Water Framework Directive, in terms of the quality of the biological community, the hydrological characteristics, and the chemical characteristics. The system is somewhat complicated, but this is inevitable given the extent of ecological variability, and the large number of parameters that must be dealt with.

Chemical Protection

Good chemical status is defined in terms of compliance with all the quality standards established for chemical substances at European level. The Directive also provides a mechanism for renewing these standards and establishing new ones by means of a prioritization mechanism for hazardous chemicals. This will ensure at least a minimum chemical quality, particularly in relation to very toxic substances, everywhere in the Community.

5.5.2 Other Uses

As mentioned above, the other uses or objectives for which water is protected apply in specific areas, not everywhere. Therefore, the obvious way to incorporate them is to designate specific protection zones within the river basin that must meet these different objectives. The overall plan of objectives for the river basin will then require ecological and chemical protection everywhere as a minimum, but where more stringent requirements are needed for particular uses, zones will be established and higher objectives set within them.

5.5.3 Coordination of Measures

There are a number of measures taken at Community level to tackle particular pollution problems. Key examples are the Urban Waste Water Treatment Directive and the Nitrates Directive, which together tackle the problem of eutrophication (as well as health effects such as microbial pollution in bathing water areas and nitrates in drinking water). The Integrated Pollution Prevention and Control Directive deals with chemical pollution. The aim is to coordinate the application of these so as to meet the objectives established above. This is done as follows:

First of all, the objectives are established for the river basin as outlined in the previous section. Then an analysis of human impact is conducted so as to determine how far from the objective each body of water is. At this point, the effect on the problems of each body of water of full implementation of all existing legislation is considered. If the existing legislation solves the problem, well and good, and the objective of the framework Directive is attained. However, if it does not, the Member State must identify exactly why, and design whatever additional measures are needed to satisfy all the objectives established. These might include stricter controls on polluting emissions from industry and agriculture, or urban waste water sources. This should ensure full coordination.

The coordination has now been overtaken by the ICPR, on February 17th, 2005, the Coordinating Committee Rhine adopted the report "International River Basin District Rhine. Characteristics, examination of the enviromental impact of human activities and economic analysis of water uses". Up to date information is given on the homepage of ICPR [15].

5.6
Combined Approach

But there is a further aspect. Historically, there has been a dichotomy in the approach to pollution control at European level, with some controls concentrating on what is achievable at source, through the application of technology, and others dealing with the needs of the receiving environment in the form of quality objectives. Each approach has potential flaws. Source controls alone can allow a cumulative pollution load that is severely detrimental to the environment, if there is a concentration of pollution sources. Quality standards can underestimate the effect of a particular substance on the ecosystem, due to the limitations in scientific knowledge regarding dose-response relationships and the mechanics of transport within the environment. For this reason, a consensus has developed that both are needed in practice – a combined approach. The Water Framework Directive formalizes this.

On the source side, it requires that as part of the basic measures to be taken in the river basin, all existing technology-driven source-based controls must be implemented as a first step. But over and above this, it also sets out a framework for further developing such controls. The framework comprises the development of a list of priority substances for action at EU level, prioritized on the basis of risk. Then the most cost-effective set of measures to achieve load reduction of those substances are designed, taking into account both product and process sources.

On the effects side, it coordinates all the environmental objectives in existing legislation, and provides a new overall objective of good status for all waters. Where the measures taken on the source side are not sufficient to achieve these objectives, additional ones are required.

5.6.1
River Basin Management Plan

All the elements of this analysis must be set out in a plan for the river basin. The plan is a detailed account of how the objectives set for the river basin (ecological status, quantitative status, chemical status, and protected area objectives) are to be reached within the timescale required. The plan should include all the results of the above analysis: the river basin's characteristics, a review of the impact of human activity on the status of waters in the basin, estimation of the effect of existing legislation and the remaining "gap" to meeting these objectives, and a set of measures designed to fill the gap. One additional component is that an economic analysis of water use within the river basin must be carried out. This is to enable rational discussion on the cost-effectiveness of the various possible measures. It is essential that all interested parties are fully involved in this discussion, and indeed in the preparation of the river basin management plan as a whole. Which brings me to the final major element of the proposal, the public participation requirements.

5.6.2 Public Participation

In getting our waters clean, the role of citizens and citizens' groups will be crucial.

There is one main reason for an extension of public participation. The decisions on the most appropriate measures to achieve the objectives in the river basin management plan will involve balancing the interests of various groups. The economic analysis requirement is intended to provide a rational basis for this, but it is essential that the process is open to the scrutiny of those who will be affected.

5.6.3 Streamlining Legislation: Seven Old Directives to be Repealed

One advantage of the framework directive approach, in its own way a significant one, is that it will rationalize the Community's water legislation by replacing seven of the first directives: those on surface water; two related directives on measurement methods and sampling frequencies and exchanges of information on fresh water quality; the fish water, shellfish water, and groundwater directives; and the directive on dangerous substances discharges. The operative provisions of these directives will be taken over in the framework directive, allowing them to be repealed.

5.7 European Drinking Water Directive in the Eyes of the Waterworks in the Rhine Area

There is one other category of uses which does not fit into the picture of a perfect water directive. It is the set of uses which adversely affect the status of water but which are considered essential on their own terms – they override policy objectives. Key examples are flood protection and, especially along the Rhine river, essential drinking water supply. The problem is dealt with by providing derogations from the requirement to achieve good status for these cases, so long as all appropriate mitigation measures are taken.

5.7.1 Groundwater

Chemical Status

The case of groundwater is most important for waterworks and is somewhat different. The presumption in relation to groundwater should broadly be that it should not be polluted at all. For this reason, setting chemical quality standards may not be the best approach, as it gives the impression of an allowed level of pollution to which Member States can fill up. A very few such standards have been established at European level for particular issues (nitrates, pesticides, and biocides), and these must always be adhered to. But for general protection, another approach has to be taken. It is essentially a precautionary one. It comprises a prohibition on direct discharges to groundwater, and (to cover indirect discharges) a requirement to monitor groundwater bodies so as to detect changes in chemical composition, and to reverse any anthropogenically induced upward pollution trend. Taken together, these should ensure the protection of groundwater from all contamination, according to the principle of minimum anthropogenic impact.

5.7.2 Quantitative Status

Quantity is also a major issue for groundwater and drinking water supply. Briefly, the issue can be put as follows: There is only a certain amount of recharge into a groundwater each year and, of this recharge, some is needed to support connected ecosystems (whether they be surface water bodies, or terrestrial systems such as wetlands). For good management, only that portion of the overall recharge not needed by the ecology can be abstracted – this is the sustainable resource, and the Directive limits abstraction to that quantity, as is the rule in integrated water resources management.

5.7.3 Waterworks Along the Rhine are Criticizing the EU Water Policy

The International Association of Waterworks in the river Rhine basin (IAWR) has three member organizations: AWBR, working in the Lake Constance/Rhine Area; ARW, the German member; and RIWA, the Dutch partner. About 120 waterworks from Austria, Switzerland, Liechtenstein, France, Germany, and Netherlands have joined. It is accepted as a non-government organizations (NGO) in the EU.

This international association has been complaining recently about the European Water Framework Directive and, in detail, about the European

Drinking Water Directive. The main point of their criticism is that in the WFD, ecological objectives are looked at instead of the more important drinking water aims. So, this directive did not take the needs of a water supply from river water into consideration. They claim that nature conservancy is not sufficient to meet the standards of water quality needed for human consumption. If, in case, the whole Rhine area is regarded as endangered, the directive did not recognize the actual condition and the big success of the river Rhine's rehabilitation in recent years. They demand equal criteria for ecological and drinking water utilization and they see no contradiction between ecology and utilization.

A "good ecological status", they claim, is unclear. It can be understood in the sense that the aim is a natural best status. This is the view of the water companies along the river. However, it can also be misunderstood as a threshold value, where the concentration of pollutants is just below a negative impact on the ecology of the river. Ecology-deduced concentrations are often higher than concentrations derived from drinking water safety. The water companies want water safe for drinking without artificial treatment, which might be necessary in case of too-high contaminant concentrations.

Not only contaminants should be considered, but also the microbiological threat. The IAWR also protests against losing the official monitoring programs when a good chemical status has been achieved. They argue that, in this case, early recognition of bad developments is impossible.

So the WFD is regarded as not perfect and has to be developed by the European officials, the NGOs, and the water companies.

6 Effects of Reduced Contamination on the Water Supply from the River Rhine

Direct Abstraction now possible

Direct abstraction of water for drinking water purposes from the Rhine was not possible in the years of heavy contamination. An example is the complicated and expensive, but efficient, waterworks combination in Wiesbaden-Schierstein. Here, two waterworks in line with soil passage were necessary to clean heavily contaminated river water. Now, after the improvement of river Rhine water quality, direct abstraction is possible with simple treatment. The question, however, is the necessary treatment in case of an accident in the river Rhine. Therefore, the decision has often been to take ground water for drinking water purposes instead of the vulnerable source of surface water, but under the concept of a sustainable water abstraction the better source should be river water than ground water.

7
Conclusion

Much progress has been made in water protection in Europe in individual Member States, but also in tackling significant problems at European level. We should continue to take up the challenge of water protection, one of the great challenges for the European Union. Let us seize the initiative generated by the present political process for the benefit of all Europe's citizens and waters, and tell this success story to other countries in the world that are still contaminating their environment.

8
Future Development

There are two models for further developments. One is described in terms of phases in pollution control and management, with interaction of natural sciences with the decision-making system, as depicted in Fig. 17.

In the pre-industrial phase the excess materials and physical impacts are low. With the start of industrial expansion the excess flow of materials increases as well as physical disturbances (phase I). The assimilative capacity of the system can, however, still cope. Once loading exceeds assimilative capacity we enter phase II, where damage impacts affecting the general public become more visible and environmental awareness increases because basic economic needs have been met. Action is taken in the form of end-of-pipe line solutions, subsequent actions include management in the form of new technology, and phase III is entered. Phase III has to deal with the mitigation of the effects of hard to control diffuse sources, such as tourism and population pressure. Phase III requires, more than any other phase, strong research and policy cooperation between the natural sciences and socio-economics in order to control the rates of change and start restoration efforts. Fast-developing countries in Asia are in phase I, "transition" economies in Eastern/Central Europe are in phase II, and the countries of Europe and North America are entering phase III.

In this scheme, the Rhine is in the middle of Phase III so it has become cleaner and control is through European Water Directives and the river basin management concept. We hope that this development remains into the future.

Another possibility might be a wave over time, with ups and downs in waste water management and river pollution. When the national income is decreasing and local governments are not able to control contamination, we cannot be sure that the achieved level of legislation and river quality can endure for long.

But, optimistically speaking, once a river in an industrialized country has recovered, the public will take care of its health. So, the river Rhine has its

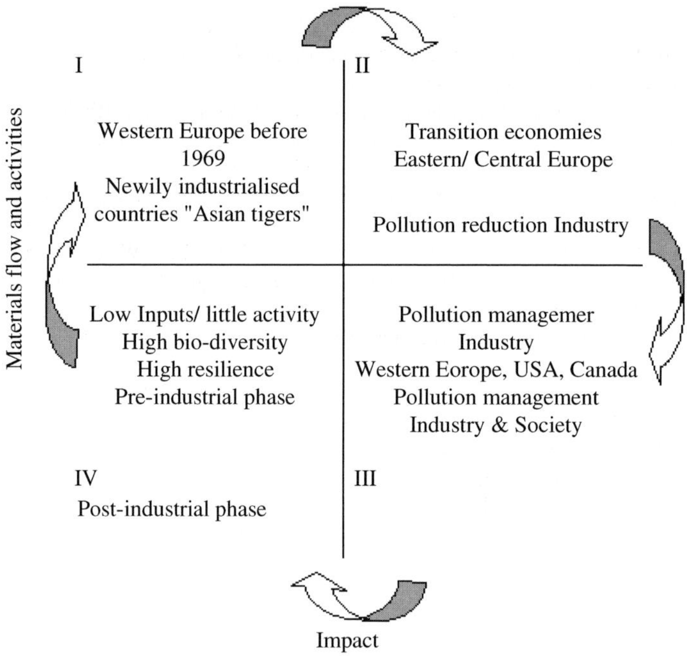

Fig. 17 Phases in pollution control and management with interaction of natural sciences with the decision-making system

success story, which is an example for all other contaminated rivers, and hopefully it will happen again in all contaminated rivers.

Acknowledgements I am very grateful for the help of my institute's staff in preparing the manuscript, especially Dipl.-Goegr. Martina Thum, who supported me both in compiling of contents but also for formatting the text. I gratefully acknowledge also the cooperation with Heike Weil and Thomas P. Knepper, who had to include my first draft into the contributions of the other authors of this book.

References

1. Cioc M (2002) The Rhine: an eco-biography, 1815–2000. Weyerhaeuser environmental books, ISBN 0-295-98254-3
2. Brockhaus FA (ed) (1998) Der Brockhaus in 15 Bänden. Leipzig, Mannheim
3. Leisering W (ed) (1993) Putzger Historischer Weltatlas. Berlin
4. www.route-industriekultur.de/geschi/lang/index.htm?ska_10.htm, 01.09.2004
5. Buchheim C (1994) Industrielle Revolutionen. München
6. Kiesewetter H (1989) Industrielle Revolution in Deutschland, 1815–1914. Frankfurt
7. Westermann (1992) Diercke Weltatlas. Braunschweig
8. Coleridge ST (1834) Cologne

9. Ministerium für Umwelt und Naturschutz, Landwirtschaft und Verbraucherschutz NRW (2000) Grundwasserbericht 2000 Nordrhein-Westfalen. Düsseldorf
10. Haberer K (1991) Geographische Rundschau 43:334
11. Haberer K (1974) Ergebnisse langjähriger Untersuchungen zur Rheinwasserqualität (Results of many years' analysis of Rhine water quality) Künstliche Grundwasseranreicherung am Rhein. Wissenschaftliche Berichte über Untersuchungen und Planungen der ESWE Stadtwerke Wiesbaden AG. Band 2. ESWE Stadtwerke Wiesbaden, p 17–42
12. EU Environment Directorate General (1998) The EU Drinking Water Directive, Council Directive 98/83/EC of 3 November 1998 on the quality of water intended for human consumption
13. DKR (2001) Deutsches Untersuchungsprogramm Rhein, 1976–2001. Landesamt für Wasserwirtschaft Rheinland-Pfalz
14. IKSR-ICPR: http://had.bafg.de:8080/iksr-zt/auswahl.asp?S=3, 25.04.2005
15. IKSR-ICPR 2000, Upstream, Outcome of the Rhine Action Programme, www.iksr.de
16. The EU – Water Framework Directive: DIRECTIVE 2000/60/EC OF THE EUROPEAN PARLIAMENT AND OF THE COUNCIL of 23 October 2000 establishing a framework for Community action in the field of water policy

Hdb Env Chem Vol. 5, Part L (2006): 89–97
DOI 10.1007/698_5_032

Published online: 9 December 2005

The Development of Water Protection Requirements from the Point of View of Waterworks, with Emphasis on 30 Years of the Memorandum of the International Association of Waterworks in the River Rhine Catchment Area (IAWR)

Klaus Lindner

RheinEnergie AG, Parkgürtel 24, 50823 Köln, Germany
k.lindner@rheinenergie.com

Abstract The International Association of Waterworks in the River Rhine Catchment Area (IAWR) defined water protection requirements for surface water that is used for drinking water supply. These requirements changed with the legal standards for drinking water quality and the scientific findings in water research. They refer to the aim of waterworks that the water used for drinking water supply should meet a quality that allows the application of natural treatment only.

Keywords Immission standards · Fixed bed bioreactor · Monitoring · Surface water memoranda · Water protection

Abbreviations

DVGW	German Technical and Scientific Association for Gas and Water
IAWR	International Association of Waterworks in the River Rhine Catchment Area
RIWA	The Dutch Association of River Waterworks
AWWR	Association of Waterworks along the River Ruhr
DOC	Dissolved organic carbon
AOS	Adsorbable organic sulphur
EDTA	Ethylenediaminetetraacetate

1
Overview on Standards for Drinking Water and Surface Water that is Used for Drinking Water Supply

Surface water that serves for drinking water supply has to meet standards that relate to those defined for drinking water. Changes in drinking water standards required the adjustment of immission standards for surface water.

The first threshold values for drinking water were defined as minimum and maximum values in 1957 in Germany. Accordingly, in 1962 the German Technical and Scientific Association for Gas and Water (DVGW) derived from these values qualification principles for surface water that can be used for drinking water supply. This technical standard W 151 was updated in 1975 and finally replaced by the technical standard W 251 in August 1996 [1].

The DVGW standard was the basis for the first IAWR memorandum in 1973 [2]. It defined immission standards for the River Rhine allowing the production of drinking water only by means of natural treatment. The steps of the natural treatment are bank filtration and slow sand filtration. The maximum allowed concentration in the surface water can be higher than the drinking water limit values, as microbiological degradation reduces the concentration during bank filtration. The surface water quality standards were defined as mean daily values instead of absolute limiting values, because short time peaks in the river are equalized during bank filtration. The IAWR standards not only covered parameters regulating drinking water, but also were extended to sum parameters that characterize river water pollution. In 1973 the quality of the River Rhine was far away from allowing it to serve as a drinking water resource where the application of natural treatment alone might be sufficient. Therefore, the IAWR introduced—similar to the DVGW standards—a second category of limiting values that considered further physical and chemical treatment in the waterworks and allowed surface water to be proportionally worse in quality.

In this respect, there is a difference from the European directive 75/440 [3] in the quality requirements for surface water being used for drinking water supply. This directive came into force on 16 June 1975 and is still valid. It differentiates three levels of quality requirements corresponding to applied treatment capacities of the waterworks: level 1 anticipates a simple physical treatment including disinfection, and thus asks for a very good surface water quality; level 2 refers to a normal physical and chemical treatment with disinfection; and level 3 requires physical and an upgraded chemical treatment to produce drinking water from surface water that only meets correspondingly poor quality standards for surface water. For each of the levels, the directive introduced both guiding and imperative values.

The first German drinking water directive came into force in 1975 [4], whereas the first European directive for drinking water (80/778), five years later in 1980 [5], introduced Europe-wide regulations for drinking water. This directive was revised in 1998 [6].

Accordingly, the IAWR memorandum was updated several times (in 1986 [7], 1995 [8] and 2003 [9]), and also took account of new experiences in river water pollution. Besides the IAWR, other waterworks developed their specific demands on river quality. The latest documents are dated from 2002 for the memorandum of the waterworks along the River Maas (RIWA-Maas) [10], and from 2005 for the standards required for the waterworks along the River Ruhr (AWWR).

2 Evaluation of Regulations and Memoranda

Table 1 gives the order of regulations and memoranda that have been developed. The grey columns indicate the standards for drinking water. By comparing these columns one can deduce the changes in drinking water standards within the last 30 years. The same applies to the changes in the IAWR memoranda and their differences from the memoranda for other river systems.

2.1 Changes in Drinking Water Directives

Of the 81 parameters listed in Table 1 there are only 48 that define drinking water quality. However, comparing the EU drinking water directives of 1980 and 1998 there remain only 13 parameters unchanged; 14 were removed from the list and 17 were added. In two cases, limiting values became less stringent: sodium and nitrite. Limiting values were tightened up for five parameters: nickel, arsenic, lead, antimony and polycyclic aromatics. The list of parameters in the EU directive increased significantly compared to the German directive from 1975.

2.2 Memoranda Stricter than Limit Values for Drinking Water

The quality standards of the memoranda faced a significant development: even though defined as mean daily values, some quality aims were stricter than the limiting values for drinking water. Chloride and sulphate are allowed in drinking water in concentrations of up to 250 mg L^{-1}, but the IAWR quality standard for surface water is 100 mg L^{-1}. The reason is that the IAWR also considered technical effects on the drinking water system, in addition

Table 1 Demands on the state of surface water from the viewpoint of drinking water supply

				IAWR 1973 daily mean values		TVO 1975 limiting values	discharge standards for surface waters that are used for drinking water supply; 75/440/EWG						EU drinking water directive 80/778/EWG		IAWR 1986 daily mean values		IAWR 1995 aiming 90 perzentiles	DVGW W251 1996 daily mean values		EU drinking water directive 98/83/EG	AWWR 2002 aiming values 90 percentiles	RIWA 2002 90 percentiles	IAWR 2003 limiting values
		Parameters		A	B		A1 G	A1 I	A2 G	A2 I	A3 G	A3 I	guiding	limiting	A	B		standard	minimum standard				
physical parameters	1	pH	-				6,5 - 8,5		5,5 - 9		5,5 - 9		6,5 - 8,5				6,5 - 8,5	6,5 - 8,5	5,5 - 9,0	6,5 - 9,5		7,5 - 8,5	7 - 9
	2	Colouring	$mg\ L^{-1}$ Pt-Skala	5	35		10	20	50	100			1	20	10	35				1)			
		Colouring (SAK 436nm)	m^{-1}														0,5	0,3	1				
	3	Turbidity	$mg\ L^{-1}\ SiO_2$										1	10									
	4	Odour factor		10	100		3		10		20						5	5	-	1)			
	5	Taste factor		5	35																		
	6	Temperature	°C				22	25	22	25	22	25	12	25			25	22	25			25	
	7	Suspended inorg. substances	$mg\ L^{-1}$				25											25	150				
		Suspended org. substances	$mg\ L^{-1}$ Pt-Skala	5	25													5	25				
	8	total content of substances in solution	$mg\ L^{-1}$	500	800													400	800				
	9	Evaporation residues	$mg\ L^{-1}$											1500									
	10	Conductivity	$\mu S\ cm^{-1}$	700	1000		1000		1000		1000		400		700	1000	700	500	1000	2500		500	700
Cations	11	Calcium	$mg\ L^{-1}\ Ca^{2+}$										100					100	-				
	12	Magnesium	$mg\ L^{-1}\ Mg^{2+}$										30	50				30	-				
	13	Ammonium	$mg\ L^{-1}\ NH_4^+$	0,2	1,5		0,05		1	1,5	2	4	0,05	0,5	0,2	1,5	0,3	0,2	0,4	0,5	0,6	0,2	0,3
	14	Sodium	$mg\ L^{-1}\ Na^+$										20	175	60	90	60	60	120	200		40	
	15	Potassium	$mg\ L^{-1}\ K^+$										10	12									
Anions	16	Chloride	$mg\ L^{-1}\ Cl^-$	100	200		200		200		200		25		100	150	100	100	200	250		75	100
	17	Sulfate	$mg\ L^{-1}\ SO_4^{2-}$	100	150	240	150	250	150	250	150	250	25	250	100	150	100	100	150	250		75	100
	18	Fluoride	$mg\ L^{-1}\ F^-$	1	1	1,5	0,7 - 1	1,5	0,7 - 1,7		0,7 - 1,7			0,7 - 1,5	1	1		1	1	1,5		1	
	19	Bromide	$\mu g\ L^{-1}\ Br^-$																			50	
	20	Bromate	$\mu g\ L^{-1}\ BrO_3^-$																	10		1	
	21	Chlorate	$\mu g\ L^{-1}\ ClO_3^-$															0,11	0,37			0,05	
	22	Phosphate	$mg\ L^{-1}\ P_2O_5$				0,4		0,7		0,7		0,4	5				0,15	0,5		0,5	0,23	
	23	Cyanide	$mg\ L^{-1}\ CN^-$	0,01	0,05	0,05		0,05		0,05		0,05		0,05	0,01	0,05	0,025	0,01	0,05	0,05		0,01	
	24	Nitrate	$mg\ L^{-1}\ NO_3^-$	25	25	90	25	50		50		50	25	50	25	25	25	25	40	50	25	25	25
	25	Nitrite	$mg\ L^{-1}\ NO_2^-$											0,1						0,5			
	26	Kjeldahl-nitrogen	$mg\ L^{-1}\ N$				1		2		3			1									
heavy metals	27	Aluminium	$mg\ L^{-1}\ Al$										0,05	0,2				0,1	0,5	0,2		0,25	
	28	Iron	$mg\ L^{-1}\ Fe$	0,1	1		0,1	0,3	1	2	1		0,05	0,2				0,2	1	0,2		0,1	
	29	Manganese	$mg\ L^{-1}\ Mn$	0,05	0,5		0,05		0,1		1		0,02	0,05				0,03	0,25	0,05		0,1	
	30	Copper	$mg\ L^{-1}\ Cu$	0,03	0,05		0,02	0,05	0,05		1		0,1		0,03	0,05		0,02	0,05	2		0,15	
	31	Zinc	$mg\ L^{-1}\ Zn$	0,5	1	2	0,5	3	1	5	1	5	0,1	1	0,5	1		0,1	0,3			0,1	
	32	Boron	$mg\ L^{-1}\ B$	1	1		1		1		1				1	1	0,2	0,5	1	1		0,1	0,2
	33	Beryllium	$mg\ L^{-1}\ Be$	0,0001	0,0002										0,0001	0,0002							
	34	Cobalt	$mg\ L^{-1}\ Co$	0,05	0,05																		
	35	Nickel	$mg\ L^{-1}\ Ni$	0,03	0,05									0,05	0,03	0,05	0,01	0,03	0,04	0,02	0,015		0,01
	36	Arsenic	$mg\ L^{-1}\ As$	0,01	0,05	0,04	0,01	0,05		0,05	0,05	0,1		0,05	0,01	0,05	0,005	0,005	0,01	0,01		0,002	0,005
	37	Cadmium	$mg\ L^{-1}\ Cd$	0,005	0,01	0,006	0,001	0,005	0,001	0,005	0,001	0,005		0,005	0,001	0,005	0,003	0,001	0,002	0,005	0,001	0,0005	0,003
	38	Chrome	$mg\ L^{-1}\ Cr$	0,03	0,05	0,05		0,05		0,05		0,05			0,03	0,05	0,025	0,03	0,05	0,05		0,002	0,025
	39	Lead	$mg\ L^{-1}\ Pb$	0,03	0,05	0,04		0,05		0,05		0,05		0,05	0,03	0,05	0,005	0,01	0,02	0,01		0,01	0,005
	40	Mercury	$mg\ L^{-1}\ Hg$	0,0005	0,001	0,004	0,0005	0,001	0,0005	0,001	0,0005	0,001		0,001	0,0005	0,001	0,0005	0,0005	0,001	0,001		0,0001	0,0005
	41	Silver	$mg\ L^{-1}\ Ag$											0,01									
	42	Barium	$mg\ L^{-1}\ Ba$	1	1			0,1		1		1	0,1		0,1	1	0,7						
	43	Antimony	$mg\ L^{-1}\ Sb$											0,01			0,002			0,005		0,0005	
	44	Selenium	$mg\ L^{-1}\ Se$	0,01	0,01	0,008		0,01		0,01		0,01		0,01	0,01	0,01	0,005	0,001	0,01	0,01		0,001	
Oxygen	45	Oxygen	$mg\ L^{-1}\ O_2$																				> 8
		Oxygen saturation	$\%\ O_2$				> 70		> 50		> 30						> 80	60	80			80	
		Oxygen deficit	$\%\ O_2$	20	40										20	40							

Table 1 (continued)

Group	No.	Parameters	Unit	IAWR 1973 daily mean values A	IAWR 1973 daily mean values B	TVO 1975 limiting values	discharge standards for surface waters that are used for drinking water supply; 75/440/EWG A1 G	A1 I	A2 G	A2 I	A3 G	A3 I	EU drinking water directive 80/778/EWG guiding	limiting	IAWR 1986 daily mean values A	IAWR 1986 B	IAWR 1995 aiming 90 perzentiles	DVGW W251 1996 daily mean values standard	minimum standard	EU drinking water directive 98/83/EG	AWWR 2002 aiming values 90 percentiles	RIWA 2002 90 percentiles	IAWR 2003 limiting values
Parameters of added groups	46	Coefficient of spectral adsorption	1/m														10						10
	47	Oxidation ($KMnO_4$)	mg L^{-1} O_2										2	5						5			
	48	Biochemical oxygen consumption (BSB5)	mg L^{-1} O_2				< 3		< 5		< 7							4	8				
	49	Chemical oxygen consumption (CSB)	mg L^{-1} O_2	10	20						30				10	20							
	50	Dissolved org. carbon (DOC)	mg L^{-1} C	4	8										4	8	3	4	8	[1]		4	3
	51	Adsorbable organic halogens (AOX)	µg L^{-1}	50	100										50	100	25	30	60			20	25
	52	Adsorbable organic sulphur compounds (AOS)	µg L^{-1}														80						80
organic substances	53	Organic halogens (IAWR 03 single)	µg L^{-1}												1	5	5	2	5			1	0,1
	54	Trihalomethane, total	µg L^{-1}														5	10	20	100			
	55	Tetrachloromethane	µg L^{-1}														1,5	1	1	100			
	56	Tetrachloroethane and Trichloroethane	µg L^{-1}																	10	1 1		
	57	Dichloroethane	µg L^{-1}																	3			
	58	Lipophile, viscous organic compounds of chloride	mg L^{-1} Cl	0,01	0,02																		
	59	Polycyclic aromatics	mg L^{-1}	0,0002	0,0003	0,00025		0,0002		0,0002		0,001		0,0002	0,0001	0,0002	0,0001	0,0001	0,0002	0,0001		0,00005	
	60	Benzo-[a]-pyrene	µg L^{-1}																	0,01			
	61	Aromatic Amines	µg L^{-1}	5	5										1	5						0,1	
	62	Hydrocarbons	mg L^{-1}	0,05	0,2			0,05		0,2	0,5	1		0,01	0,1	0,2		0,05	0,2				
	63	Fleeting aromatic hydrocarbons (FAK)	µg L^{-1}																			0,1	
	64	Benzol	µg L^{-1}																	1			
	65	Methylene blue active substances	mg L^{-1}				0,2		0,2		0,5		0,2										
	66	anionic tensides	mg L^{-1}	0,1	0,3										0,1	0,3		0,1	0,3				
	67	synthetic complexing agent (EDTA)	mg L^{-1}															0,05	0,01		0,01	0,05	0,05
	68	Nitrilo-triacetic acid (NTA)	mg L^{-1}															0,01	0,02		0,01		
	69	Substances repressing chlolinesterase	mg L^{-1}	0,03	0,05																		
	70	Endokrine effective substances	µg L^{-1}																				0,1
	71	Pharmaceutics (incl. Antibiotics)	µg L^{-1}																				0,1
	72	Persistent pollutants	µg L^{-1}																				1
	73	Pesticides (Single Subst.)	µg L^{-1}											0,1	0,1	0,5	0,05	0,1	0,1	0,1	0,05	0,05	0,1
		Pesticides total	µg L^{-1}					1		2,5		5		0,5						0,5			
	74	Phenols	mg L^{-1} C_6H_5OH	0,005	0,01			0,001	0,001	0,005	0,01	0,1		0,5	0,001	0,005			0,001				
	75	Vinyl chloride	µg L^{-1}																	0,5			
	76	Epichlorohydrin	µg L^{-1}																	0,1			
	77	Acrylic amide	µg L^{-1}																	0,1			
	78	Chlorophyll	µg L^{-1}																			25	
Radioac	79	Rest of beta radioactivity	Bq L^{-1}																			0,1	
	80	Total alpha radioactivity	Bq L^{-1}																			0,1	
	81	Tritium	Bq L^{-1}																	100		20	

I = imperative value
G = guiding value
A = intromission standards that allow the production of drinking water by natural treatment steps only.
B = surface water quality standards considering further physical and chemical treatment in waterworks

[1] without unexpected changes

A1 = anticipates simple physical treatment including disinfection
A2 = refers to a normal physical and chemical treatment with disinfection
A3 = physical and upgraded chemical treatment is needed

to toxicological criteria. An increase in salt content can support corrosion in the distribution network causing secondary impairments of drinking water. For other parameters like heavy metals, the IAWR standards focused on the ban of quality deterioration and took the actual river quality as the desired standard. The general ban does not accept a deterioration of the river quality even if the immission is below the drinking water standards. This is one of the reasons why memoranda for different river systems differ from each other.

2.3 The Character of the Standards

There are only five parameters that have kept the same values throughout the development of the four memoranda: conductivity, chloride, sulphate, nitrate and mercury. Despite the constant figures their meaning has changed a great deal, because mean daily concentrations changed in 1995 to 90 percentiles and in 2003 the IAWR standards became absolute limiting values. The reasons for changing to 90 percentiles were on the one hand the increasing practice of comparing annual mean values with the memorandum standards, and on the other hand state agencies had switched to the 90 percentile evaluation of their official monitoring results. In order to keep pace with the official statements on the river quality development, the waterworks along the Rhine also decided to change to the percentile evaluation [11]. The 90 percentile defines the fact that the value in a row of monitoring results is not exceeded by 90% of the measurements. Defining a quality aim by means of 90 percentiles allows 10% of the monitoring measures to fail the aim. The correlation between a strict limiting value and the 90 percentile is best described by taking half the limiting value as the new 90 percentile to guarantee that the allowed exceeding monitoring results keep within the range of the original limiting value. Accordingly, all memorandum values should have been given half the original figure when the memorandum changed to 90 percentiles. This was actually done for many of the parameters: boron, nickel, arsenic, chromium, lead, selenium, pesticides, dissolved organic carbon and absorbable organic halogens. For other parameters the derivation rule resulted in higher values: ammonium, cadmium, cyanide, barium, organic halogens. Another reason won more supporters for the 90 percentiles: after the Sandoz accident in 1986, waterworks started a research project to investigate the processes of bank filtration. The results proved that peak concentrations in the river need several days before they arrive at a bank filtration well, and that peak concentrations in the well are reduced for at least 95% of the peak in the river. Thus, waterworks abstracting bank infiltrated water are actually more dependent on long-term quality development than on short-time accidents. In this respect percentile evaluations are quite helpful.

2.4
The In-Between Aims

Another significant change in the memoranda was discarding those values that IAWR referred to for 22 years as originally tolerated in-between aims. These so-called B values defined a surface water quality that asked for highly sophisticated techniques to produce drinking water from that surface water.

2.5
The New Memorandum 2003

Why did the IAWR introduce a new memorandum 8 years later? Up to 2000, peak concentrations for the pesticides diuron and isoproturon were repeatedly discovered in the river. The concentrations forced the Dutch waterworks to stop the intake of Rhine water. The official monitoring programs, good for describing long-term quality development, did not register that peak pollution. This was one reason for changing back to limiting values in the memorandum. However, the halving of the values for the standards resulting from the percentiles was not taken back in every case. A special need for the new memorandum resulted from the required input of the waterworks for participating in the water framework directive of the EU. New, problematic pollutants were added to the memorandum: endocrine active substances and pharmaceuticals. This is the first time that a risk acceptance of 0.1 $\mu g\,L^{-1}$ has been documented for such substances in the Rhine. Moreover, all potential pollutants are covered by the respective sum parameters in the memorandum.

2.6
Summarized Requirements on Surface Water Quality

As a conclusion it can be stated that the requirements on surface water quality used for drinking water supply should be derived from the drinking water standards. In addition, the quality should not affect the drinking water supply systems. Also, those substances that are not regulated in directives but may pollute drinking water should be limited on a precautionary level. Since not every single substance in surface water can be monitored, sum parameters have to be added to the monitoring programmes. Concerning the quality aims, there are several variations of limiting values, from daily mean values to percentiles. Whatever is chosen has consequences for the monitoring strategy and legal input in the transposition of the EU water framework directive. All the variations correspond when describing the same water quality, that is, being in a constantly good state—and that is the aim of the waterworks.

2.7 Reality and Quality Aims

So far, the required quality standards of the waterworks have not been reached in the Rhine at Cologne, in 2003 for the following parameters: oxygen content, conductivity, chloride, lead, DOC, AOS, pesticides, some pharmaceuticals, X-ray contrast compounds and EDTA [12]. Even though this does not endanger the drinking water supply along the River Rhine, there is a need for action to reduce pollution that exceeds the limiting values. Waterworks feel that this demand is in accordance with the aim of the EU water framework directive, which asks for a good ecological quality of the water bodies. Man-made substances that cannot be degraded biologically should not enter the water cycle, because otherwise they would endanger the ecological quality.

3 Research with Fixed-Bed Bioreactors

A very successful cooperation developed with major chemical industries along the Rhine. In several joint research projects, waterworks and the chemical industry [13–15] developed and standardized fixed-bed bioreactors that simulate biological degradation and thus enable the detection of pollutants in industrial wastewater, which eventually can become relevant for drinking water by passing through all the treatment steps. The latest research project allowed knowledge to be gained about biological degradation effects for micro-level concentrations of pollutants, as they are present in the environment. One result is that there are minimum concentrations for individual substances below which biological degradation cannot start, whereas beyond such concentrations degradation starts and continues, even when the concentration falls below that threshold value. Another experience is that adaptation of microorganisms in the natural water cycle supports the degradation potential of a substance. Due to the cooperation of laboratories of waterworks, which are used for developing analytical means for the detection of micro concentrations, and the chemical industries, which have knowledge of their wastewater processes, it was possible to find drinking water relevant substances! It has to be gratefully acknowledged that the industry prevented detected substances from entering the water cycle either by stopping production or by applying changes to production processes.

4
Input to Water Safety Plans

Waterworks care greatly about the protection of water resources, even though water resources management is the responsibility of the state authorities. But the best guarantee for safe drinking water is good raw water. Therefore waterworks are the pressure group that pushes the stakeholders of the water resources for effective water protection, and they participate themselves in monitoring critical control points, in terms of water safety plans for sources of old and new pollution. Moreover, waterworks are involved in cooperation with farmers, in order to keep agricultural non-point sources of pollution under control. This is far more than is required by the modern approach of water safety plans, which only tries to bring processes under control. This cooperation based on the coexistence of water users is the input of waterworks to water resources management that will carry on into the future.

References

1. DVGW (1996) Eignung von Fließgewässern für die Trinkwasserversorgung. DVGW Regelwerk, Technische Mitteilung, Merkblatt W 251, Bonn, p 43
2. IAWR (1973) Pollution of the Rhine and the abstraction of drinking water; IAWR-Memorandum, Amsterdam, p 24
3. European Directive on discharge standards for surface waters that are used for drinking water supply (1975) 75/440/EWG
4. Verordnung über Trinkwasser und über Brauchwasser für Lebensmittelbetriebe (Trinkwasser-Verordnung) (1975). In: Bundesgesetzblatt Nr 16, Bonn, pp 453–461
5. European Directive on the quality of water for human consumption (1980) 80/778/EWG. In: Amtsblatt der Europäischen Gemeinschaften Nr L 229, pp 11–29
6. European Directive on the quality of water for human consumption (1998) 98/83/EG
7. IAWR (1986) Rhine Memorandum. IAWR, Amsterdam, p 24
8. IAWR (1995) Rhine Memorandum. IAWR, Amsterdam, p 20
9. IAWR (2003) Rhine Memorandum. IAWR, Amsterdam, p 24
10. RIWA (2002) Maas Memorandum. RIWA-Maas/Meuse, Werkendam/Bruxelles, p 17
11. Lindner K, Brauch H-J, Fleig M (1995) Das IAWR-Memorandum 1995. In: Jahresbericht 1994 der ARW. ARW, Köln, pp 93–121
12. ARW (2004) Bericht 2003, Karlsruhe, p 42
13. Lindner K, Knepper TP, Karrenbrock F, Rörden O, Brauch H-J, Lange FT, Sacher F, (1996) Erfassung und Identifizierung von trinkwassergängigen Einzelsubstanzen in Abwässern und im Rhein. In: IAWR Rhein-Themen 1, Köln, p 180
14. Lindner K, Knepper TP, Müller J, Karrenbrock F, Rörden O, Brauch H-J, Sacher F (2000) Entwicklung von Verfahren zur Bestimmung und Beurteilung der Trinkwassergängigkeit von organischen Einzelstoffen. In: IAWR Rhein-Themen 3, Köln, p 104
15. Lindner K, Knepper TP, Müller J, Karrenbrock F, Rörden O, Juchem H, Brauch H-J, Sacher F (2003) Bestimmung und Beurteilung der mikrobiellen Abbaubarkeit von organischen Einzelstoffen bei umweltrelevanten Konzentrationen in Gewässern. In: IAWR Rhein-Themen 5, Köln, p 114

Hdb Env Chem Vol. 5, Part L (2006): 99–124
DOI 10.1007/698_5_015

Published online: 25 October 2005

Early Warning Strategies and Practices Along the River Rhine

Peter Diehl[1] (✉) · Thomas Gerke[2] · Ad Jeuken[3] · Jaqueline Lowis[2] · Ruud Steen[4] · Jaap van Steenwijk[3] · Peter Stoks[5] · Hans-Günter Willemsen[2]

[1]Rhine Water Control Station Worms, State Environment Agency Rhineland-Palatinate, Am Rhein 1, 67547 Worms, Germany
peter.diehl@luwg.rlp.de

[2]State Environment Agency North Rhine-Westphalia, P.O.B. 102363, 45023 Essen, Germany
thomas.gerke@lua.nrw.de, jaqueline.lowis@lua.nrw.de, hans-guenter.willemsen@lua.nrw.de

[3]Institute for Inland Water Management and Waste Water Treatment (RIZA), P.O.B. 17, 8200 AA Lelystad, The Netherlands
a.jeuken@riza.rws.minvenw.nl, j.steenwijk@riza.rws.minvenw.nl

[4]Het Waterlaboratorium, P.O.B. 734, 2003 RS Haarlem, The Netherlands
ruud.steen@hetwaterlaboratorium.nl

[5]Association of Rhine Water Works RIWA, P.O.B. 402, 3430 AK Nieuwegein, The Netherlands
stoks@riwa.org

Abstract In 1986, the International Commission for the Protection of the Rhine (ICPR) installed a Warning and Alarm Plan (WAP) in order to avert the dangers and to detect and investigate the causes of incidental pollution (spills, industrial or shipping accidents). Since then, the number of reported incidents has decreased considerably. The WAP approach is both emission- and immission-based. Within the former, incidents are directly reported by the discharger; within the latter, early warning monitoring systems provide (semi-)continuous measurements of water quality from which unreported spills can be deduced and traced. Both the water authorities and drinking water companies operate such systems, in which various chemical screening techniques are used in combination with biological early warning systems. This two-way strategy, using sensitive equipment, is considered to be best for a river such as the Rhine, as it has a large flow and there is an endless spectrum of chemicals potentially being used and produced within its catchment area. Future improvements in the WAP will focus on information exchange and the use of internationally available "expert" knowledge, the embedding of biological alarms, and a new system of guidance values. The future challenges for early warning monitoring systems include a broadening of analytical and effect windows, and standardization with a better understanding of relations between substance and effect. In a good early warning system, a balance should be sought between effectiveness for the specific river situations, manageability of the WAP-type system, robustness, and cost. Along the Rhine this message is well understood.

Keywords Biomonitoring · Early warning · International cooperation · Screening of organic micropollutants · Warning and Alarm Plan Rhine

Abbreviations

BEWS	Biological early warning systems
BTEX	Benzene, toluene, ethylbenzene, xylene
CAS	Chemical Abstracts Service
EOA	Emission-oriented approach
FID	Flame ionization detector
GC	Gas chromatography
HPLC	High-performance liquid chromatography
ICPR	International Commission for the Protection of the Rhine
IMBL	International Monitoring Station Bimmen-Lobith
IMWC	International Main Warning Center
IOA	Immission-oriented approach
LC-DAD-UV	Liquid chromatography with diode-array detector and UV detector
MS	Mass spectrometry
NRW	North Rhine-Westphalia
PAH	Polycyclic aromatic hydrocarbon
PCB	Polychlorinated biphenyl
SPE	Solid-phase extraction
UV	Ultraviolet extinction/detector
WAP	Warning and Alarm Plan Rhine
WFD	Water Framework Directive

1 Introduction

The Rhine is one of the major rivers in Europe. Its course of 1320 km takes it through six different countries. The entire catchment area of 185 000 km^2 is home to approximately 50 million people. Previously, the riverside was characterized by fishermen's huts and barges. Today these have given way to towns and cities, industry, and cargo vessels. Fifty percent of all Europe's chemical industry is now located on the Rhine. At the same time, the Rhine is a source of re-treated drinking water for approximately 20 million people, 5.5 million of them getting their drinking water directly from the flowing river. About 11 000 ships carry approximately 175 million tons of goods every year, many of them loaded with hazardous chemical substances. For the protection of the river's ecology and water quality, which means protection of the people living by and from the river, intensive international cooperation is essential.

The problems of water body protection become apparent by looking at the many ways in which the Rhine is used: as a waterway, for leisure and recreation, as a habitat for plants and animals, as a reservoir of drinking and cooling water, for hydropower, and for the discharge of wastewater. The intensive use of the river has almost led to its biological death. In the mid-1970s certain sections were considered highly polluted, but various efforts helped it to recover by and large. Today it is the habitat of many different forms of life. The improvements were only possible with intensive international cooperation, and continuing cooperation in the future is an essential condition for the implementation of the European Directive 2000/60/EC. This European Water Framework Directive (WFD) demands of the member states that "basic measures" shall consist of "...any measures required ... to prevent and/or to reduce the impact of accidental pollution incidents ... including through systems to detect or give warning of such events..." (Article 11, paragraph 3 (l)) [1].

The continuous chemical and biological monitoring at the River Rhine dates back to the 1960s and 1970s, beginning with conventional techniques for obtaining water samples to monitor trends through collected data. The major reasons for improvements in river monitoring in the Rhine basin are the establishment of an international monitoring network along the river and its tributaries, improvements in the regulatory policies of regional and national governments, and strict regulations on industrial and municipal wastewater emissions. Consequently, from the 1960s until the 1980s an off-line and online monitoring network was established along the Rhine, as had been the case along most of the major river systems in Europe. Today the Rhine is investigated using sophisticated technologies, extending from detailed chemical investigations of water (more than 300 parameters are routinely analyzed) to sediment, biota, and comprehensive biological surveys [2].

In addition to the water monitoring programs of the state authorities, many of the users of Rhine water run their own monitoring programs. This was especially true of the waterworks, which developed a sophisticated monitoring program in order to get an overview of the common loads as well as information about compounds not commonly analyzed in the routine programs of the authorities (e.g., [3]).

A particular strain on water quality stems from industrial companies and from the high level of shipping activity. Industrial incidents or shipping accidents can lead to spills of substances that are harmful to the water. If the contamination is detected at an early stage, the alarm can be triggered and measures can be taken in order to protect the drinking water supply and the ecosystem of the Rhine. The earlier that information about the source of the spill is available, the sooner efforts to stop the discharge can be successfully implemented.

There was a major chemical incident in Basel, Switzerland, on 1 November 1986 [4], when the River Rhine was polluted due to a fire in an industrial storage building. This gave serious impetus to the development and installation of early warning systems, especially in Germany and the Netherlands [2, 5–11]. During this accident, pesticides were released into the Rhine in large quantities (nearly 30 tons) along with fire water, which resulted in serious damage of a large portion of the river biotope. Consequently, fish suffered very badly; the eel population (specialized benthos feeder) was completely eradicated in the upper Rhine. Even 500 km down, in North Rhine-Westphalia (NRW), the dynamic daphnia tests in operation registered an "alarm" situation. These experiences gave a signal to the environmental protection organizations on the importance of biological early warning systems (BEWS) [2].

Chemical monitoring was introduced in the Rhine river basin during the last three decades, with varying degrees of sophistication. The earliest monitoring was limited to easily measurable physicochemical parameters such as temperature, pH, conductivity, and oxygen. Today, it has evolved into automated laboratories that measure a large number of parameters, including organics [9], which need elaborate sample preparation procedures. These procedures are achieved through the development of efficient online methodologies for screening of a large number of organic micropollutants, which include halogenated hydrocarbons, herbicides and pesticides, nitroaromatics, phosphoric esters, and other priority substances (screening with online techniques using GC, GC–MS, HPLC–UV, and HPLC–MS/MS). The results are obtained after a few hours of laboratory analysis (see [5, 12]) in time to inform the downstream water supplies. Since the early 1970s, the chemical monitoring of the Rhine has been supplemented with continuous biomonitoring; the organisms used at the beginning were fish [7, 13]. Later a number of biomonitors were deployed for continuous monitoring, either in flow-through systems or in situ (e.g., daphnids, algae, mussels, fish, and photobacteria) [6]. Biomonitoring, as an alternative to chemical tests, is capable of rapidly detecting acutely toxic conditions in river water [2, 11].

Whether it is the information by the discharger (emission principle) or the detection of spills by monitoring techniques (immission principle), protecting the river's ecology and the drinking water supplies will only be successful if any unusual findings are reported right away under a multitier alarm system, and notified through the International Warning and Alarm Plan for the Rhine, whenever required. This article will introduce the strategies for early warning at the River Rhine, the main techniques used for early warning, and experiences with these techniques, including statements, problems, and constraints.

2 The International Warning and Alarm Plan Rhine (WAP)

In spite of all the action intended to prevent polluted water from flowing into the River Rhine, this may well still happen. If and when it does, time is of the essence to address the pollution. Above all, those concerned downstream must be informed, especially the authorities responsible for the struggle against pollution and the waterworks that take Rhine water for drinking water supplies. Therefore, the International Commission for the Protection of the Rhine (ICPR) has installed the International Warning and Alarm Plan "Rhine" (WAP Rhine). The first version was officially introduced in 1986; the recent version dates from 2003 [14]. The goals of the WAP Rhine are: averting danger; detection of causes; investigation into the cause; and taking measures to remove damage, to avoid or reduce damage, and to avoid consequential damage. In addition, damage events that are expected to lead to higher public interest should be passed on.

Between Basel and the German–Dutch frontier, seven International Main Warning Centers (IMWCs) "share" the Rhine (Table 1, Fig. 1). Two more such warning centers are located on the main tributary Moselle, forming the spine of the Warning and Alarm Plan "Moselle".

Each IMWC is responsible for a certain part of the Rhine or for the tributaries. The IMWC on whose territory the spill has happened sends a

Table 1 International Main Warning Centers according to the International Warning and Alarm Plan Rhine (2003)

Country	Name	Operator
Switzerland	R1	Office for Environment and Energy (Amt für Umwelt und Energie) of Canton Basel-Stadt
France	R2	Prefecture for the Departement Bas-Rhin (Préfecture du Bas-Rhin), Strasbourg
Baden-Württemberg, Germany	R3	State Water Guard Head Office (Wasserschutzpolizeidirektion), Mannheim
Hessen, Germany	R4	Water Guard Station (Wasserschutzpolizeistation) Wiesbaden
Rheinland-Pfalz, Germany	R5	Water Guard Station (Wasserschutzpolizeistation) Koblenz
Nordrhein-Westfalen, Germany	R6	District Government (Bezirksregierung), Düsseldorf
The Netherlands	R7	Water Authority Head Office (Rijkswaterstaat directie), Oost-Nederland, Arnhem
	S	Secretariat of the ICPR

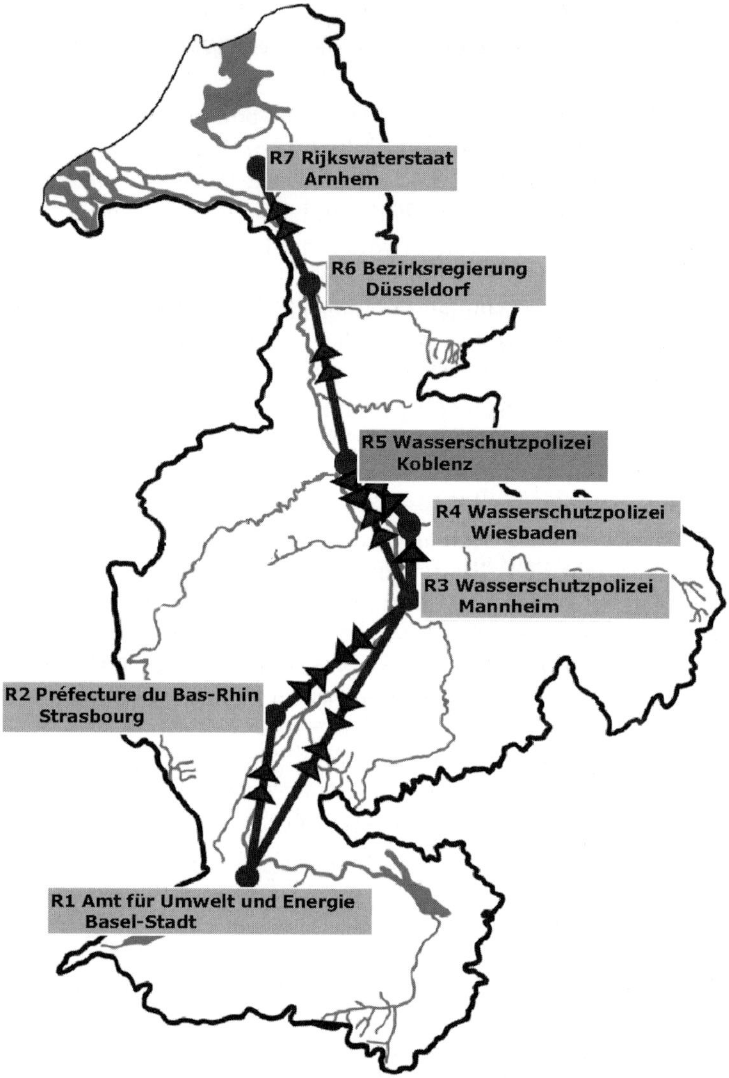

Fig. 1 Location of International Main Warning Centers and directions of communication according to the International Warning and Alarm Plan Rhine (2003)

"first report" to all centers downstream as well as to the ICPR secretariat in Koblenz.

The WAP Rhine [14] regulates in detail the procedures of reporting: language, forms, technical means, time windows, distribution, addresses and telephone/fax numbers, definitions of the "all clear" signal, guidance values for loads and concentrations, regulations for special cases, and so forth. In the following paragraphs the most important regulations are described.

Language: The WAP Rhine consistently uses the three main languages of the river basin area in parallel: French, German, and Dutch. All forms are available in the three languages; most of them are in fact trilingual.

Forms: The reports have to be written by using special trilingual forms, which are also available on an electronic basis. The paragraphs of the main report form are numerically coded so that—independently from a knowledge of the language—it is possible to add all relevant information to the form at the correct place.

Staff and material: The WAP Rhine demands of the IMWCs that they are manned 24 hours a day, and that they are well-equipped with handbooks, databases of hazardous goods, and lists of substance-identifying numbers (CAS), as well as with the computer-based flow–time model, the Rhine Alarm Model [15].

Protocol: For each alarm a chronological diary has to be kept.

Cause of triggering an alarm: In order to facilitate the decision as to whether or not an alarm should be triggered, the ICPR defined "guidance values" for loads and concentrations for several groups of hazardous substances (Table 2). The guidance values are related to discharges of compounds that are not allowed and are caused by accidents, misperformances, or illegal acts. Normally, the authorities get information on such spills from the dischargers themselves. In those cases, the guidance values for loads (central column in Table 2) allow a rapid estimatation of the importance of triggering an alarm or not (emission-oriented approach, EOA). Sometimes, however, spills are detected only by chemical monitoring techniques. In these cases the guidance values for an increase of concentration in a 24-h sample at the International Monitoring Station Bimmen-Lobith (IMBL, reference measuring site at the German–Dutch frontier) are used to estimate the implications (immission-oriented approach, IOA). Starting from measured concentrations it is possible to calculate loads. When the guidance values are exceeded, an "information", the lower of the two levels of alert, is regularly triggered. Additional information about the danger resulting from the spill, or other public interest, may lead to the "warning" level.

The number of WAP alarms caused by the chemical industry has decreased within the last 15 years (1989, $n = 33$; 2002, $n = 3$; Fig. 2), whereas the spills caused by ships have remained more or less constant (1989, $n = 6$; 2002, $n = 4$; Fig. 2). Therefore the IOA, i.e., the network of monitoring stations, has become more and more important.

Between Lake Constance and the Rhine estuary, a chain of early warning monitoring stations is equipped with sampling equipment, and with modern online monitoring and analyzing systems (Fig. 3). Some of the stations are even able to distinguish between spills originating from different sources across the river section, which allows a much quicker assignment of the origin. The reason for this lies in the fact that lateral inflow disperses only slowly over the cross section due to laminar flow.

Table 2 Guidance values of the International Warning and Alarm Plan Rhine

Substance	Guidance values Daily load kg	Resulting increase in concentration measured in daily composite water samples at Lobith water monitoring station $\mu g\ L^{-1}$
Arsenic	500	5
Beryllium	100	1
Cadmium	300	3
Organic micropollutants (individual substances)	300	3
PAH (individual substances)	50	0.5
PCB (individual substances)	10	0.1
Pesticides (individual substances)	50	0.5
Mercury	100	1
Selenium	500	5
Cyanide	500	5
The following applies to radioactivity		
Parameter	Activity GBq	 $Bq\ L^{-1}$
Total alpha	20	0.2
Total beta	200	2.0
Tritium	10 000	100

Information transfer: The reports on special forms are regularly sent as faxes to those concerned. If the location of the spill is known, the report is sent to all IMWCs downstream of this location. If the location is unknown, in addition the upstream IMWCs are asked to help in investigating the source (searching report). Additional regulations according to the involvement of the ICPR secretariat, the procedures of answering a report, the procedures of telephone communication, and special regulations for the complicated circumstances at the upper Rhine, are also part of the WAP Rhine documents. Finally, the WAP Rhine gives advice on triggering the "all clear" signal.

Figure 4 summarizes the WAP Rhine as a schematic overview. It makes clear that the WAP Rhine is based on two main pillars: the emission-oriented

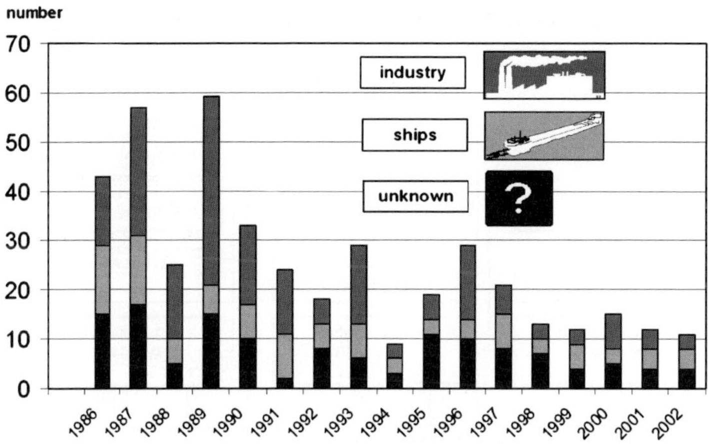

Fig. 2 Number of reports according to the International Warning and Alarm Plan Rhine

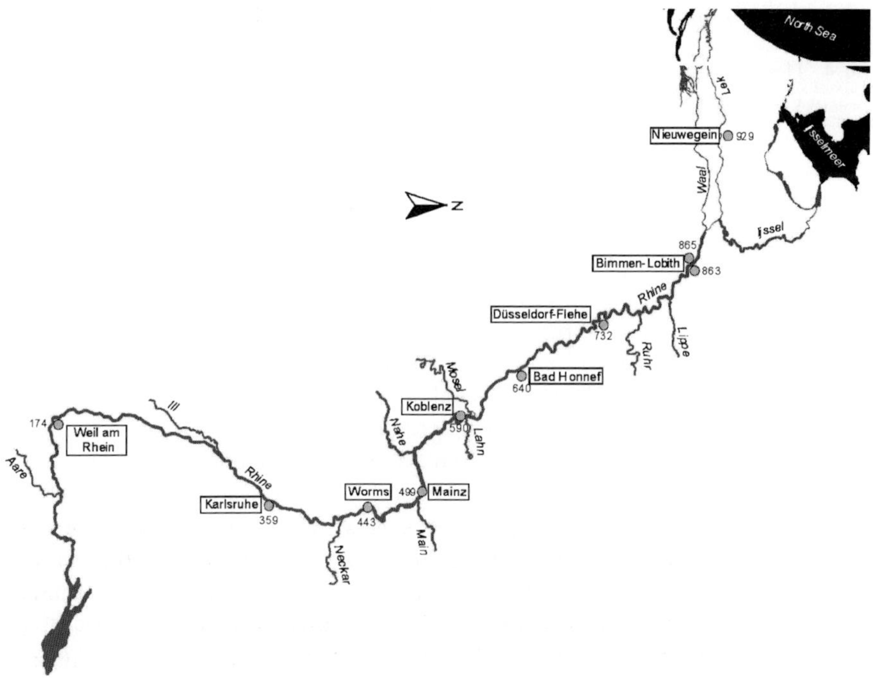

Fig. 3 Monitoring stations along the River Rhine equipped for early warning

approach (EOA, first report is given by the emittent), and the immission-oriented approach (IOA, first report is given by laboratories or water monitoring stations). The guidance values are well defined for the EOA and the chemical monitoring aspects of the IOA. The biological monitoring aspects,

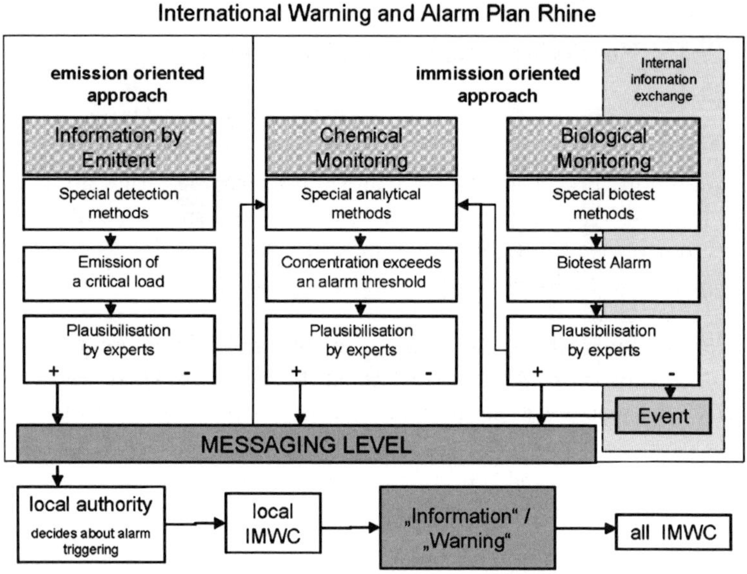

Fig. 4 Flow chart of decisions leading to activation of the International Warning and Alarm Plan Rhine

however, are still not as well integrated into the WAP Rhine, although first assumptions have been made [16]. Biotest alarms should only be taken into account if they are confirmed by experts to be plausible, as well as by indications from chemical–physical results.

The functionality of the WAP Rhine is tested every year with an international practice alarm. The results always show that the system works well, but also show that there is still a need for improvement. One has to regard what the European Directive 2000/60/EC [1] demands of the member states in Article 11 (program and measures), paragraph 3 (l) (see also the Introduction), which means that early warning systems should be further developed.

The recently developed computer-generated alarm system, which contains a flow–time model of the river, will be improved further, so that the relevant authorities can predict more and more rapidly and reliably the passage of a wave of pollution in the Rhine and foresee the expected contaminant concentration. Preliminary ideas to modify the guidance values so that their hazardous potential is taken into better account were discussed [17]. Finally, the World Wide Web may give the opportunity to exchange information in a quicker and more reliable way. Here, the first steps to introduce such a web-based application within the WAP are made. Comparable developments in other river basin areas (e.g., Elbe, Danube) are observed.

3
Early Warning Monitoring: Practices and Experiences

In the previous section the International Warning and Alarm Plan for the Rhine and its underlying principles were introduced. In this section the early warning methods practiced at the various monitoring stations along the river are highlighted. The advantages and constraints of the methods are given. Methods in use with both governmental authorities and drinking water companies are presented together with practical experiences, and are illustrated by real alarm cases.

The monitoring stations along the Rhine use multiple techniques (Table 3), from online sensors that measure temperature, pH, oxygen, and electrical conductivity to trace-level analysis of organic compounds by chromatography

Table 3 Physical–chemical and biological early warning systems in use along the Rhine

Location	Physical–chemical Online sensors	Volatile	GC–MS	LC–UV	LC–MS
Weil am Rhein	x	x	x	x	x
Karlsruhe	x				
Worms	x		x		
Mainz	x				
Koblenz	x				
Bad Honnef	x	x	x	x	
Düsseldorf	x	x	x		
Bimmen/Lobith	x	x	x	x	x
Nieuwegein	x	x	x	x	x

Location	Biological Fish	Daphnia	Mussels	Algae	Bacteria
Weil am Rhein					
Karlsruhe	x	x		x	x
Worms		x		x	
Mainz		x			
Koblenz				x	
Bad Honnef		x	x		
Düsseldorf		x			
Bimmen/Lobith		x	x		
Nieuwegein	x	x		x	x

and mass spectrometry (Sect. 3.1) and to an effect-oriented approach using continuous biotest systems (Sect. 3.2). Explanations as to why these systems are thought to be the most effective for early warning along the Rhine will be provided.

3.1 Chemical Monitoring and Backtracing

The essential task of the monitoring stations is the detection of extraordinary loads of pollutants. Additional amounts of organic micropollutants (50 to 300 kg, Table 2) cause short-term elevated concentrations in the $\mu g\ L^{-1}$ or $ng\ L^{-1}$ range (see Table 2). They should be detected in order to alert users of Rhine water (such as waterworks) and to trace back compounds to their sources. Due to the large discharge of the River Rhine (1000–10 000 $m^3\ s^{-1}$), these small amounts are only detectable with specific, sensitive, chemical analytical methods at high sampling frequencies.

Gas chromatography (GC) and high-performance liquid chromatography (HPLC) are used for the separation of organic compounds, and usually mass spectrometry (MS) or spectral absorption (UV) is used for detection (Table 4). The results are interpreted by experts, typically daily. Whenever the findings exceed the alarm thresholds, a message is sent to the IMWCs, even at weekends. Information about conspicuous concentrations below the alarm thresholds is exchanged between the experts of the monitoring stations along the river.

Many substances (e.g., PAHs, PCBs) are adsorbed onto solid materials (suspended solids, sediments). Filtered water is usually analyzed at the monitoring stations. Sample preparation of solids is too time-consuming for early warning purposes. Furthermore, solids are readily eliminated prior to drinking water production. However, compounds attached to these solids could be

Table 4 Screening methods for the detection of organic micropollutants

Targets	Highly volatile compounds	Nonpolar and medium-polar evaporable compounds	Nonpolar and medium-polar compounds
Sample preparation	Headspace or purge-and-trap	RP-SPE continuous adsorption, desorption in the lab	RP-SPE
Separation	GC	GC	HPLC
Detection	MS, FID, ECD	MS	UV, MS/MS
Calibrated compounds	BTEX, chlorinated hydrocarbons, MTBE	Industrial organics, nitrogen, phosphorus, and halogenated compounds, pesticides	Pesticides

an indication of forbidden emissions. This is partly compensated by continuous biotests (see Sect. 3.2).

A common approach to the detection of organics is solid-phase extraction (SPE) followed by full-scan GC–MS of nonpolar and medium-polar compounds, which can be extracted from water by reversed-phase material and evaporated without disintegration. Most of the monitoring stations use GC–MS systems for the detection of nonpolar and medium-polar evaporable compounds. The adsorption step of SPE can be used to prepare composite samples (e.g., of 12, 24 hours or more) directly from the pipes of the monitoring stations, in closed systems to prevent contamination and deficits due to evaporation.

Full-scan mass spectrometry in GC–MS accomplishes simultaneous quantification of calibrated compounds and detection, or even the identification of unknown substances. A rough estimation of the concentrations of these compounds is possible with regard to internal standards, assuming similar efficiency of sample preparation and ionization. If pure material is available early, mass spectral identification is confirmed by matching retention times. Subsequent calibration by applying the complete process of sample preparation and analysis accomplishes quantitation of the total load.

At some monitoring stations, special instruments employing headspace or purge-and-trap GC with FID or MS detection are used for the analysis of highly volatile compounds like benzene, toluene, ethylbenzene, xylenes (BTEX), and halogenated hydrocarbons. Any contact of the samples with the laboratory's atmosphere has to be avoided because of potential contamination and evaporation of substances. Usually spot samples are analyzed once a day or more (Bad Honnef, Düsseldorf). The International Monitoring Station Bimmen-Lobith (IMBL) at the German–Dutch border is equipped with a gas-tight online purge-and-trap GC–MS system for the continuous monitoring of volatiles. The station at Weil am Rhein uses a gas-tight sampling system for 24-hour composite samples mixed from five intakes across the Rhine, which are finally analyzed at the Office for Environment and Energy of Canton Basel-Stadt [18]. At these two monitoring stations selected compounds, which cannot be evaporated, are additionally analyzed by SPE and HPLC. The limited resolution of HPLC (compared to capillary GC) requires specific detection by MS/MS or UV spectral absorption. Estimation of the concentrations of unknown compounds by the specific signals is impossible here.

Full-scan spectra generated by the LC–MS systems are highly dependent on the method of ionization and the instrument type. Moreover, the few special libraries are restricted to several hundreds of spectra. Thus, high-resolution MS is necessary for analytical identification, which is not available at the monitoring stations. The GC and HPLC methods are based on reversed-phase materials. Polar compounds escaping the SPE process cannot be detected at all.

Sampling frequencies should be optimized with regard to reliable detection of unusual loads. Preparing composite samples over time is a common approach to continuous monitoring, but maximum concentrations tend to be suppressed. Moreover, the preparation of composite samples requires special equipment (e.g., for continuous filtration). Composite aqueous samples from the common bulk sampling systems are often unsuitable for trace-level analysis. Evaporation of analytes out of the open flasks and cross-contamination are major problems. However, the analysis of fresh spot samples requires high sampling frequencies for quasi-continuous monitoring and delicate equipment for unattended operation overnight.

Some concentrations change slowly, e.g., pesticides from diffuse sources with seasonal rural application. Analyzing a single spot sample a day or a composite sample collected over 24 hours might be sufficient to monitor slightly elevated concentrations over weeks. Material discharged directly into the river requires higher sampling frequencies. Moreover, there will be no even distribution in the cross section of the river.

A central task of monitoring is to locate and remove the sources of loads. Within a tight-knit monitoring network it is possible to trace materials back to their sources by specific patterns (e.g., mass spectra) even without complete analytical identification. To begin with, the region of a source is determined by comparing analytical results of the main monitoring stations. Samples from automatic sampling systems at the mouths of major tributaries and from further upstream are then analyzed. Additional samples from cross sections of the river can be taken by laboratory ships, or by boats of the water police. Finally, samples from the effluents or the wastewater treatment plants of potential dischargers will be checked. Upon application of this strategy, problems can be recognized and traced back to their source in short order.

The monitoring station at Worms is situated at a bridge over the Rhine. Four distinct intakes are mounted to the pillars of the bridge and at the riverbanks. The water sampling site at the left riverbank is influenced by the wastewater plume of a large chemical factory 10 km upstream. The sampling site at the right riverbank, in contrast, enables the station to measure the influence of several dischargers there. Suspicious results at only one of the sites help to identify or to exclude the potential dischargers.

In 2000, at the monitoring station in Worms, 31 suspicious results exceeding the lowest alarm threshold of 1 $\mu g\,L^{-1}$ were detected in the intake, which is influenced by the treated wastewater of the chemical factory [19]. In September 2000, *p*-methylanisole, identified by GC–MS, was detected. This pollutant could be detected downstream throughout the entire expanse of the river up to and abutting the German–Dutch border (Fig. 5). At the right bank of the river, however, only one suspicious result higher than 1 $\mu g\,L^{-1}$ was registered in the intake in 2000. This incident was due to the *p*-methylanisole alert on the left bank.

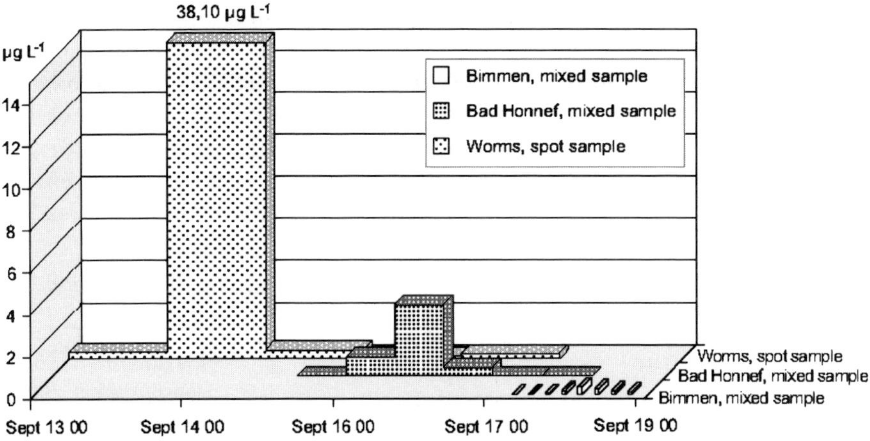

Fig. 5 *p*-methylanisole concentration in September 2000 measured at three consecutive monitoring stations

In February 2003 elevated concentrations of cyclohexanone were detected first at the monitoring station at Bad Honnef at Rhine km 640 near the right bank. At the same time a chemical factory about 200 km upstream unloading a tanker discovered that some thousand kilograms of cyclohexanone were missing. The captain of this tanker eventually reported to have grounded at Rhine km 540 near the left bank. Figure 6 (left) shows the concentrations of cyclohexanone detected at the IMBL at the German–Dutch border. After the wave had traveled more than 300 km, the concentrations changed by more than 10 $\mu g\,L^{-1}$ within 4 h (composite samples). On the left bank still higher concentrations were observed than on the right bank. The bulk passed the border within 24 hours.

A much broader wave was observed when about 1200 kg of nitrobenzene was discharged over several days from a wastewater treatment plant on the left bank approximately 80 km upstream of the IMBL in July 2003 (Fig. 6 (right)). During this period, the concentrations of nitrobenzene next to the

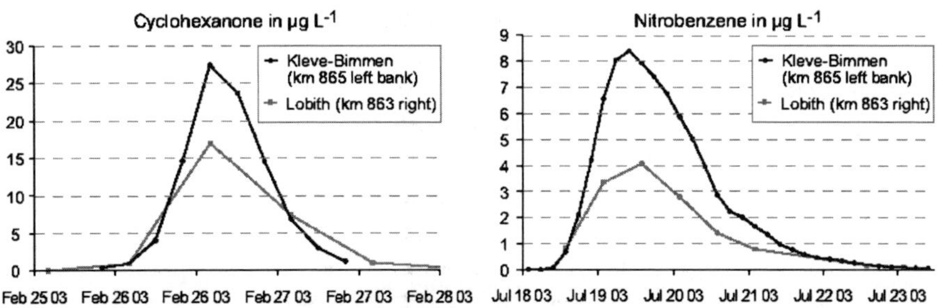

Fig. 6 Elevated concentrations of cyclohexanone (*left*) and nitrobenzene (*right*) at the International Monitoring Station Bimmen-Lobith (IMBL) in 2003

left bank were double those next to the right bank, with intermediate concentrations in the middle of the Rhine.

By far the most extraordinary loads detected at the IMBL consist of volatile compounds. Typically 100 to 3000 kg of BTEX, 1,2-dichloroethane, or especially the fuel additive methyl *t*-butyl ether (MTBE) pass the border as sharp waves within 12 to 24 hours [20]. The sources of these waves remain undetected. Backtracing is difficult, because the volatile compounds readily

Table 5 Selected maximum concentrations of some events reported by monitoring stations

Station	Date	Compound	Maximum concentration ($\mu g\ L^{-1}$)	Load (kg)
Weil	03.01–05.01.2002	Metolachlor	0.13	11
Weil	09.01–10.01.2002	Toluene	1.2	114
Weil	26.04.2002	Metabenzthiazuron	0.13	> 13
Weil	13.09–24.09.2002	Triethylene glycol dichloride	1.4	2000
Weil	02.11–06.11.2002	Isoproturon	0.16	79
Weil	01.12–05.12.2002	Dimethylaniline	0.54	150
Worms	22.01–23.01.1999	Isophoronenitrile	24	[a]
Worms	18.08–31.08.2000	1,2-Dichlorobenzene	9.0	[a]
Worms	14.09–15.09.2000	*p*-Methylanisole	38	[a]
Worms	27.11–28.11.2000	Tributyl phosphate	3.1	[a]
Worms	19.–20.07.2003	Hydrodehydrolinaloole	24	[a]
Bimmen-Lobith	17.01.2003	*m*-, *p*-Xylenes	4.3	230
Bimmen-Lobith	18.01.2003	MTBE	9.4	500
Bimmen-Lobith	26.02.2003	Cyclohexanone	28	2500
Bimmen-Lobith	25.03.2003	MTBE	25	1000
Lobith	09.04.2003	1,2-Dichloroethane/ -dichloromethane	25/27	[a]
Bimmen-Lobith	27.05.2003	Benzene/toluene/ *m*,*p*-xylenes	8.5/1.6/1.5	390
Bimmen-Lobith	29.08.2003	MTBE	30	550
Nieuwegein	06.04–13.04.1999	Isoproturon	0.32	–
Nieuwegein	16.11–21.12.2001	Isoproturon	0.48	–
Nieuwegein	16.11–21.12.2001	Chlorotoluron	0.21	–
Nieuwegein	05.02–03.03.2002	Isoproturon	0.19	–
Nieuwegein	05.02–03.03.2002	Chlorotoluron	0.34	–

[a] The load could not be calculated because the alert was detected at only one side of the river (not mixed)

evaporate from the open flasks of the automatic fraction collectors upstream, and elevated concentrations cannot be detected beyond a few hours after sampling.

Table 5 lists some of the classified maximum concentrations of events that have been reported by monitoring stations to the environmental authorities. The detection of unusual loads by measuring global parameters such as temperature or pH is nearly impossible. Even the controlled spill of 1800 tons of concentrated nitric acid from a tanker near Krefeld, which had caught fire on 26 November 2002, decreased the pH for only a few kilometers within the plume [21]. Critical concentrations of oxygen below 4 mg L^{-1} have not been observed since the 1980s, even in periods of low discharge and elevated temperature, such as those in summer 2003.

3.2 Biological Early Warning Systems

Given the fact that many more pollutants may be present in the water than are detected by the existing analytical equipment, and given the fact that among any occurring unknown pollutants those possessing toxic properties will be of primary concern, several online biological early warning systems (BEWS) are installed along the River Rhine.

Biotest methods have been used for several decades in a number of areas for the monitoring of water courses. They assess the effect of biologically harmful or biologically stimulating substances on selected test organisms. The aim of biological effect monitoring is to determine the effects of as many pollutants as possible. Due to the multitude of specificities of action of pollutants, test methods with various kinds of organisms must be deployed.

In the biotests, a distinction is made between static biotests (those deployed in the laboratory) and dynamic biotests (those deployed in various fields of application). Static biotests with aquatic organisms play a significant role in central Europe, as exemplified by the approval (or disapproval) of chemicals. Generally, static biotests determine the effects on test organisms exposed to a test medium (e.g., wastewater, surface water) or to a chemical for a certain test period (e.g., 24 or 48 h). As a rule, in the course of these acute tests, no exchange of the test medium occurs.

As opposed to the static biotests, the dynamic biotests expose the test organisms to the test medium in a continuous flow, or in semicontinuous systems in which the cycles are of several minutes. Changes in metabolism, or of behavioral/physiological parameters induced by sublethal effects, are measured using automatic detection systems. The designation "biomonitor" characterizes the main function of the system as a monitoring and warning system for the identification of combined pollutant effects. Continuous biotests to examine quantities of immissions and emissions offer the possibility of ongoing water monitoring. Biotests are primarily deployed as BEWS in the monitoring of immissions. For instance, they might indicate elevated concentrations caused by accidental discharges above the usual background concentration of the water course. In conjunction with chemical analysis and more detailed biological examinations, dynamic biotests assist in securing evidence of illegal discharges and accidents. Applied to the monitoring of emissions, they can supplement the data with standardized static biotests as a means of continuous quality assessment [10].

Changes in the behavior or properties of the organisms used for BEWS may indicate the sudden occurrence of a pollutant not detected in conventional analytical warning systems [22]. It should be noted that BEWS are capable only of signaling that something is wrong, but not the exact cause. Thus, the BEWS detect the effects of pollutants, rather than the types or the concentrations thereof. The number and type of these systems may differ among the different monitoring stations. Since biotests with organisms of distinct trophic levels are not interchangeable, efforts are made to install a battery of tests with organisms of different trophic levels at each suitable monitoring station. Thus, monitoring stations generally install one or more continuous biotest methods using organisms such as fish, mussels, daphnids, algae, and luminescent bacteria.

All BEWS include some mathematical algorithms to detect "alarm" situations. Although static alarm thresholds are sufficient for some, other methods need an evaluation of the data using dynamic alarm thresholds. The latter are important when natural changes or drifts that are not due to contaminations of the water occur in the measured parameters. In addition to the results of mathematical evaluation, the interpretation of the results from continuous biotest systems includes evaluation of several other pieces of online data. This pertains to both alarm and nonalarm situations. Such a "hierarchy" of BEWS results can be stated, as shown in Fig. 7. Depending on whether or not the alarm threshold is reached, or whether or not chemical–physical online data confirm changes in water quality, the following alarm levels are defined: "irregularity", "indication", "event", and "messaging level" [16, 23].

As already mentioned, the BEWS are incorporated in the immission-oriented approach of the WAP Rhine. Furthermore, both BEWS and the chemical screening methods play a prominent role in protection of the drinking water intake. For instance at Nieuwegein, several BEWS safeguard the intake of Rhine raw water which is used for the production of drinking water for well over 2 million people, or around one-seventh of the Dutch population. Once a contamination of the river water is detected, certain measures can be taken to control the intake of raw water. These measures include, but are not limited to, dilution with groundwater, mixing with the treated product from the other treatment plant at the delivery point, or interruption of the extraction [24]. Table 6 gives an overview of the BEWS that are installed along the Rhine, although some of them are still in a testing phase.

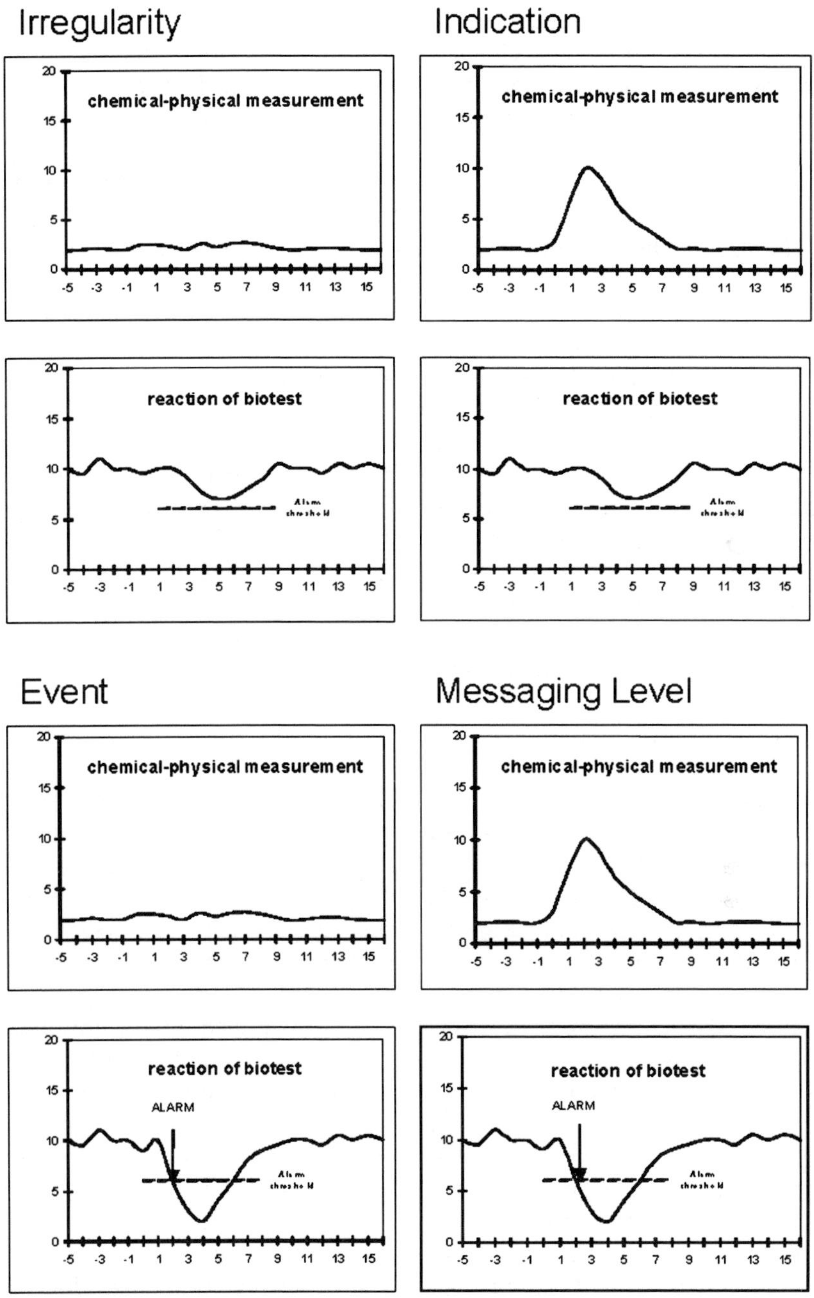

Fig. 7 Alarm levels of BEWS defined according to whether or not the alarm threshold is reached, or whether or not chemical–physical online data confirm changes in water quality

Table 6 Overview of BEWS along the Rhine

Biotest system	Organisms	Principle of measurements	Evaluation	Producer test
Kerren Fish Test	Golden ide *Leuciscus idus*	Behavioral parameters as touching sensors	Dynamic alarm thresholds	Kerren, Viersen, Germany
bbe Fish Toximeter	*Leuciscus idus* or other fish	Behavioral parameters as video images	Dynamic alarm thresholds	bbe Moldaenke, Kiel, Germany
bbe Daphnia Toximeter	*Daphnia magna*	Behavioral parameters as video images	Dynamic alarm thresholds	bbe Moldaenke, Kiel, Germany
Dynamic Daphnia Test	*Daphnia magna*	Swimming behavior by IR light sensor	Dynamic alarm thresholds	Electron GmbH, Krefeld, Germany
Dreissena-Monitor	*Dreissena polymorpha*	Shell movement by electromagnetic fields	Statistics over all mussels, static alarm thresholds	Envicontrol, Frechen, Germany
Mosselmonitor®	*Dreissena polymorpha* or other mussels	Shell position by electromagnetic fields	Statistics for each individual mussel, static alarm thresholds	Delta Consult b.v. AC Kapelle, Netherlands
DF Algae Test	Algae (various species)	Variable delayed fluorescence (DF)	Static: comparison of fading curves, sample vs. control; dynamic: comparison with preceding data	Regensburg University, Regensburg, Germany
bbe Algae Toximeter	Algae (various species)	Variable chlorophyll *a* fluorescence	comparison sample vs. control, Adaptive Hinkley Detector	bbe Moldaenke, Kiel, Germany
Regensburg Luminescent Bacteria Test	*Photobacterium phosphoreum*	Variable bioluminescence	Comparison sample vs. control, static alarm thresholds	Regensburg University, Regensburg, Germany

Figure 8 gives an example of an alarm generated by a Fish Test on 23 July 2003 from 16:30 to 23:50 h at the monitoring station at Nieuwegein. A composite sample (16:00 to 24:00 h) was subsequently analyzed by both GC–MS and LC-DAD-UV. These techniques detected high concentrations of 2,4-dichlorophenol with an estimated concentration of 4.3 $\mu g\,L^{-1}$ (a compound used for the production of pesticides).

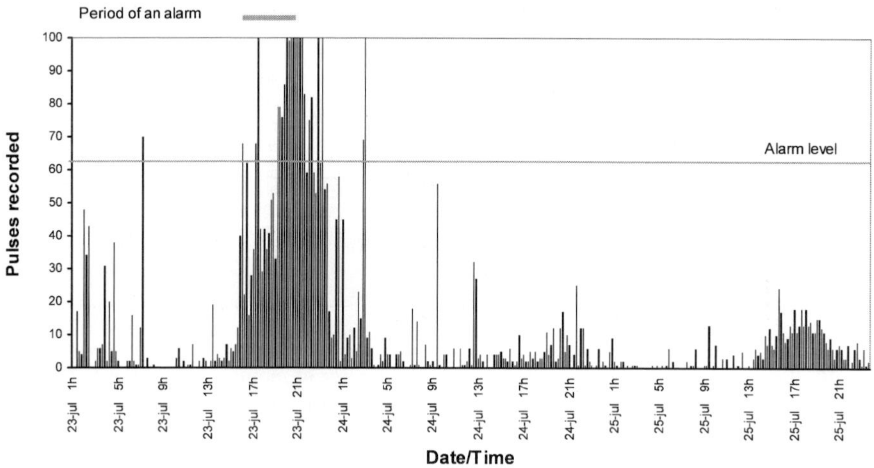

Fig. 8 Biotest alarm generated by a fish test at Nieuwegein in July 2003 caused by 2,4-dichlorophenol

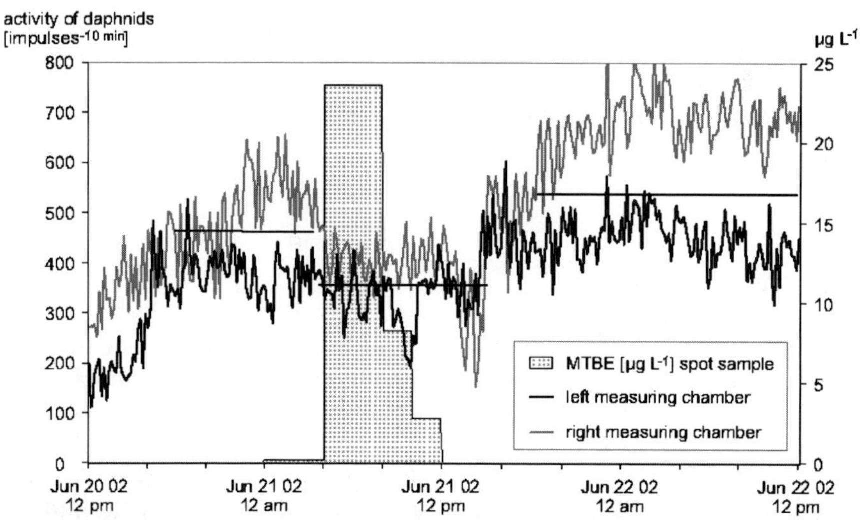

Fig. 9 Reaction of the daphnia in the Dynamic Daphnia Test at monitoring station Bimmen-Lobith (river km 865, left bank) to increasing concentrations of methyl *t*-butyl ether (MTBE)

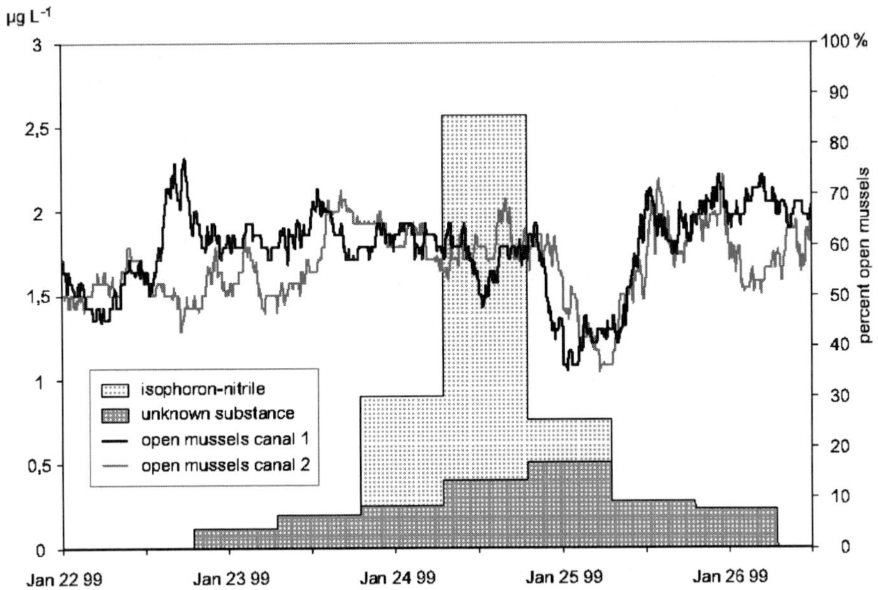

Fig. 10 Reaction of the mussels in the Dreissena-Monitor at monitoring station Bad Honnef (river km 640, right bank) to increasing concentrations of isophorone nitrile and an unknown compound

Figure 9 shows the reaction of the daphnia in the Dynamic Daphnia Test to increasing concentrations of MTBE. There was a discernible decrease in the daphnia activity. Afterwards, the activity showed irregularities on at noticeably higher level. Figure 10 shows the reaction of the mussels in the Dreissena-Monitor to increasing concentrations of isophorone nitrile and an unidentified substance. The percentage of open mussels hence decreased significantly.

4 Perspective

Within the ICPR good progress has been made with the WAP Rhine throughout the years. The plan is functioning without major problems. Improvements that can be made are initiated in dedicated workshops with members of all participating IMWCs, experts, and representatives from the waterworks.

The good cooperation between the different countries along the Rhine, in addition to the fact that the river is regarded as a whole, is in the spirit of the European Water Framework Directive [1]. From the perspective of this directive (Article 11, see Sect. 1) the WAP Rhine and the monitoring stations will continue to play an important role. In addition, the broad chemical screening that is carried out on a daily basis at various monitoring stations can help to determine which substances are really relevant for the Rhine and which should be tackled to obtain the requisite "good ecological status" that is required. For substances at regulatory level (priority substances), the river authority should identify their emission sources and implement an action program to reduce emissions in order to achieve the goals set out. If this approach is translated into incidental pollution events, this means that the identification of sources, e.g., by backtracing, will gain importance in the perspective of the WFD. Inventories of potential threats over the whole river basin can be helpful in this respect as well. As the ecological status of the river improves, the vulnerability with respect to incidental pollution will increase, underscoring the importance of early warning systems.

For an ideal early warning monitoring system, the requirements can include the following [25]:

1. Provides warning in sufficient time for action
2. Cost is affordable
3. Requires low skill and training
4. Covers all potential threats
5. Is able to identify the source
6. Is sensitive to quality changes at regulatory levels
7. Gives minimal false-positive or -negative responses
8. Is robust
9. Is reproducible and verifiable
10. Allows remote operation
11. Functions year-round

Although it is impossible to meet all these criteria in one system, they can be used to evaluate the systems in use along the Rhine. The intention of setting up a monitoring station in the first place is to forewarn, doing so all year round, which requires that systems are robust. Therefore, for none of the stations along the Rhine do criteria 1, 8, and 11 pose any major problems. In some cases, the requirements are conflicting. For example, it is nearly impossible to design a system that covers all threats and is at the same time able to identify the source, or which requires little skill training. This means that choices have to be made. Monitoring stations along the Rhine have one clear common strategy: a two-way approach of using chemical screening and biological monitoring techniques. With this approach many potential threats will be covered, since a broad view on the water quality is obtained. Techniques (especially the chemical ones) are very sensitive at regulatory levels. As substances can be identified, hints on possible sources can be obtained. Continuous biotests can pose problems with respect to reproducibility for two reasons: first, because standard procedures are still in the developing stages, and second because the response of the biotests depends highly on the whole matrix of the river water. These dynamics will change when an observation is confirmed by additional experiments.

With this strategy the challenges for early warning monitoring along the Rhine are:

- Broadening and opening new analytical windows
- Broadening effect windows (see Sect. 3)
- Standardization
- Better understanding of relations between substance and effect (e.g., toxprint techniques) [26]

This is the direction chosen by water authorities and drinking water companies for the development of new techniques with the limited budget available (due to the improvement of Rhine water quality, political attention has been shifted). With this strategy, skilled personnel are needed; automated and remote operation is complex, and costs may be considerable. Thus, the fifth challenge is to maintain balance among all elements.

Joint international cooperation in the coming years will ensure continued improvement of the WAP Rhine. Improvements are to be expected in:

- The exchange via the internet of "expert" knowledge on substance toxicology etc.
- The real-time availability of a Rhine-wide expert network in case of an alarm
- Better embedding of biological alarms in the WAP
- A new system of guidance values
- Passing through alarms via a web-based application [27]

The characteristics of the river determine which early warning system is most suitable [2]. Therefore different strategies will be chosen for different river basins. Along the River Rhine the riparian states have chosen to collaborate and to organize themselves in a very pragmatic and effective way, which was recognized by the 2002 Rhine Award issued by a Dutch drinking water company for the international cooperation at the Dutch–German border.

Acknowledgements The authors would like to thank Anke Lauer (Worms), Dr. Jan Mazacek (Basel), and Harald Teicher (Düsseldorf) for contributions to the manuscript. Special thanks to Heather Prescott (Mercersburg Academy, Pennsylvania) for correcting the language.

References

1. European Union (2000) Directive 2000/60/EC of the European Parliament and of the Council, establishing a framework for Community action in the field of water policy, Brussels
2. Gunatilaka A, Diehl P (2001) In: Butterworth FM, Gunatilaka A, Gonsebatt Bonaparte ME (eds) Biomonitors and biomarkers as indicators of environmental change, vol 2. Kluwer/Plenum, New York, p 9

3. ARW (ed) (2003) 59 Jahresbericht 2002, Cologne, ISSN 0343-0391
4. LWA (1986) Sonderbericht: Brand bei Sandoz und Folgen für den Rhein in NRW. Eigen, Düsseldorf
5. Brusske A, Willemsen HG (1990) Vom Wasser 74:393
6. Kramer KJM, Botterweg J (1991) In: Jeffrey DW, Madden B (eds) Bioindicators and environmental management. Academic, London, p 95
7. Hendriks AJ, Stouten M (1994) Ecotoxicol Environ Saf 26:265
8. Irmer U (ed) (1994) Continuous biotests for water monitoring of the river Rhine: summary, recommendations, description of test methods. Umweltbundesamt Texte 58/94, Berlin
9. Diehl P, Krauß-Kalweit I, Lüthje S (1997) Wasser Boden 49:25
10. LAWA (1998) Recommendations on the deployment of continuous biomonitors for the monitoring of surface waters. Kulturbuch, Berlin
11. Gunatilaka A, Diehl P, Puzicha H (2001) In: Butterworth FM, Gunatilaka A, Gonsebatt Bonaparte ME (eds) Biomonitors and biomarkers as indicators of environmental change, vol 2. Kluwer/Plenum, New York, p 29
12. Alberti J, Willemsen HG (1989) Forum Städte-Hyg 4:118
13. Poels CLM (1977) In: Cairns J et al. (eds) Biological monitoring of water and effluent quality. ASTM, STP 607, p 85
14. IKSR (2003) Internationaler Warn- und Alarmplan Rhein, IKSR-Bericht http://www.iksr.org/bilder/pdf/bericht_nr_137d.pdf
15. Sreafico M, van Mazijk A (1993) Alarmmodell Rhein, ein Modell für die Vorhersage des Transportes von Schadstoffen im Rhein, KHR, Bericht Nr I-12
16. DK Rhein (ed) (2003) Handbuch für die Vereinheitlichung von kontinuierlichen Biotestverfahren und die Einbindung in den Warn- und Alarmplan Rhein, Teil A: Allgemeiner Teil. Worms
17. IKSR (2003) Empfehlungen des 3. IKSR-Workshops Warn- und Alarmplan Rhein (Bonn, 30.09/01.10.2003), IKSR-Document S 43-03
18. Dolf R, Mazacek J (2002) Rheinüberwachungsstation Weil am Rhein (RüS), Jahresbericht 2001, Amt für Umwelt und Energie Basel-Stadt
19. Rheingütestation Worms (2001) Tätigkeitsbericht 2000, Worms
20. Gerke T, Brandt P, Willemsen HG (2003) In: Bilitewski B, Werner P (eds) Proceedings of the first European conference on MTBE, Dresden, 8/9 September 2003. Dresden University of Technology, DTU Technical University of Denmark, IIQAB Barcelona, p 128
21. Mentfewitz J, Kremeskötter U, Bonny-Wefers M, Friesel J (2002) Jahresbericht 2001 Landesumweltamt Nordrhein-Westfalen, Düsseldorf
22. Umweltbundesamt (1995) Kontinuierliche Biotestverfahren zur Überwachung des Rheins (Continuous biotest systems for the surveillance of the Rhine). UBA Berichte 1–95. Schmidt, Berlin
23. DK Rhein (ed) (2003) Handbuch für die Vereinheitlichung von kontinuierlichen Biotestverfahren und die Einbindung in den Warn- und Alarmplan Rhein, Teil B: Spezieller Teil Dynamischer Daphnientest. Worms
24. Stoks P (1994) Water quality control in the production of drinking water from river water. In: Adriaanse M, Kraats J, Stoks PG, Ward R (eds) Proceedings of the international workshop Monitoring Tailor-made, Beekbergen, 20–23 September 1994, p 118

25. Brosnan TM (1999) Early warning monitoring to detect hazardous events in water supply. ILSI risk science institute workshop report. ILSI, Washington DC
26. Bobeldijk I, Stoks P, Emke E, Muilwijk B, Brandt A, van Leerdam T, Noij Th (2002) Proceedings of the 223rd American Chemical Society Conference, Orlando, FL
27. Stoks P (2002) In: Fleischmann N, Langergraben G, Haberl R (eds) Proceedings of the international IWA conference on automation in water quality monitoring, Vienna, 21–22 May 2002, p 445

Part II
Water Quality in the Alpine Lakes

Hdb Env Chem Vol. 5, Part L (2006): 127–138
DOI 10.1007/698_5_018

Published online: 23 November 2005

Water Quality of Lake Constance

Michael Petri

Zweckverband Bodensee-Wasserversorgung, Betriebs- und Forschungslabor,
78354 Sipplingen-Süßenmühle, Germany
michael.petri@zvbwv.de

Abstract Lake Constance is used as a drinking water reservoir for about 5 million people. Because of its glacial origin from the Alps, the amount of minerals, heavy metals, and trace elements in the lake water is very low. During the 20th century, a rising population in the catchment area of Lake Constance, which was accompanied by growing industries and intensified agriculture, led to an increasing pollution by wastewater and runoff from agriculture areas. Subsequently, in the last century Lake Constance entered a phase of eutrophication.

Currently, Lake Constance is approaching a tropic state, which may not be far from the beginning of the 20th century. After the phase of eutrophication, the re-oligotrophication was a great success and an encouraging example for an international environmental management of an aquatic system.

Most of the wastewater that is entering Lake Constance is now purified to a high technical standard, but the treated wastewater still contains organic xenobiotics (e.g., pesticides, chelating agents, pharmaceuticals), that cannot be removed efficiently by wastewater purification. The contaminations are still rather low, but must not be neglected. Since it is not possible to assume the ecotoxicological effects of these persistent organic pollutants, their long-term behavior and their ecotoxicological effects should investigated in more detail.

Keywords Heavy metals · Lake Constance · Persistent organic pollutants · Trace elements · Water quality

Abbreviations

AWBR Arbeitsgemeinschaft Wasserwerke Bodensee-Rhein (Association of Waterworks Lake Constance-Rhine)
DDT Dichlorodiphenyl trichlororethane

EDTA Ethylenediamine tetraacetic acid
IGKB Internationale Gewässerschutzkommission für den Bodensee (International Commission for the Protection of Lake Constance)
NTA Nitrilotriacetic acid

1 The Alpine Rhine and Lake Constance

The Rhine, the longest and most important river of Germany, has its origin in the Alps of Switzerland. The tributaries Hinter-Rhine and Vorder-Rhine, which come from two different Alpine springs, combine at Reichenau (Switzerland) and flow as the Alpine Rhine to Lake Constance, which is the second largest lake of Europe at the northwestern edge of the Alps. Lake Constance has a volume of about 48.4 km^3, a surface of 535 km^2 and a maximum depth of 253 m. Lake Constance has two basins: the deep Upper Lake Constance with its fjord-like appendix called Lake Überlingen and the shallow Lower Lake Constance [1] (Fig. 1). More morphometric data of Lake Constance and some details of its inflows and outflow are given in Table 1.

The catchment area of Lake Constance, which is about 11 500 km^2, is located in Germany (24%), in Switzerland (48%) and in Austria (28%). More than the half of the catchment area is in the Alps above an altitude of 1500 m. The mean annual inflow is about 12 km^3, of which more than 80% are glacial origin and comes from the Alps by the inflows Alpine Rhine, Bregenzerach and Dornbirnerach [2]. The water of these inflows enter the Upper Lake in the

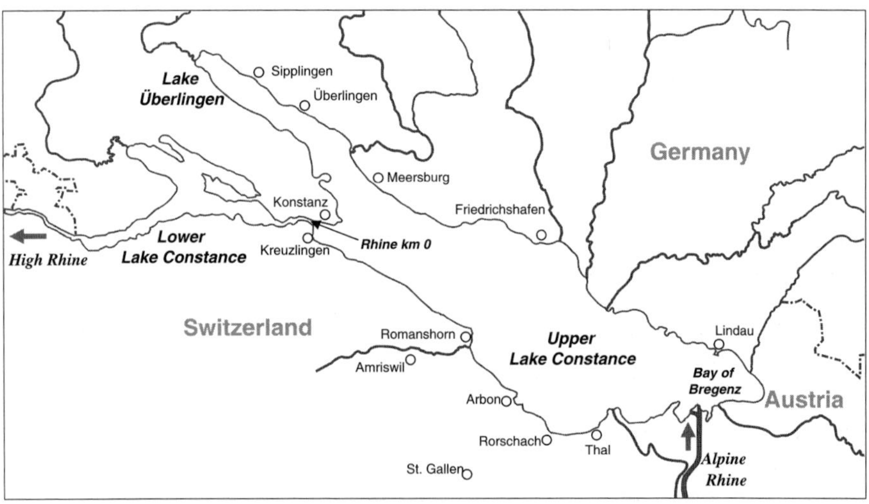

Fig. 1 Map of Lake Constance with Upper Lake Constance, Lake Überlingen, and Lower Lake Constance

Table 1 Morphometric data of Lake Constance and its inflows and outflow [2]

	Total	Upper Lake Constance	Lower Lake Constance
Surface area [km^2]	535	473	62
Volume [km^3]	48.4	47.6	0.8
Max depth [m]	–	253	40
Mean depth [m]	–	101	13
Length of shoreline [km^3]	273	186	87
Catchment area [km^2]	11 474	10 919	555
Inflow by rivers [$km^3\ year^{-1}$]	11.2	10.7	0.52
Inflow by rain [$km^3\ year^{-1}$]	0.45	0.40	0.05
Inflow by difuse sources[$km^3\ year^{-1}$]	0.28	0.20	0.08
Inflow by sewage water [$km^3\ year^{-1}$]	0.10	0.08	0.02
Mean outflow [$km^3\ year^{-1}$]	12.0	–	–
Residence time of the water [year]	–	4.3	0.08

eastern part near Bay of Bregenz. At Konstanz, the water flows from the Upper Lake Constance into the Lower Lake Constance (Fig. 1) and then into the High Rhine to Basel, including the waterfalls at Schaffhausen. The Alpine Rhine has a typical runoff characteristic of high mountain rivers with a pronounced water regime: during winter, a minimum and high water in spring and summer from snowmelt and heavy rains in the Alps. These variations in flow are not completely evened out by Lake Constance but strongly damped. As a sink, Lake Constance accumulates all the coarse and fine-grained sediments of the Alpine Rhine. Up to 1.4 million tons of mud and gravel is transported each year by the Alpine Rhine into the lake [2]. Most of it is deposited close to the mouth of the Alpine Rhine and in the main basin of the Upper Lake Constance.

2 Drinking Water from Lake Constance

Lake Constance has been used as a drinking water reservoir for a long time. In 1885 the municipality of St. Gallen started to pump water from Lake Constance to its drinking water utility. Nowadays 17 drinking water treatment plants withdraw an annual amount of about 170 million m^3 of water from Lake Constance to serve more than 5 million people with drinking water. The biggest drinking water utilities using water of Lake Constance are the municipality of St. Gallen (about 10 million m^3 per year) and the Zweckverband Bodensee-Wasserversorgung (about 135 million m^3 per year).

The raw water is taken from a depth of 40–60 m, where the temperature of the water is about 5 °C throughout the year. The lake water is of such a good quality that it can easily be treated to obtain drinking water by

Table 2 Water quality of Lake Constance – Sampling site: Zweckverband Bodensee-Wasserversorgung in Sipplingen at Lake Überlingen (Fig. 1), annual means of 2004, raw water from a depth of 60 m

Main anions and cations			
Cations	mg L^{-1}	Anions	mg L^{-1}
Na^+	4.4	HCO_3^-	150
K^+	1.3	SO_4^{2-}	33
Ca^{2+}	47	NO_3^-	4.5
Mg^{2+}	8.3		
Sr^{2+}	0.46		

Halogenated compounds					
	mg L^{-1}		mg L^{-1}		mg L^{-1}
F^-	< 0.10				
Cl^-	5.3				
Br^-	0.0089	BrO_3^-	< 0.0005	Br_{total}	0.010
I^-	0.0002	IO_3^-	0.0002	I_{total}	0.002

Trace elements and heavy metals			
	mg L^{-1}		mg L^{-1}
Al	0.0045	Li	0.0023
Ag	< 0.000025	Mn	0.00074
As	0.0012	Mo	0.001
B	0.019	Ni	0.00071
Ba	0.026	P	0.008
Be	< 0.000025	Pb	< 0.000050
Bi	< 0.000050	Rb	0.001
Cd	< 0.000025	Sb	0.00014
Co	0.00008	Se	0.00013
Cr	0.00028	Th	< 0.000025
Cu	0.0011	Tl	< 0.000025
Fe	0.006	U	0.001
Ga	< 0.00001	V	0.00025
Hg	0.000003	Zn	0.0015

using natural treatment methods like microstraining, ozonation and rapid sand filtration. Drinking water of Lake Constance complies all drinking water regulations of Germany, Austria, Switzerland and European Union (Table 2). Because of its glacial origin, the amount of minerals in Lake Constance is low. But the minerals and compounds are not mixed homogeneously in the whole waterbody of Lake Constance. Local and seasonal fluctuations in the concentration of the constituents are not strictly due to the variations in the rates of precipitation throughout the whole catchment area of Lake Constance [3]. Consequently, the concentrations of the constituents in the lake water are not necessarily proportional to the mean concentration of the total inflows. Elements involved in some biogeochemical cycles such as phosphorous, silicon, nitrogen, aluminium and calcium can also differ significantly from its mean concentration of the total inflows. The cations of sodium, potassium, calcium, magnesium and strontium covers more than 95% of the positive charge, whereas hydrogen carbonate, sulphate, chloride and nitrate are the four main anions in the ionic balance. The contents of all other cations and anions are very low (Table 2).

The complexity and time-dependent variation of biogeochemical cycles of most trace elements in water of Lake Constance are not well understood. Whereas aluminium, iron and manganese are mainly bound to suspended matter, other trace elements may preferably be dissolved or complexed. However, the elemental speciation like oxidation state, the forming of organic or inorganic complexes and their transition and conversion are in most cases completely unknown [3].

3
Eutrophication and Re-oligotrophication of Lake Constance

Like in many other European lakes, the increasing population during the 20th century accompanied by growing industries and intensified agriculture in the catchment area of Lake Constance led to an increased discharge of treated and untreated wastewater and to an increased runoff from agriculture areas. Until the beginning of the 1960s, Lake Constance was an oligotrophic lake [4]. Because of the intensified usage of phosphorus-containing fertilizers and detergents in the catchment area of Lake Constance, the concentration of phosphorus in the lake water increased dramatically during the 1960s and 70s (Fig. 2). With the increasing phosphorus concentration, significant changes in the biotic communities of Lake Constance were observed [2]. Water administrations of the states around Lake Constance were concerned about the development of eutrophication, which led to the foundation of the International Commission for the Protection of Lake Constance (IGKB) in 1959. In 1964 the IGKB decided to implement an extensive wastewater purification program for the whole catchment area. After the implementation and opti-

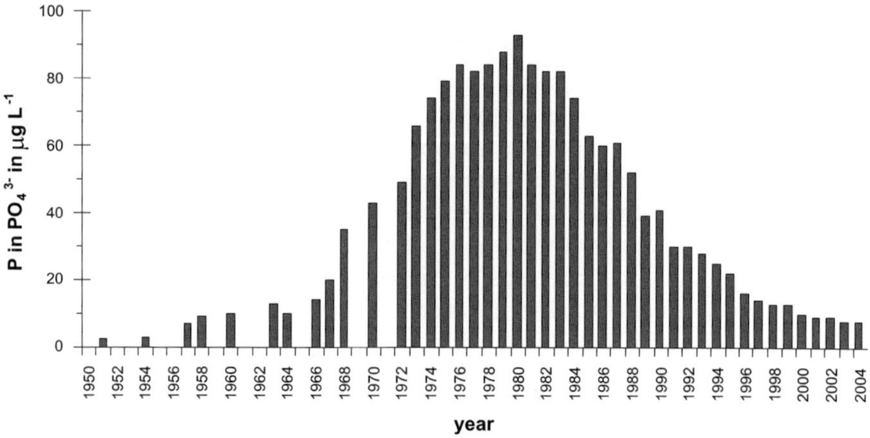

Fig. 2 Long-term course of phosphorus concentration in Lake Überlingen. PO_4^{3-} concentration given in P in water from a depth of 60 m after spring-turnover. Sampling site: Zweckverband Bodensee-Wasserversorgung in Sipplingen at Lake Überlingen (Fig. 1)

mization of sewage treatment plants and due to the reduction of phosphorus compounds in detergents, the concentration of phosphorus in water of Lake Constance peaked at 90 $\mu g\,L^{-1}$ P at the end of the 1970s and decreased steadily to a phosphorus concentration of 8 $\mu g\,L^{-1}$ P in 2004. Nowadays, Lake Constance is approaching a tropic state, which may be not far from the beginning of the 20th century.

Whereas the annual mean concentrations of hydrogen carbonate and sulphate have remained stable for decades, the chloride concentration has increased gradually from 3.8 $mg\,L^{-1}$ in 1977 to 5.3 $mg\,L^{-1}$ in 2004 in Lake Überlingen (Fig. 3). The IGKB reported a chloride concentration of 2.5 $mg\,L^{-1}$ in 1961 in Upper Lake Constance [5]. This increase is probably due to a rise in the anthropogenic emission of inorganic chlorine by households, industry, agriculture and the usage of sodium chloride for removing of ice and snow in winter. At the same time, the nitrate concentration in the water of Lake Überlingen has increased from 3.55 $mg\,L^{-1}$ in 1974 to 4.5 $mg\,L^{-1}$ in 2004 (Fig. 3). The IGKB reported a nitrate concentration of 2.2 $mg\,L^{-1}$ in 1961 for the Upper Lake Constance [5]. Stabel [3] suggested the following reasons for the increase of nitrate: agriculture runoff, long-distance transport in the atmosphere and wastewater. Nitrate is the main chemical nitrogen compound found in water of Lake Constance. Ammonia and nitrite are not always detectable, but they are important parts in the biochemical nitrogen cycle of lake water. The detection of both compounds points to a microbiological decomposition of protein-containing organic substances. Consequently, there are seasonal and local fluctuations in the concentration of ammonia and nitrite in water of Lake Constance (ammonia up to 90 $\mu g\,L^{-1}$ and nitrite up to 60 $\mu g\,L^{-1}$) [5].

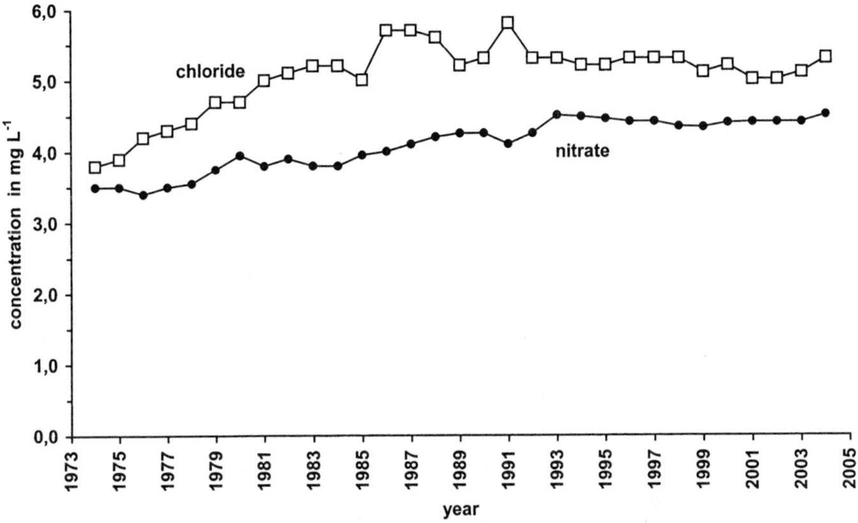

Fig. 3 Long-term course of chloride (□) and nitrate (•) concentration in Lake Überlingen – Sampling site: Zweckverband Bodensee-Wasserversorgung in Sipplingen at Lake Überlingen (Fig. 1), annual means of raw water from a depth of 60 m

4
Trace Elements and Heavy Metals

The concentration of most trace elements and heavy metals in Lake Constance is very low (Table 2). Since the end of the 1970s it has been possible for the first time to measure trace elements in the lake water with atomic absorption spectrometry in such a low concentration level and to get reliable results [6–8]. With the further developments and increasing sensitivities of atomic absorption spectrometric methods in the 1980s, and with inductive plasma mass spectrometry in the 1990s, the array of detectable heavy metals and other trace elements was considerably enlarged. Before this time, only by measuring the lake sediments was it possible to get information about heavy metal contamination (e.g. Cd, Cu, Cr, Hg, Pb, Zn) in Lake Constance [9].

Except for arsenic, no significant long-term trend for heavy metals in Lake Constance was noticeable. Since the beginning of low-level metal determination pollution by heavy metals leading to enhanced concentrations in the main water body of Lake Constance has not been observed. The seasonal and local variations in their contents is due to environmental conditions, meteorological and hydrodynamical aspects and due to biogeochemical cycles [3].

During the period of re-oligotrophication in Lake Überlingen the arsenic concentration gradually decreased from 3.0 $\mu g\,L^{-1}$ in 1977 to 1.3 $\mu g\,L^{-1}$ in 1996. Since 1996 the arsenic concentration has remained stable between 1.2 and 1.3 $\mu g\,L^{-1}$ (Fig. 4). Stabel [10] assumed that during the vegetative seasons

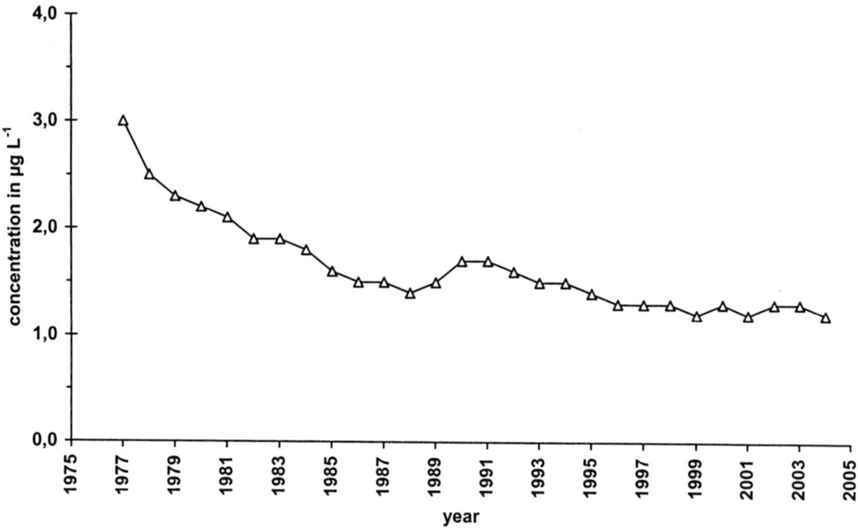

Fig. 4 Long-term course of arsenic concentration in Lake Überlingen – Sampling site: Zweckverband Bodensee-Wasserversorgung in Sipplingen at Lake Überlingen (Fig. 1), annual means of raw water from a depth of 60 m

arsenic is actively taken in by phytoplankton followed by sedimentation of As-containing dead cells in the depths of the lake. Subsequently, the release of soluble arsenic from sediments due to diminished oxygen concentration in the deep water layers during the period of eutrophication and redox-controlled biogeochemical cycles led to higher arsenic concentrations in the lake water.

5 Organic Pollutants

The pollution of Lake Constance with organic xenobiotica (e.g., pesticides, chelating agents, pharmaceuticals) is fortunately relatively low [2, 3]. In the 1980s, pesticides like atrazine, simazine and the atrazine-metabolite desethyl-atrazine were continuously found in Lake Constance in concentrations between 0.01 and 0.05 $\mu g\,L^{-1}$ [3]. The sources of the pesticide pollution were little streams and rivers entering Lake Constance, the runoff from agriculture areas, and the wet and dry depositions on the water surface [2, 11]. Due to the prohibition of the usage of atrazine and other triazine-herbicides in Germany and other laws and regulations for the usage of pesticides, the concentration of triazine-herbicides and other pesticides decreased slowly below detection limits during the 1990s. In the sediments of Lake Constance persistent organochlorine pesticides like DDT, endosulfane and lindane, which were used in the 1960s and 1970s, are still detectable in low concentrations [2].

In 1985 chelating agents like EDTA and NTA were detected in Lake Constance and its inflows by Rossknecht [12] for the first time. The EDTA concentrations of the inflows fluctuated between 0.4 μg L^{-1} and 145 μg L^{-1}. In the offshore area of Upper Lake Constance the EDTA concentration was about 4.0 μg L^{-1} in 1985 [12] and decreased gradually to 1.0 μg L^{-1} in 2003 [13]. Since NTA is biodegradable in sewage plants and in surface water, the NTA concentration in Lake Constance is normally below the detection limit of 0.5 μg L^{-1} [3, 13].

Human pharmaceuticals cover a very diverse group of compounds. The most important groups that have been investigated in waste, surface and ground water are blood lipid regulator, β-blocker, antiepileptics, analgesics, cytostatics, antibiotics, antidepressants and iodinated X-ray contrasting agents [14–16]. Most of the detectable pharmaceuticals, especially the iodinated X-ray contrasting agents, have a very low biodegradability. In tests with bacteria, algae, water fleas and fishes, acute toxicological effects were not observed for pharmaceutical quantities below 10 μg L^{-1} [15]. So it can be assumed, that the low concentrations of the detectable pharmaceuticals will not generate any acute risks for water organisms in Lake Constance. Only speculations can be made about possible environmental risks in the long term, because there are no data available about ecotoxicological effects for pharmaceuticals at so low concentration levels.

In offshore water of Lake Constance the following have been detected: clofibric acid (blood lipid regulator), carbamazepine (antiepileptic) and the iodinated X-ray contrasting agents iopromide, iopamidol and diatrizoate. The

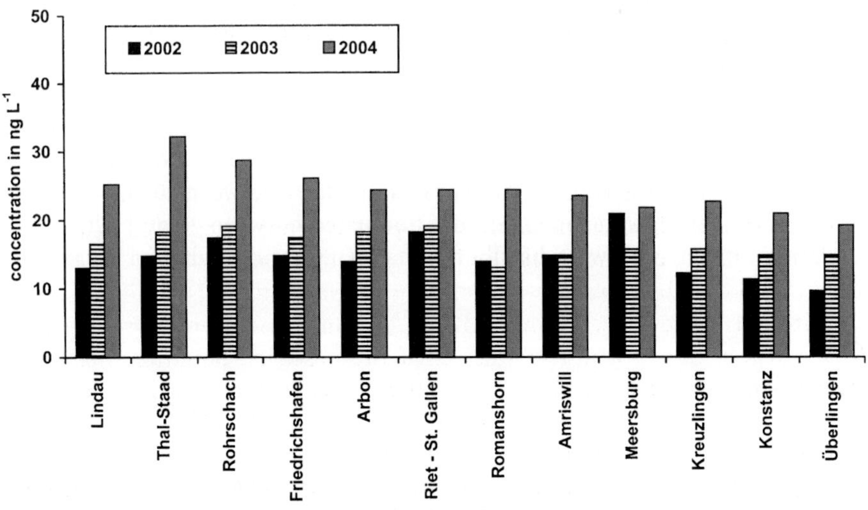

Fig. 5 Iopromide concentration in water from Lake Constance 2002–2004. Raw water from the drinking water supplies around Lake Constance [13]. Sampling sites listed in east–west direction

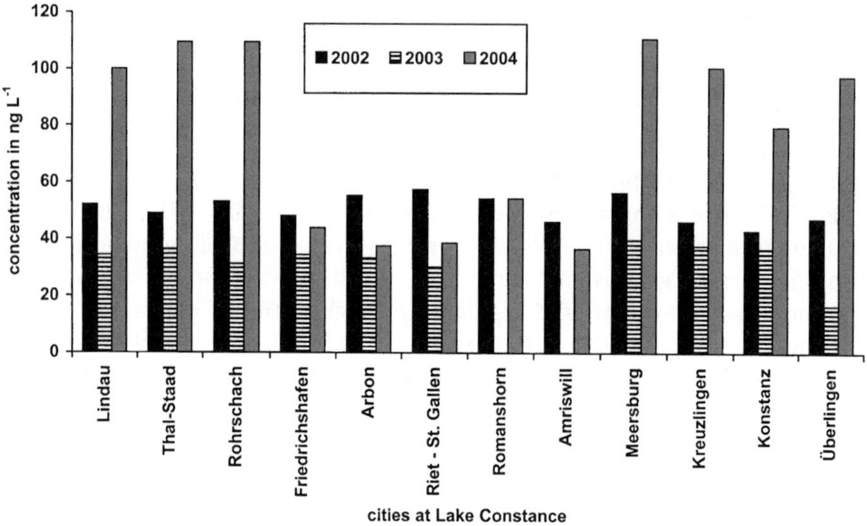

Fig. 6 Iopamidol concentration in water from Lake Constance 2002–2004. Raw water from the drinking water supplies around Lake Constance [13]. Sampling sites listed in east–west direction

concentrations fluctuated between the detection limit of 0.01 $\mu g\,L^{-1}$ and up to 0.12 $\mu g\,L^{-1}$ [2, 13]. In the incoming rivers and little streams the concentration of the detected pharmaceuticals are much higher and can exceed 1 $\mu g\,L^{-1}$. In estuaries, where the incoming rivers mix into the lake water, the concentration decreases due to dilution. In sampling campaigns in 2002–2004, organized by the Association of Water Works of Lake Constance and Rhine (AWBR), raw water of the drinking water supplies around lake Constance was analyzed for iodinated X-ray contrasting agents. In 2004 the concentrations of iopromide (Fig. 5) and iopamidol (Fig. 6) were higher than in the years before [13]. Iopromide seemed to be distributed more or less regularly between 20 and 30 $ng\,L^{-1}$ in Lake Constance, whereas Iopamidol fluctuated between 40 and 120 $ng\,L^{-1}$. But no significant east–west distribution pattern was noticeable for iopamidol.

6 Conclusions

Lake Constance is approaching a tropic state which may be not far from the beginning of the 20th century. The re-oligotrophication between 1980 and 1990 was a great success and an encouraging example for an international environmental management of an aquatic system. The implementation of improved wastewater treatment, the policy of restricted usage of chemical compounds (e.g., pesticides) by national regulations, and the enhanced consciousness for environmental issues in the population living in the catchment area were the fundamentals for the improving of water quality and the re-oligotrophication of Lake Constance.

Currently, over 95% of the domestic and industrial wastewater [2] is treated by intensified purification on a high technical standard before entering the lake. But the treated wastewater still contains organic xenobiotics (e.g., pesticides, complex-forming agents, pharmaceuticals, and personal care products), that cannot be efficiently removed by wastewater purification. At this time, it is not possible to assume the ecotoxicological effects of the especially persistent pollutants at the low concentration levels usually observed. The contaminations are still rather low, but must not be neglected. In the future, the long-term behavior of these pollutants and their ecotoxicological effects should be investigated in more thoroughly. After the success of re-oligotrophication, the reduction of point and diffuse sources of these pollutants will be the next challenge for the international environmental management of Lake Constance.

Preventing the pollution of Lake Constance with heavy metals, pesticides, pharmaceuticals, and other organic pollutants is undoubtedly more cost effective than a time-consuming clean up of a contaminated and damaged system. It should be in the interest of all countries along the Rhine River to secure and improve the water quality of the Alpine Rhine and Lake Constance by decreasing the amount of pollution. This will ensure a reliable and steady supply of clean fresh water in the outflow of Lake Constance and downstream the High Rhine.

References

1. Kiefer F (1972) Naturkunde des Bodensees. Thorbecke, Sigmaringen
2. Mührle U, Ortlepp J, Rey P (2004) Der Bodensee Zustand – Fakten – Perspektiven. IGKB, Bregenz
3. Stabel H-H (1998) Arch Hydrobiol 53:13
4. Vollenweider RA (1968) Scientific fundamentals of the eutrophication of lakes and flowing waters, with particular reference to phosphorus and nitrogen as factors of eutrophication. In: OECD Techn Rep DAS/CSI/68 27, OECD, Paris, p 1
5. Rossknecht H (1998) Langjährige Entwicklung chemischer Parameter im Bodensee-Obersee. IGKB-Bericht Nr. 48
6. Maier D, Sinemus HW, Wiedeking E, Fresenius Z (1979) AAS-Bestimmung gelöster Spurenelemente im Bodenseewasser des Überlinger Sees. Anal Chem 296:114
7. Sinemus HW, Maier D, Radziuk B (1987) Zehn Jahre Spurenelementbestimmung im Bodenseewasser. In: Welz B (ed) Atomspektrometrische Spurenanalytik. Bodenseewerk Perkin-Elmer, Überlingen, p 607
8. Stabel H-H, Kleiner J, Merkel P, Sinemus HW (1991) Stoffkreisläufe ausgewählter Spurenelemente im Bodensee. Vom Wasser 76:73
9. Wessels M, Schröder H-G (1998) Lead and zinc in the sediments of Lake Constance, SW Germany. Arch Hydrobiol 53:335

10. Stabel H-H, Kamperdick P, Sinemus HW (1988) Arsengehalte im Bodenseewasser. Vom Wasser 71:163
11. Wendtlandt, Stabel H-H, Wieland K (1990) Atmosphärischer Herbizid-Eintrag in ein Wasserschutzgebiet am Bodensee. In: Jahresbericht der AWBR 21:125
12. Rossknecht H (1998) Die Entwicklung der NTA- und EDTA-Konzentration im Bodensee und einigen Bodensee-Zuflüssen von 1985–1990. IGKB-Bericht Nr. 41
13. Fleig M, Brauch H-J, Kühn W (2003) Ergebnisse der AWBR-Untersuchungen 2003. In: Jahresbericht der AWBR 35:91
14. Ternes TA (1997) Occurence of drugs in German sewage treatment plants and rivers. Wat Res 32:3245
15. Derksen JGM, van Eijnatten GM, Lahr J, van der Linde P, Kroon AGM (2002) Environmental effects of human pharmaceuticals – the presence and risks. RIZA Report 2001.051, RIWA-RIZA, Nieuwegein
16. Ternes TA, Meisenheimer M, McDowell D, Sacher F, Brauch H-J, Haist-Gulde G, Preuss G, Wilme U, Zulei-Seibert N (2002) Removal of pharmaceuticals during drinking water treatment. Environ Sci Technol 36:3855

Hdb Env Chem Vol. 5, Part L (2006): 139–151
DOI 10.1007/698_5_017

Published online: 18 November 2005

MTBE in Lake Zurich

Marcel Leemann (✉) · Richard Forster

Wasserversorgung Zurich, Hardhof 9, CH-8023 Zurich, Switzerland
Marcel.Leemann@wvz.stzh.ch

Abstract Methyl tert-butyl ether (MTBE) is found in Lake Zurich at quite low concentrations between 40 ng L^{-1} in deep water or during the winter season, and up to 1250 ng L^{-1} in surface water on sunny summer days. Investigations showed that on sunny weekends an increase of the MTBE concentration in the upper water layers takes place within a few hours. It can be assumed that this pollution is mainly caused by recreational boat traffic on the lake. Investigations in 2001, followed by monitoring starting in 2002, confirmed this hypothesis. The MTBE concentration starts to increase in early summer, reaches a maximum in late summer, and falls to a low level in the winter season. Most of the MTBE is mixed into water layers at depths above 20 m, which is washed out in the summer within a few weeks when a stratification of the lake is observed. In deep water, on the other side, an unpleasant background concentration of 40 to 50 ng L^{-1} is already established.

Keywords Environment · Methyl tert-butyl ether · MTBE · Water quality

Abbreviations

MTBE methyl tert-butyl ether
BTEX benzene, toluene, ethylbenzene, xylenes
GC gaschromatograph
FID flame ionization detector
MS mass spectroscopy
SPME solid-phase micro extraction
PDMS Polydimethysiloxane
IS internal standard

1 Introduction

1.1 Methyl Tert-butyl Ether

With the introduction of unleaded gasoline, methyl tert-butyl ether (MTBE) was added as an octane enhancer to fuels, replacing toxic organo-lead compounds. Together with the replacement of harmful substances, MTBE also leads to a more complete combustion of gasoline in engines. MTBE is added to fuels at concentrations of up to 9%. It is one of the most commonly produced organic compounds (about 20 million t per year worldwide). With approximately 100 000 t a year it is one of the most used chemicals in Switzerland [1].

Studies in the USA showed that with the widespread use of MTBE as a combustion enhancer the contamination in the surface and groundwater also rose [2].

MTBE has a high water solubility (48 g L^{-1}), a low Henry coefficient (61.34 Pa m^3 mol^{-1}) and low microbial degradability. It is persistent and highly mobile in the environment. Studies from Finland showed that there are no toxicity or health risks to humans [3], even though MTBE showed carcinogenic effects in animal tests. Acute toxicity could be observed for freshwater organisms in another study [4], but only at levels that are significantly higher than the concentrations commonly found in the environment.

Most important sources for the release of MTBE into the environment are losses at refuelling, leakages of gasoline tanks and emission of unburned fuels from engines.

Because MTBE is not a substance occurring naturally in the environment, it is undesirable in surface and groundwater, especially if such water serves as the source for the production of drinking water.

If present in drinking water, MTBE leads to problems with taste and odor starting from concentrations of 5 to 180 μg L^{-1} [5].

1.2 Lake Zurich

Lake Zurich is a deep, perialpine lake in a densely populated region. It is used as the main source for the production of drinking water for nearly one million people. Outside of this, the lake is mostly used for recreation activities.

The lake is usually monomictic, but holomixis is rather exceptional [6]. In warm winters, or in such conditions with poor wind activities, mixing depth reaches only 60 to 100 m. Inverse stratification is observed only in cold winters, mostly during a short period at the beginning of March. Complete ice cover occurs (statistically) once in every 25 years, the last one dating back to 1963. From May until October a stable thermocline develops between 8 and

Table 1 Morphometric data of Lake Zurich

Surface area of lake A_0	65.06 km^2
Catchment area	1740 km^2
Volume of lake V	3.30 km^3
Maximum depth	136 m
Mean depth $V\ A_0^{-1}$	51 m
Average outflow	89.2 $m^3\ s^{-1}$
Theoretical residence time	428 days
Altitude	406 m above sea level
Ratio of catchment area to lake surface	27

50 m, with maximum gradients of 3.5 °C m^{-1}. The temperature in the epilimnion exceeds 20 °C for usually two months and reaches maximum values of 25 °C at the surface.

The lake has been intensively monitored for many decades. In the period from 1937 to 1975, chemical and physical measurements have been made routinely at several sites along the length axis of the lake by the Cantonal Laboratory Zurich [6]. Since then the investigations are done by the Zurich Water Supply at various sampling sites. S. Gammeter et al. summarized the results in 1997 [7].

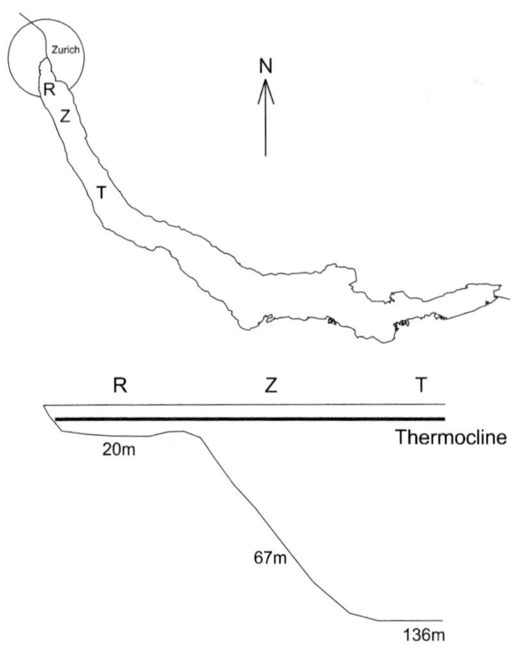

Fig. 1 Lake Zurich with sampling sites and depth profile (R = Riesbach; Z = Zollikon; T = Thalwil)

Theoretically, the water is interchanged in 1.2 years. In reality, the water interchange during the summer stratification happens only in the upper 10 m of the lake. Hence this water has a residence time of about three months. Pollution from the surface during stratification has therefore a lesser effect on the whole lake than during winter time. On the other side, the water in the hypolimnion gets no exchange during stratification. As a result, the residence time of the hypolimnion is about six years. From October to March influx is intermixed into deeper layers than during summer stagnation. Consequently, pollution from the surface is likely mixed into lower layers during full circulation. Over the years, pollution will accumulate slowly, if the input of a pollutant is regular and at a relatively high concentration.

2
Experimental Section

2.1
Analytical Methods

The analytical method applied for the analysis of MTBE allows the detection of benzene, toluene, ethylbenzene and xylenes (BTEX) with one enrichment procedure in one step.

Analyses were carried out in 2001 on a Varian GC 3800 gaschromatograph with split/splitless injector 1177 and flame ionization detector (FID). For the separation of the analytes a GC-column CP-Select 624, 30 m $\times$ 0.32 mm, df 1.8 μm film (Varian) was used.

Since 2002 quantification has been performed on a Varian Saturn 2200 GC/MS, using a GC-column CP-Select 624, 30 m $\times$ 0.25 mm, df 1.4 μm (Varian).

On both instruments headspace solid-phase micro extraction (SPME) on a Combi-PAL autosampler was used for enrichment and injection. SPME enrichments were done with a 75 μm Carboxen/PDMS-fibre (Supelco).

GC- and GC/MS-conditions for analysis are listed in Table 2.

Quantifications were done for both methods with an external calibration curve after correction by the internal standards. Characteristics of the analytical methods listed for GC-FID method in Table 3 and for GC/MS method in Table 4, respectively. The reliability of the methods were validated in two round-robin tests for the determination of MTBE at low concentrations.

2.2
Chemicals

As a calibration standard, a mixture of MTBE, benzene, toluene, ethylbenzene, m-, p- and o-xylene (Dr. Ehrenstorfer, Augsburg, Germany) was used. This mixture contains all analytes at a concentration of 1 g L^{-1} in methanol.

Table 2 GC and GC/MS conditions

	GC 3800	GC/MS Saturn 2200
Carrier gas	Helium, constant flow 2.0 mL min^{-1}	Helium, constant flow 1.0 mL min^{-1}
Make up gas	Nitrogen, 25 mL min^{-1}	–
FID-gas	Hydrogen, 30 mL min^{-1} Synthetic. Air, 300 mL min^{-1}	– –
MS-parameters	–	Segment Low Mass: 50 m z^{-1} Segment High Mass: 160 m z^{-1} Max. Ionization Time: 65 000 μs Quant Ions: MTBE-d^3 (IS) 76 MTBE 73 Benzene 78 Toluene 91 Ethylbenzene 91 m/p-Xylene 91 o-Xylene 91 1,2-Dichlorbenzene (IS) 146
Analyt enrichment	15 min at 20–25 °C, in Vial Headspace	15 min at 20–25 °C, in Vial Headspace
Analyt desorption	Fiber for 20 min in injector, 1 min split OFF, then split ON at 25 mL min^{-1}	Fiber for 20 min in injector, 1 min split OFF, then split ON at 25 mL min^{-1}
Temperatures	Injector 280 °C Detector 300 °C	Injector 280 °C Transfer line 280 °C
Temperature program	5 min at 40 °C 5 °C min^{-1} to 150 °C 40 °C min^{-1} to 250 °C 5 min at 250 °C	5 min at 40 °C 5 °C min^{-1} to 150 °C 40 °C min^{-1} to 250 °C 5 min at 250 °C

Table 3 Analytical method parameters for GC-FID

Substance	Detection limit [ng L^{-1}]	Working range [ng L^{-1}]	Measurement uncertainty [%]	Recoveries [%]
MTBE	100	100–10 000	15	110
Benzene	100	100–10 000	10	110
Toluene	500	500–10 000	10	105
Ethylbenzene	100	100–10 000	10	105
m/p-Xylene	200	200–10 000	10	105
o-Xylene	100	100–10 000	10	108

Table 4 Analytical method parameters for GC/MS

Substance	Detection limit [ng L^{-1}]	Working range [ng L^{-1}]	Measurement uncertainty [%]	Recoveries [%]
MTBE	20	20–1000	15	96
Benzene	50	50–1000	10	105
Toluene	500	500–1000	10	105
Ethylbenzene	50	50–1000	10	106
m/p-Xylene	100	100–1000	10	101
o-Xylene	50	50–1000	10	99

Methanol and water used for standard preparation was HPLC-grade from Baker (Stehelin, Basel, Switzerland). 1,4-Dichlorobenzene (purum, Fluka, Buchs, Switzerland) and Methyl tert-Butylether-d_3 (purum, Fluka (Aldrich), Buchs, Switzerland) were used as internal standards.

3 Results and Discussion

3.1 Influence of Motorboat Traffic on MTBE Concentration

In 2001, the first investigations were carried out to confirm the hypothesis that boat traffic on Lake Zurich is mainly responsible for the occurrence of MTBE.

Two series of samples were taken at two sampling sites (Riesbach, Thalwil, Fig. 1) on weekends with good weather (June 24, 2001 and July 27, 2001). Each series covered one sampling in the morning and one sampling in the afternoon to study the influence of little and intensive recreational boat traffic, respectively.

The detailed results in Table 5 (Riesbach) and Table 6 (Thalwil) show a steep increase of MTBE in the most upper layers at both dates of sampling, but also that the basic pollution in the morning is remarkably higher in July than in June.

Furthermore, a trend to an intermixing of MTBE into deeper layers can be detected at both sampling sites within one month. The weighted means of the MTBE concentrations of the depths from 0 to 7.5 m undergo an increase, particularly at sampling site Riesbach.

Figures 2 and 3 show the results for the sampling sites Riesbach (Fig. 2) and Thalwil (Fig. 3), respectively. For both sampling sites and dates, the difference between morning and afternoon is obvious. In the afternoon, the concentration of MTBE in the upper layer of the lake is clearly higher than in the morning. MTBE is mixed only in the epilimnion but not below the metalimnion. Higher concentrations are found at Riesbach, because most traffic

Table 5 MTBE results for sampling site Riesbach

MTBE ng L^{-1}	24.06.01 Morning	Afternoon	29.07.01 Morning	Afternoon
0 m	410	1250	550	840
1 m	390	405	585	865
2.5 m	160	265	540	645
5 m	90	150	420	320
7.5 m	110	140	130	445
10 m	110	125	95	145
20 m	75	85	80	120
Weighted mean 0 m–8.75 m	173	273	397	538
Weighted mean 8.75 m–20 m	89	101	86	131

Table 6 MTBE results for sampling site Thalwil

MTBE ng L^{-1}	24.06.01 Morning	Afternoon	29.07.01 Morning	Afternoon
0 m	265	385	330	530
1 m	240	325	340	365
2.5 m	245	275	305	330
5 m	120	140	265	255
7.5 m	90	120	135	150
10 m	95	120	100	135
20 m	90	80	70	120
Weighted mean 0 m–8.75 m	164	205	252	273
Weighted mean 8.75 m–20 m	90	96	83	125

passes this sampling site coming from the harbors in Zurich. At Thalwil, the traffic is distributed already on a greater area on the lake.

The lower concentrations in the morning samples point to a partial evaporation of MTBE overnight, but due to the good solubility in water, the evaporation is not complete. There seems to be an accumulation during the summer period. The higher levels in the second sample series indicate such an accumulation of MTBE in the epilimnion of the lake.

3.2 Monitoring MTBE in Lake Zurich

The results of the first investigations led to a monitoring program for MTBE, which began in 2002. The analytical method was changed from GC-FID to GC/MS, resulting in a more sensitive detection. The sampling site Thalwil was

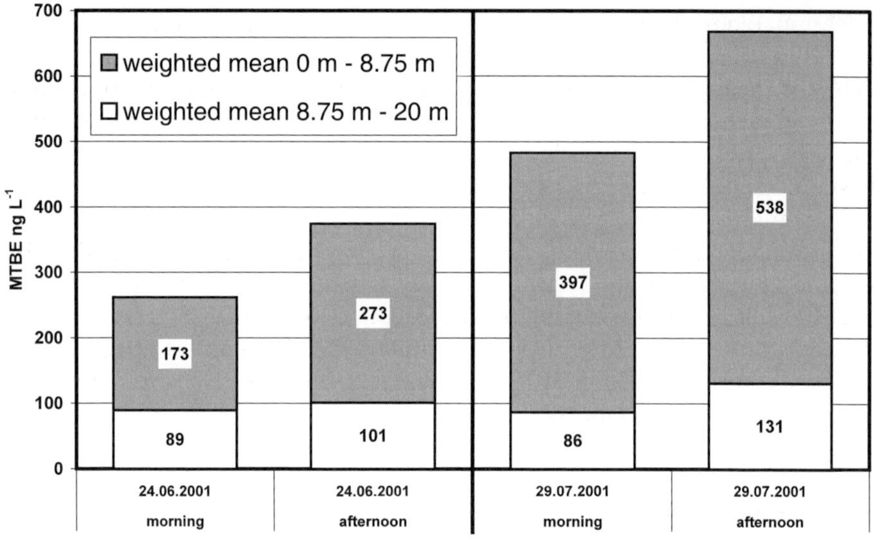

Fig. 2 Average MTBE concentrations at Riesbach, campaign 2001

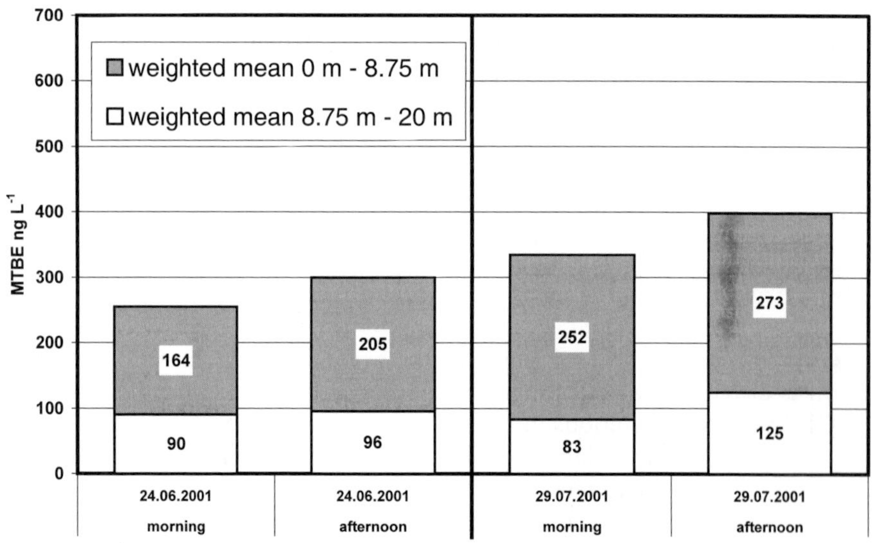

Fig. 3 Average MTBE concentrations at Thalwil, campaign 2001

kept, and the sampling site Riesbach was changed to Zollikon (Fig. 1), which is regularly used for taking water samples by the Zurich Water Supply.

Sampling site Zollikon has less motorboat traffic than Riesbach, but still more than Thalwil. It is situated between the intakes for the two lake water plants of Zurich Water Supply.

Compared to the first analytic series the sampling program was extended and there were also samples taken from layers below the metalimnion. The depth of the lake at site Zollikon is 67 m; the Thalwil site is at the maximum depth of Lake Zurich with 136 m. The goal of this procedure was to find out if the MTBE reaches the hypolimnion, and a change of the concentration in the lower levels of the lake could be observed.

The analytical results show a similar characteristic of MTBE concentrations for both sampling sites, with higher maximum values at Zollikon in summer. Figures 4 and 5 show (for Zollikon and Thalwil, respectively), that MTBE concentrations reach relatively high values in the upper layers (0–5 m), dropping to lower values at depths around the metalimnion. At depths below 20 m the concentrations are stable all year round with concentrations between 30–50 ng L^{-1}. This is important for the water plants at Lake Zurich, because water for the production of drinking water is mostly taken from depths between 30 and 40 m.

At the beginning of the boating season, the MTBE concentrations rise quickly and reach maximum values in late summer. Because the vertical stratification of the lake is still stable by the end of the summer, MTBE intermixed in the upper layers can flow off quite fast. With the beginning of the turnover in autumn the concentration falls sharply within a month.

In 2003 concentrations at both sampling sites reached higher values than in 2002. A reason for this is certainly the extraordinary warm summer in 2003, which led to more recreational activities on the lake.

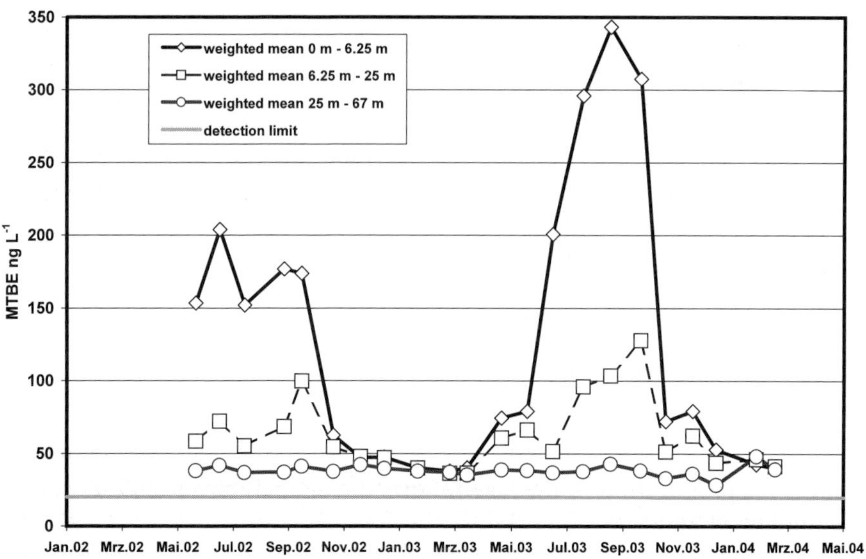

Fig. 4 Zollikon, temporal course of MTBE concentration

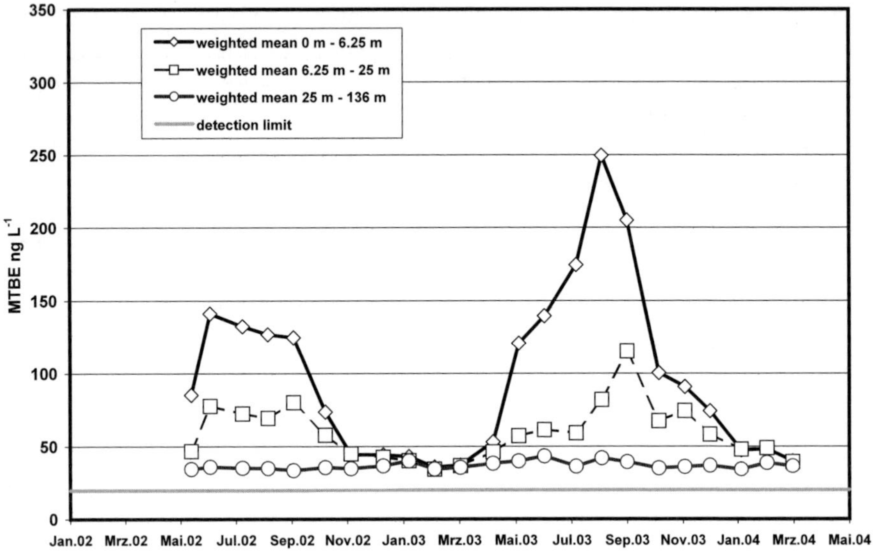

Fig. 5 Thalwil, temporal course of MTBE concentration

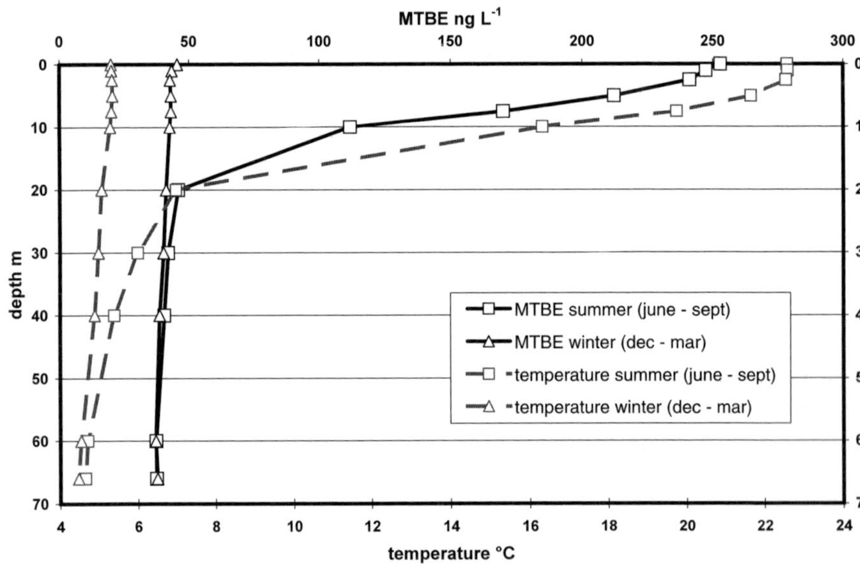

Fig. 6 Temperature and MTBE concentration at Zollikon (summer/winter averages)

MTBE concentration in the lake follows the temperature profile as can be seen in Figs. 6 and 7.

Maximum concentrations of MTBE reached 352 ng L^{-1} at Zollikon and 317 ng L^{-1} at Thalwil, respectively (Table 7).

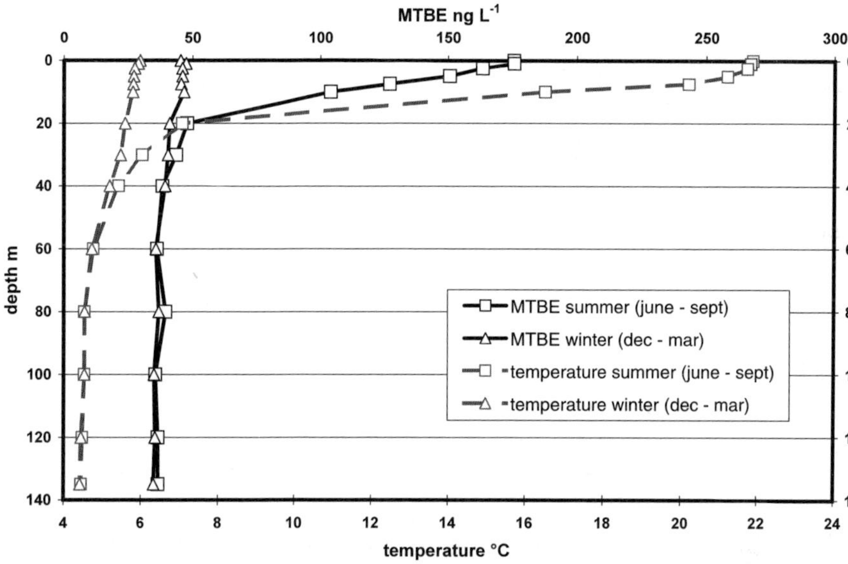

Fig. 7 Temperature and MTBE concentration at Thalwil (summer/winter averages)

Table 7 MTBE concentrations 2002–2003

Depth [m]	Maximum conc. [ng L^{-1}] Zollikon	Thalwil	Minimum conc. [ng L^{-1}] Zollikon	Thalwil
0–6.25	352	317	37	35
6.25–25	299	207	35	32
25–ground	56	51	27	30

Similar behavior can be found for other substances contained in gasoline, such as BTEX. Due to the less solubility, these substances stay in the water for a shorter time and are much less intermixed into the lake than MTBE. BTEX can be detected occasionally in the epilimnion but never in the hypolimnion.

4 Conclusions

The investigations showed that MTBE is accumulated during the summer in the upper layers of Lake Zurich and is caused mainly by boat traffic. Even when changes in the concentration of MTBE can occur within a few days – depending on weather conditions and boat traffic – the pollution rises to a maximum value by the end of the boating season in autumn.

According to today's knowledge, MTBE is not yet a problem in Lake Zurich. The detected maximum concentrations are not of acute toxicity to organisms [4]. Concentrations in the hypolimnion, where all water plants located at Lake Zurich extract water for the production of drinking water, are at levels far beyond thresholds for smell and odor. On the other side, MTBE is a persistent substance that does not occur naturally in surface water and does not belong into lakes and rivers or even drinking water. So it is quite worrying to find concentrations of 40 to 50 ng L^{-1} MTBE in the hypolimnion after only about 10 years of intensive use.

As the observation time was relatively short, it is not yet possible to make statements for future developments of MTBE concentrations in Lake Zurich. Due to the long residence time of the water in the hypolimnion, the MTBE concentrations in these depths are still low. An intermixing of gasoline components happens mainly in summer until autumn, when the lake is in stratification. The contamination with MTBE is not likely mixed into the hypolimnion at these times because the epilimnion water is exchanged within a few months. Nevertheless, an accumulation of MTBE in water layers below the metalimnion is observed at very low levels. A continuation of accumulation can be expected, if recreation boat traffic stays at the same level. If small personal water crafts for high-speed boating with high power two-stroke engines would become common on Lake Zurich, even an increase of MTBE influx could be expected in the next years.

More information about the accumulation and developments of MTBE concentrations is given in the work of Schmidt et al. [8], where model calculations for epilimnion and hypolimnion in Lake Zurich were done. The modelling of Schmidt et al. confirm an accumulation of MTBE in the hypolimnion, but measured MTBE concentrations are about four times higher than the predicted ones.

To be aware of future developments, a monitoring of MTBE is essential. Zurich Water Supply plans to keep the examination program for a few years to get enough information for making prognosis of future developments.

References

1. Lottenbach B, Munz C (2002) Abschätzung der Altlastenrelevanz von Methy-tert-butylether (MTBE). Umweltmaterialien Nr. 151, BUWAL, Bern
2. Gullik RW, LeChevallier MW (2000) Occurrence of MTBE in drinking water sources. Journal AWWA Volume 29, January
3. Saarinen L, Hakkola M, Pekari K et al. (1998) Exposure of gasoline road-tanker drivers to methyl tert-butyl ether and methyl tert-amyl ether. Int Arch Occup Environ Health 71:143
4. Hernando MD, Ejerhoon M, Fernández-Alba R, Chisti Y (2003) Combined toxicity effects of MTBE and pesticides measured with *Vibrio fischeri* and *Daphnia magna* bioassays. Water Res 37:4091

5. Schmidt TC, Haederlein SB (2000) Does the use of MTBE pose environmental problems on Switzerland? Results of a pilot study carried out at EAWAG March–August
6. Oern CG (1980) Die Sauerstoffverhältnisse im Zürichsee (Untersee) von 1937–1975 und ihre Beeinflussung durch meteorologische Faktoren. Vierteljahresschrift der Naturforschenden Gesellschaft in Zürich, pp 259-364
7. Gammeter S, Forster R, Zimmermann U (1997) Limnologische Untersuchung des Zürichsees 1972–1996. Eigenverlag Wasserversorgung, Zurich
8. Schmidt TC, Haederlein SB, Pfister R, Forster R (2004) Occurrence and fate modeling of MTBE and BTEX cpmpounds in a Swiss Lake used as drinking water supply. Water Res 38:1520

Part III
Analysis and Monitoring of Organic Micropollutants and Radioactivity in the River Rhine, Its Tributaries and its Estuary

Hdb Env Chem Vol. 5, Part L (2006): 155–175
DOI 10.1007/698_5_016

Published online: 2 December 2005

Trends in Pesticide Transport into the River Rhine

M. Peschka[1] · J. Müller[1] · T. P. Knepper[1] (✉) · P. Seel[2]

[1]Europa University of Applied Sciences Fresenius, Limburger Straße 2, 65510 Idstein, Germany
Knepper@fh-fresenius.de

[2]Hessian Agency for the Environment and Geology, Rheingaustraße 186, 65203 Wiesbaden, Germany

Abstract The occurrence of relevant pesticides in the River Rhine and two of its tributaries is presented over a period of ten years. Trace determinations of 66 target pesticides and their metabolites in water from the River Main and the River Nidda were performed on continuously sampled wastewater and surface water utilizing different solid phase extraction protocols and detection by gas chromatography mass spectrometry, directly or after derivatization. The transport rates of pesticides in municipal wastewater treatment plant (WWTP) effluents and surface waters were determined from data obtained in 1994, and these show that WWTPs contribute significantly to the pesticide pollution in the surface water. A trial education program providing improved methodology, spraying equipment and support to farmers living close to a single WWTP lead to a drastic reduction (more than 90%) in the total pesticide transport caused by this WWTP.

During two extensive sampling campaigns in 1999 and 2000, mixed samples from a total of 106 (for 1999) and 35 (for 2000) WWTPs in agricultural used areas from Hesse (Germany) were investigated for selected priority pesticides and metabolites. In this case, the mitigation measures mentioned above were found to be unsuccessful overall, which is most likely attributable to less interaction with the pesticide users as compared to projects in small villages with high public attention.

Keywords Monitoring · Pesticides · Rhine · Tributaries · Water analysis

Abbreviations

50-P/90-P	50-percentile/90-percentile
AIPA	Anthranilic acid isopropylamide
AMPA	Aminomethylphosphonic acid
GC	Gas chromatography
GLY	Glyphosate
IPU	Isoproturon
LOD	Limit of detection
MS	Mass spectrometry
NPD	Nitrogen phosphor detector
n.d.	No data
PAC	Phenoxyalkane carboxylates
PU	Phenyl ureas
Rt	Retention time
SIM	Single ion monitoring
SPE	Solid phase extraction
WWTP	Wastewater treatment plant
LC	Liquid chromatography
GC	Gas chromatography
MS	Mass spectrometry
2,4-D	2,4-Dichlorophenoxy acid
MCPA	(4-Chloro-2-methylphenoxy)acetic acid
MCPP	Mecoprop; (*RS*)-2-(4-chloro-2-methylphenoxy)propionic acid
2,4,5-T	2,4,5-trichlorophenoxy acid

1 Pesticides

One of the greatest threats to the environment comes from modern crop protection methods, since an unacceptable aftereffect of pesticide use is the contamination of water sources, including river systems used for domestic supplies of drinking water. The yearly output of pesticides in Germany is about 28 600 tons [1]. Organic synthetic pesticides have been used for more than 70 years to protect crops against damage from organisms such as insects, fungi or herbs [2]. Herbicides are the pesticides sold in the largest quantities in Germany [1], and they are also the most commonly used pesticides in the area of the Rhine catchment. There are three main groups of herbicides (Fig. 1):

- chlorinated phenoxy acids
- triazines
- urea derivatives

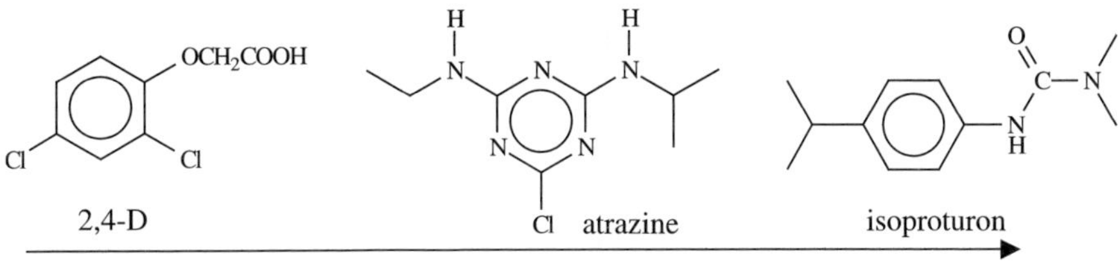

Fig. 1 Examples of the three main chemical classes of herbicides in order of their chronological development

The first organic synthetic herbicides were nitrophenoles [2], which were first used in the 1930s. Chlorinated phenoxy acids became popular after the Second World War, and due to its effectiveness and low cost of production, 2,4-D-((2,4-dichlorophenoxy)acetic acid) is still one of the most popular herbicides in the world [2]. About 40 years ago, herbicides based on triazines and urea derivatives were synthesised by a Swiss company in Basel.

The properties of the herbicides favored for use have changed over the last few decades, shifting from being nonpolar and highly adsorbed by sediment to being polar substances that can easily reach surface water. Finally, the extremely high effectiveness of new herbicides based on sulfonyl ureas mean that that only require a very low application rate, so they tend to be present in very low concentrations in the environment.

2
Fate, Regulations and the Environment

In Germany there are currently 252 licensed pesticides [3]. These can be found in products based upon a single active compound or in products that contain combinations of pesticides. Each active compound needs to pass a licensing procedure. One important criterion of this procedure is that, when applied in at intended and adequate levels, no substance should exceed a concentration of 0.1 $\mu g L^{-1}$ in groundwater. Since 1989, European drinking water regulations have stipulated that potable water should not contain any individual pesticide at levels higher than 0.1 $\mu g\, L^{-1}$, and the concentration of total pesticide should be below 0.5 $\mu g\, L^{-1}$. The establishment of this threshold has prompted the development of advanced analytical methods and instrumentation. It has made industry, farmers, waterworks and authorities take action in order to monitor and minimize the entry of pesticides into surface waters.

Nevertheless, some herbicides leach into groundwater due to their high polarities, and so they exceed allowed concentration levels. Such herbicides include bentazone, isoproturon (IPU) and diuron. Also, even though it has been prohibited for over ten years in Germany, atrazine (which has been superceded by the more adsorbable terbuthylazine) can still be detected in some groundwaters [4]. These facts illustrate the importance of a suitable monitoring program to evaluate pesticides, assess their risks to the aquatic environment, and to trace them back to their sources.

The huge catchment of the River Rhine is subject to the influence of agriculture and industry. In the past few years the Rhine has experienced several environmental accidents; for example, in 1986 several tons of the pesticides disulfotone and thiometone were flushed out into the river with fire-fighting water after a major fire in a chemical plant (Sandoz) in Switzerland. This resulted in the deaths of large numbers of fish 100 km along the river. However, the water quality of the river is still impaired by pesticides; not only due to the risk of industrial accidents, but also due to the pesticides that enter the river from agricultural use each year.

There are several routes that allow pesticides to reach surface waters: run-off, leaching, dispersion, point-source pollution and via wastewater treatment plants [5, 6]. Polar substances that can easily reach the water phase need to be rapidly degradable in soil and in water in order to prevent their dispersion thoughout the whole aquatic environment.

3
Analysis and Monitoring

The high dispersions of pesticides and reduced application rates pose new challenges for analysts. The changing nature of pesticides over the last 40 years has resulted in the continuous development of sample pretreatment, separation and the detection techniques intended for herbicide monitoring.

When pesticide analysis first began (during the 1960s), the most important methods were thin layer chromatography with infrared detection and gas chromatography coupled with an electron capture or a nitrogen phosphor detector (NPD) [7]. In most cases the samples were prepared by liquid–liquid extraction. During the 1970s, gas chromatography became the main analytical method used to separate pesticides. Of these techniques, the NPD is still used today for several families of pesticides, such as carbamates, organophosphorus pesticides or urea derivatives. One disadvantage of gas chromatography (GC) is that the analytes need to be volatile and stable at elevated temperatures, a property which is not associated with polar pesticides. As well as the enrichment step, a second sample pretreatment step is required, a derivatization step. Several methods have been developed for measuring polar analytes with GC [8]. Because it minimizes the pretreatment steps required (and so minimize analyte loss), liquid chromatography (LC) has become more and more popular [9–11]. Nowadays, chromatographic methods coupled to mass spectrometric detectors are widely used for the analysis of pesticides [8]. They provide very high sensitivity and selectivity. Due to the high mass accuracies of modern MSD techniques, they also can be used to determine the structures of unknown compounds [12, 13].

Preconcentration is still generally required, however. Solid phase extraction (SPE) is now an established technique for the preconcentration and clean-up of organic pollutants from a variety of matrices. The most commonly used material for SPE is chemically bonded silica. Numerous attempts have been made to develop new materials that can be used for solid phase extraction of polar compounds; carbon-based sorbents are more efficient than C_{18} silica at trapping polar pesticides [14]. The new generation of polymers is designed for use in the extraction of acidic, basic and neutral compounds, that can be polar or nonpolar.

Other methods for extracting pesticides from environmental samples derive from immunochemistry, which is based on the use of antibodies. Antibodies raised against a specific target pesticide can be immobilized on a solid phase, and these will selectively retain the pesticide, resulting in both preconcentration and clean-up [15].

An overview of the bioanalytical tools used to monitor polar pollutants has been given by C. Nistor and J. Emnéus [16].

The application rate of sulfonylurea is about 10 to 100 $g\,ha^{-1}$ instead [17] of the rate of 1 to 3 $kg\,ha^{-1}$ used for other pesticides classes like triazines or phenylureas, which corresponds to an initial soil concentration of 0.1–1 ppm. Various methods for the determination of sulfonylureas in soil and water have been published [17–25]. The proposed techniques are mainly based on liquid chromatography with UV or MS detection – the limits of detection ranging from 20 to 50 ngL^{-1}.

The reason for huge financial investment in pesticide monitoring is mainly the occurrence of pesticides in surface and ground water, especially that used for drinking water, as well as the risk of direct toxicity to aquatic communities and ecosystems. Moreover, some ecotoxicological effects are still not yet well understood, with some pesticides considered to be probable endocrine disrupters.

It is therefore important to understand the fate of pesticides from their application through to their final appearance in surface and ground waters, which requires that specific regional and local conditions are considered. Knowledge of the most significant sources, the transport pathways and the fates of the different pesticides provides a basis for developing regional guidelines for sustainable agricultural practices with respect to pesticides and water quality.

Agricultural activities, and in particular herbicide use on grain and maize, are the most important diffuse sources of pesticide contamination of natural waters. However, this loss of pesticides from agricultural fields is not the only risk to surface and groundwater. Sewage plants may also play an important role as input pathways of pesticides into surface waters. This has been shown in extensive studies carried out since 1990 (see Chap. 4), which investigated pesticide pollution of the Nidda river (one of the tributaries of the Main river, itself one of the main tributaries of the Rhine), in Hesse, Germany.

Table 1 Application rates, classification and use of herbicides and fungicides in Germany in 1998 [a]

Application rates	Compound	Classification	Use
> 1000 t	IPU	H	grain
	Metamitron	H	beets
500 t – 1000 t	Metazachlor	H	rape
	2,4-DP	H	grain
200 t – 500 t	Bentazone	H	maize/grain
	MCPA	H	grain
	MCPP	H	grain
	Terbuthylazine	H	maize
	Tebuconazole	F	grain
	Epoxiconzazole	F	grain
	Fenpropimorph	F	grain
100 t – 200 t	Fluoxypyr	H	grain
	n-Chloridazon	H	beets
	Ethofumesate	H	beets
	Diuron	H	total

H: Herbicide, F: Fungicide [a] Sourced from [1]

The pesticides selected were those that were used in large amounts in Germany (Table 1), and those that produced known degradation products such as anthranilic acid isopropylamide (AIPA, from bentazone), terbutylazine-desethyl (from terbutylazine) and atrazine-desethyl and -desisopropyl (from atrazine).

Although the use of atrazine has been restricted in Germany since 1991, it remains important to monitor its persistence and the occurrence of its metabolites due to its high mobility and poor biodegradability.

Upon analyzing the pesticide transport into rivers, it became evident that (contrary to former suppositions) the main method of entry is not surface run-off from pesticide-treated fields after heavy rainfalls or by drainage, but via local wastewater treatment plants (WWTPs).

A study was carried out to reveal the sources of pesticide entry into surface waters, to reveal the long-term trends for pesticide pollution in rivers, and to investigate the contribution of WWTPs to pesticide levels in these rivers, utilizing state-of-the-art analytical techniques. The effects of mitigation measures (such as giving farmers feedback and education about the proper use of pesticides) on the final levels of contamination (in order to ascertain possible preventative measures) were also studied.

3.1 Selection of Pesticides and Chemicals

The 62 selected pesticides (fungicides, herbicides, insecticides) were divided into three groups according to their chemical structures: (a) nitro-

Table 2 The pesticides and metabolites investigated, their classification and use, retention times (Rt) and characteristic ions used for GC/MS quantification, and LOD and recoveries ($n = 3$) obtained in surface water and WWTP effluent

No.	Substance	Application	Retention time (min)	Ion 1 (m/z)	Ion 2 (m/z)	LOD Surface water (μg/l)	LOD WWTP-effluent (μg/l)	Recovery Surface water (%)	Recovery WWTP-effluent (%)
	N- and P-Pesticides								
1	AIPA	M	20.8	120	178	0.05	0.06	90	109
2	Alachlor	H	33.8	160	188	0.06	0.12	100	92
3	Ametryn	H	34.1	212	227	0.05	0.1	92	97
4	Atrazine	H	26.6	200	215	0.03	0.06	100	100
5	Atrazine-desethyl	M	22.5	172	187	0.08	0.08	92	104
6	Atrazine-desisopropyl	M	21.9	158	173	0.04	0.16	92	107
7	Azinphos-ethyl	I	62.9	132	160	0.06	0.2	92	102
8	Chlorfenvinphos	I	41.7	267	323	0.03	0.08	88	95
9	Chlorpyrifos	I	37.7	197	199	0.03	0.08	96	110
10	Desmetryn	H	32.0	198	213	0.05	0.1	102	101
11	Dichlobenil	H	10.5	171	173	0.03	0.06	84	120
12	Dimethoate	I	25.4	87	125	0.06	0.16	94	107
13	Epoxiconazole	F	56.3	192	194	0.06	0.06	96	116
14	Ethofumesate	H	36.2	161	207	0.03	0.06	96	102
15	Fenoxaprop-ethyl	H	63.4	268	361	0.03	0.08	98	95
16	Fenpropimorph	F	37.6	128	303	0.03	0.1	98	84
17	Fluchloralin	H	29.6	306	326	0.12	0.24	94	81
18	Furalaxyl	F	42.3	95	242	0.03	0.08	102	105
19	Hexazinone	H	54.7	128	171	0.05	0.08	102	111
20	Iprodione	F	57.7	314	316	0.03	0.08	92	96
21	Metalaxyl	F	34.5	206	249	0.03	0.08	80	83
22	Metamitron	H	47.5	104	174 c)	0.24	0.32	96	114
23	Metazachlor	H	40.4	133	209	0.04	0.08	98	102
24	Methidathion	I	42.9	85	145	0.03	0.08	110	93
25	Methoprotryne	H	47.8	256	271	0.07	0.14	102	96
26	Metolachlor	H	37.1	162	236	0.03	0.06	100	100
27	Metribuzin	H	34.9	198	199	0.03	0.08	78	87
28	n-Chloridazon	H	54.1	77	221 d)	0.20	0.32	86	97
29	Parathion-ethyl	I	37.8	139	291	0.12	0.24	94	107
30	Pendimethalin	H	40.7	252	281	0.20	0.4	96	103
31	Prometryn	H	34.5	226	241	0.05	0.1	88	91
32	Propazine	H	27.0	214	229	0.03	0.06	98	94
33	Propham	H	13.8	93	179	0.03	0.08	116	97
34	Propikonazole	F	53.2 a)	173	259	0.06	0.08	96	111
35	Sebutylazine	H	29.8	200	229	0.05	0.1	100	98
36	Simazine	H	26.1	186	201	0.03	0.06	96	109
37	Tebuconazole	F	54.6	125	250	0.06	0.08	114	109
38	Terbutylazine-desethyl	M	23.3	186	201	0.03	0.06	100	99
39	Terbutryn	H	35.6	226	241	0.03	0.06	106	93
40	Teerbutylazine	H	27.0	214	229	0.03	0.06	98	93
41	Trifluralin	H	23.4	264	306	0.06	0.12	100	86
42	Vinclozolin	F	33.1	212	285	0.03	0.1	92	83
	PAC-Pesticides								
43	2,4,5-T	H	28.8	233	268	0.04	0.08	112	110
44	2,4-D	H	29.8	175	234	0.04	0.08	106	121
45	2,4-DB	H	31.9	101	162	0.05	0.1	114	106
46	2,4-DP (Dichlorprop)	H	20.8	162	248	0.03	0.06	106	123
47	Bentazone	H	33.4	212	254	0.03	0.06	108	107
48	Bromacil	H	26.2 b)	219	221	0.04	0.08	108	86
49	Dicamba	H	16.9	203	234	0.05	0.06	106	92
50	Fluazifop	H	39.9	254	282	0.03	0.06–0.08	96	88
51	Fluoxypyr	H	31.8	209	268	0.04	0.06–0.08	108	97
52	Haloxyfop	H	44.5	316	375	0.03	0.06	106	89
53	MCPA	H	18.8	155	214 e)	0.04	0.08	100	123
54	MCPB	H	28.7	59	101 f)	0.05	0.1	118	110
55	MCPP (Mecoprop)	H	18.1	169	228	0.03	0.06	98	129
56	Triclopyr	H	24.8	210	269	0.03	0.08	102	116
	PU-Pesticides							6	
57	Chlorotoluon	H	25.6	72	226	0.03	0.06	96	97
58	Diuron	H	30.1	72	246	0.04	0.06	100	98
59	Fenuron	H	14.3	72	178	0.1	0.08	98	100
60	Fluometuron	H	13.4	72	246	0.03	0.08	98	101
61	IPU (Isoproturon)	H	24.1	72	220	0.03	0.06	96	99
62	Linuron	H	29.3	174	202	0.05	0.06	108	97
63	Metobromuron	H	26.1	164	212	0.1	0.08	106	103
64	Metoxuron	H	32.8	72	242	0.08	0.16	110	105
65	Monolinuron	H	21.8	140	168	0.05	0.06	108	98
66	Monuron	H	21.9	72	212	0.05	0.06	98	100

a,b) Retention times of isomeric compounds: a) 53.8, b) 35.9, c–f) further ions selected for quantification: c) 202, d) 223, e) 216, f) 242; F: fungicide, H: herbicide, I: insecticide, M: metabolite

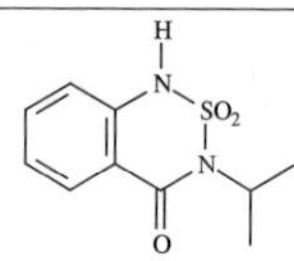	bentazone	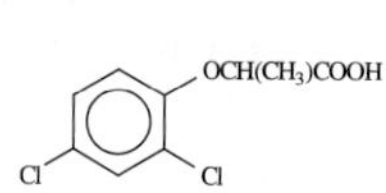	2,4-DP
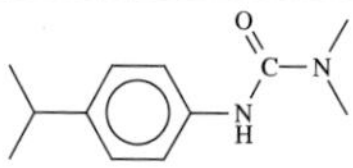	iso-proturon	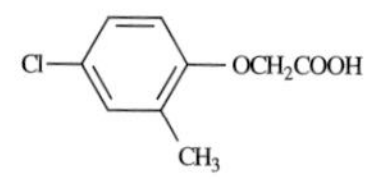	MCPA
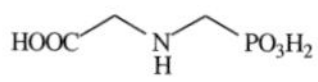	glyphosate		AMPA
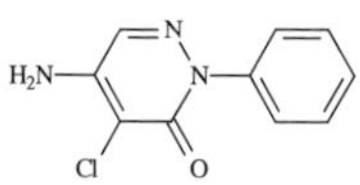	n-chlori-dazon	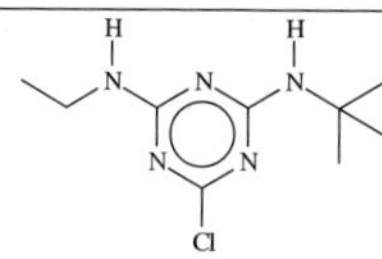	terbuthyl-azine
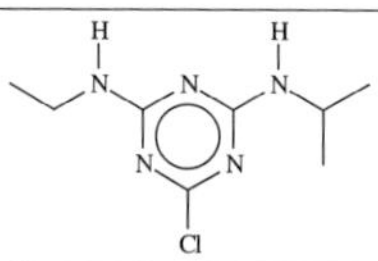	atrazine	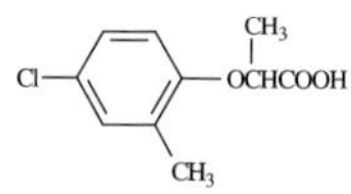	MCPP
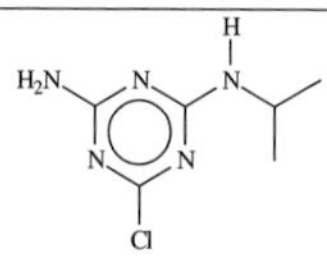	desethyl-atrazine	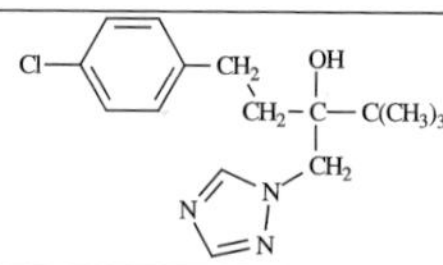	tebu-conazole
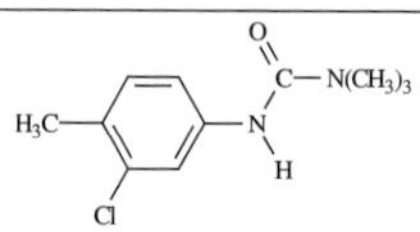	chloro-toluron	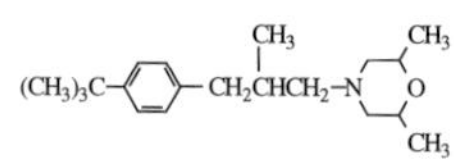	fenpropi-morph
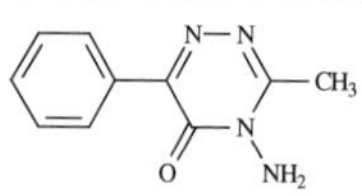	metamitron	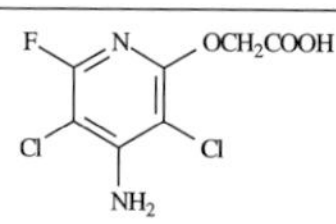	fluoxypyr
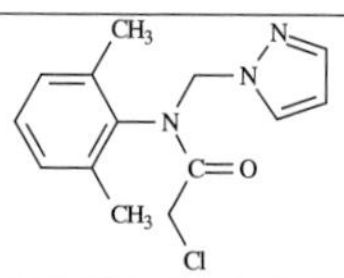	metaza chlor	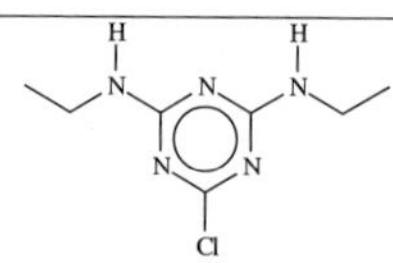	simazine
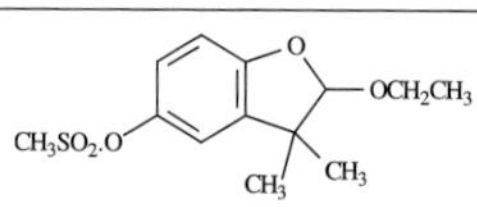	etho-fumesate	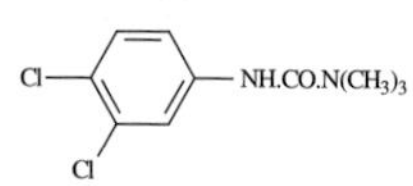	diuron

Fig. 2 Names and formulae of monitored pesticides

gen and phosphorous pesticides (N- and P-pesticides), the two atrazine metabolites atrazine-desethyl and -desisopropyl, the terbutylazine metabolite terbutylazine-desethyl, and the bentazone metabolite AIPA; (b) phenoxyalkane carboxylates (PAC) and (c) phenylureas (PU). These priority pesticides were selected according to their production, use and regulation in the respective region. The names of the pesticides and metabolites investigated as well as their application types (insecticide, fungicide and herbicide) are shown in Table 2. The names and chemical structures of several important pesticides are given in Fig. 2.

3.2 Sampling

Receiving streams (the Main and the Nidda) and WWTP located in agricultural areas were chosen for study in Hesse, Germany. Sampling points are given in Fig. 3.

Fig. 3 Schematic map of sampling points. 1: Bischofsheim (Main), 2: Frankfurt-Nied (Nidda), 3: Cologne (Rhine), • waste water treatment plant

Mainly grain and maize, but also rape and sugar beet are grown over the catchment area of the River Nidda. No companies discharging industrial waste containing pesticides were located on the river. At Frankfurt-Nied, the Nidda joins with the River Main, which subsequently joins with the River Rhine close to Bischofsheim. The Main receives discharges from many chemical industries, including those producing pesticides. Therefore, both Frankfurt Nied and Bischofsheim were included as sampling points.

The period from April to May was selected for sampling, as this time frame reflects the peak period for pesticide application. All of the selected pesticides are applied in the spring in Germany, except for IPU, which is applied in spring and late autumn. In general, a total of 100 mL of WWTP effluent was sampled each day, giving a total sampling volume of 2.1 L.

River samples were taken twice a week from the Rhine (Rhine-km 685.8 (left side)). Sampling and analysis was carried out by GEW RheinEnergie AG. Data from a period of ten years (1993–2003) will be presented.

In the period from 6th April to 17th May 17 1999, a total of 106 WWTP effluent samples were collected twice as three-week mixed samples. Following this sampling, an intensive education and information program for the farmers living in the areas close to the most polluted WWTPs investigated (a total number of 35 WWTPs) was conducted. The sites found to be most polluted in 1999 were then sampled again over the same period in 2000. During the same time period, mixed samples from the WWTP at Woelfersheim, Hesse, Germany were also taken daily from 1994 to 1998. Mixed weekly surface water samples were collected automatically from the Main during pesticide application time (April to June), and collected as two-week mixed samples for the rest of the year.

3.3 Sample Preparation and Analysis

Rhine: Samples were filtered if necessary and then enriched over C-18 cartridges [26].

Main, Nidda, WWTP: All samples were passed through glass fiber filters (0.45 μm) prewashed with methanol and Milli-Q water. Solid phase extraction (SPE) was performed on 1 L of the surface water and 0.5 L of the WWTP effluent samples. SPE of group (a) (N- and P-pesticides) and (c) (PU) pesticides was performed in the neutral pH range, whilst samples of group (b) (PCA) were adjusted to pH 2 by adding 3.5 M sulfuric acid prior to enrichment. The samples were filtered under vacuum (20 mL min^{-1}) through glass cartridges filled with 0.1 g Lichrolute EN (Merck) and 0.25 g Isolute C18ec (IST). Prior to extraction, the cartridges were conditioned with 6 mL *n*-hexane, 6 mL methanol and 10 mL pristine ground water for the group (a) and (c) analyses, and with 10 mL ground water adjusted to pH 2 for group (b); 100 ng atrazine-D5 (for group A) and 200 ng MCPP-D3 (for group B) serving as internal standards were added to the samples. After application, the enriched compounds were dried under a gentle stream of nitrogen gas for 60 min before they were eluted and prepared for analysis by the following methods. Each solvent extraction was allowed 15 min, after which the resulting extract was eluted by passing compressed air through the cartridge.

Group (a), N- and P-pesticides: After eluting with 3× 1.5 mL acetone/ethylacetate (1:1 v:v), the extracts were evaporated under nitrogen gas to 100 μL, external standard was added (fluazifop-butyl, final concentration: 0.7 μg mL^{-1}), and the extract was made up to a final volume of 200 μL.

Group (b), PCA: After eluting with 2 × 1.5 mL methanol, the extracts were evaporated under nitrogen gas to dryness. The samples were filled up to 700 μL and then derivatized using 150 μL diazomethane in diethylether (in excess) at 20 °C, with the reaction being terminated after 60 min through the addition of two droplets of acetic acid in acetone (10% v/v). External standard (heptadecanoic nitrilo acid, final concentration: 1 μg/mL) was added and the extract was made up to a final volume of 1 mL with hexane.

Group (c), PU: After eluting with 2 × 1.5 mL acetone, the extracts were evaporated to dryness under nitrogen gas. Derivatization was then performed by adding 1 mL dioxane, a spatula tip of sodium hydride, and 100 μL methyliodide and allowing the reaction to occur for 70 min at 60 °C. After cooling to room temperature, 10 mL phosphate buffer (0.2 mol L^{-1} Na_2HPO_4 and 0.5 mol L^{-1} KH_2PO_4), 1 mL *n*-hexane and the external standard (heptadecanoic nitrilo acid, final concentration: 1 μg mL^{-1}) were added. The solutions were then mixed, frozen, and after approximately 2 h the organic phase was removed. All samples were stored at –18 °C until GC/MS analysis.

The samples were analyzed with a GC equipped with a mass spectrometer (MS). Qualitative analyses of the individual substances, 10 μg each, were performed in the scanning mode over the mass range m/z 60 to 550 amu. Quantification was performed in selected ion monitoring mode (SIM), based on retention times (Rt) and two or three characteristic ions (see Table 2). Prior to each measurement of a real sample series, a five-point calibration was performed for each of the ions selected for quantification.

4 Results and Discussion

The results obtained demonstrate that GC/MS analysis following SPE is an extremely sensitive and almost matrix-independent method of analyzing a wide range of chemically different priority pesticides (total sum: 62, Table 2) as well as four major metabolites. Derivatization of PAC with diazomethane and PU with sodium hydride and methyl iodide led to detection with almost no background interference.

Linear correlation coefficients between 0.990 and 1.000 were obtained for the calibration curves. The limit of detection (LOD) was set as twice the value of the lowest calibration point analyzed. The lowest responses were observed for metamitron, pendimethalin and *n*-chloridazon, which gave LODs in the range of 0.2 $\mu g\,L^{-1}$ in surface water and between 0.32 and 0.4 $\mu g\,L^{-1}$ in wastewater. The LODs of all of the other compounds investigated were between 0.03 and 0.1 $\mu g\,L^{-1}$ in surface water and between 0.06 and 0.2 $\mu g\,L^{-1}$ in WWTP effluents (Table 2).

Excellent recoveries for the selected pesticides (exhibiting a broad range of chemical polarities) were obtained from spiked surface water samples, with values between 78 and 118%. The recovery rates obtained in spiked wastewater effluents were slightly higher, with values between 83 and 129%, due to stronger matrix effects (Table 2).

4.1 Rhine 1993–2003

Since the end of the 1980s, the concentrations of several important pesticides have decreased significantly in the Rhine. This can be traced back to prohibitions (for atrazine for instance), and strict official conditions for industrial companies. Figure 4 shows the trends for 11 important herbicides

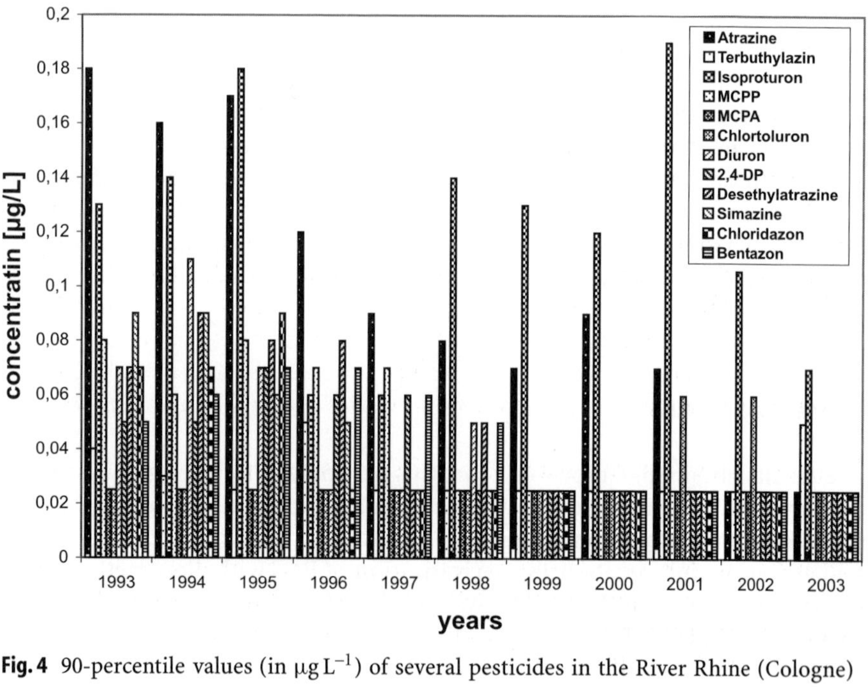

Fig. 4 90-percentile values (in $\mu g\,L^{-1}$) of several pesticides in the River Rhine (Cologne)

over a period of ten years in the Rhine. Data were taken from ARW reports [26–36]. Values < LOD were set at half LOD in Fig. 4.

The concentration of isoproturon (IPU) is the one that stands out in recent measurements. This pre- and post-emergence herbicide usually exceeds $0.1\ \mu g\,L^{-1}$ in the Main (Fig. 5) during its application time and shows no significant reduction in levels in the Rhine over this measurement period (Fig. 4). IPU is comparatively cheap and provides highly efficient weed control. The amount applied exceeds 1000 t per year (Table 1). Increased concentrations can be observed in spring and autumn [32].

The detection of high concentrations of the herbicides isoproturon and chlortoluron led to an interruption in the use of Rhine water to recharge groundwater in the Netherlands [35]. As a result of this, a monitoring program was set up with the aim of investigating the occurrence of IPU, chlortoluron and diuron in the River Rhine and its tributaries, as well as to determine their input pathways. The program found that the three compounds reach the Rhine predominantly via the Mosel. A calculated amount of 2.2 tons of IPU entering the Rhine every two months came mainly from the Mosel catchment area (30%) and from the North Rhine-Westphalia area (30%).

4.2
Main and Nidda

4.2.1
Nidda, Spring 1991–2000

The concentrations of the selected pesticides were also determined in one-week mixed samples from the River Nidda from 1991 until 2000, excluding 1993 (data not shown). The calculated transport rates during their main application time (1st April to 30th June) showed a measurable reduction of up to 50%, for all compounds, over the 1991–2000 period. The maximum transport rate during the investigated period of application was 37 kg, for IPU. Despite its ban in 1991, the level of atrazine did not immediately drop dramatically. The most significant reduction, from 14 to 2 kg, was only observed after 1994. This time lag may be due to the consumption of excess and stored material, as well as the purchase in neighboring countries where the substance was still allowed.

In the Nidda, approximately two-thirds of the total amount of pesticide found in the river can be attributed to input from local WWTPs. To demonstrate this, the transport rates of MCPP atrazine, IPU, diuron and 2,4-DP from the sum of all WWTP effluents investigated were compared with that seen in the Nidda where it joins to the River Main. Similar to all of the other pesticides investigated, MCPP shows almost identical transport patterns in the sewage effluents and the river water.

4.2.2
Main, Spring 1993–2003

The River Nidda river is a tributary of the River Main; they join at Frankfurt-Nied, Germany (Fig. 2). Thus, the transport of the selected priority pesticides was also investigated in the River Main during the years 1993–2003. Figure 5 presents the trends for nine priority pesticides during this period (values < LOD were set at half LOD).

The transport rate patterns (Fig. 6) of the selected pesticides IPU, MCPP and 2,4-DP in the River Main are similar to those observed in the much smaller River Nidda. However, the values obtained for the Main are about one order of magnitude higher, which can be attributed to the significantly higher water flow. Peak transport rates were observed in 1994, 1995 and 1999, mainly related to higher than average water flow in these years. The maximum values determined during the investigated period of time were 435 kg for IPU in 1995, 200 kg for MCPP in 1994 and 180 kg for 2,4-DP in 1994. In 2000, the transport rates were 159 kg for IPU, 63 kg for MCPP and 24 kg for 2,4-DP.

Those pesticides known for high application rates and low elimination rates in WWTP, which are therefore present in WWTP effluents, were also detected in the rivers investigated. Since the Nidda has a much lower water flow rate compared with the Main, and a higher share of wastewater, the amounts and concentrations of pesticides detected were higher than in the Main river. During the main pesticide application time, maximum concentrations of 0.70 $\mu g\,L^{-1}$ and 0.28 $\mu g\,L^{-1}$ were found for IPU in the Nidda and Main

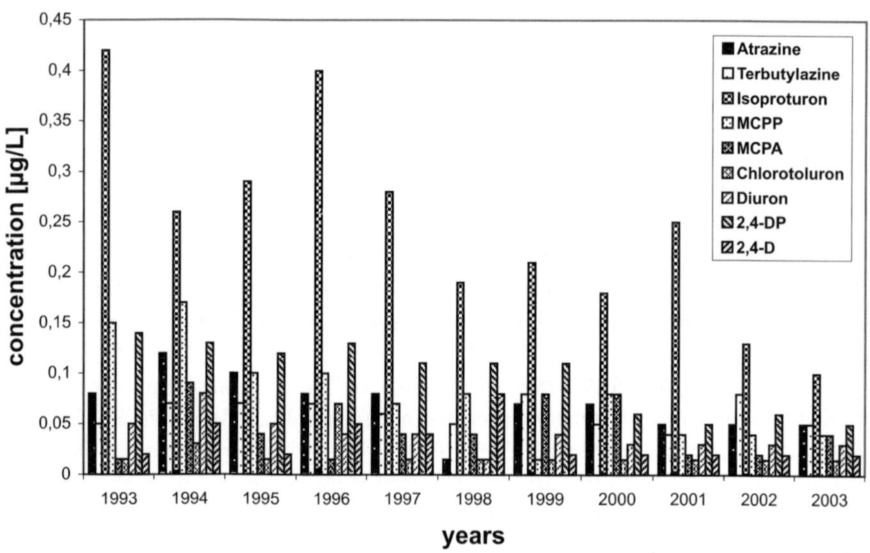

Fig. 5 90-percentile values (in $\mu g\,L^{-1}$) of several pesticides in the River Main (Bischofsheim)

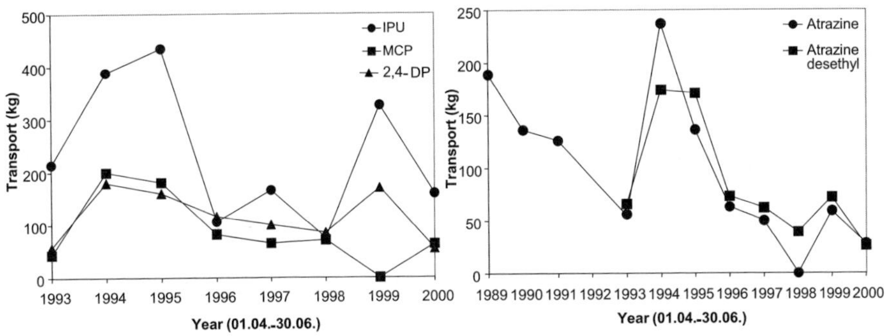

Fig. 6 Transport rates of the selected pesticides 2,4-DP, MCPP, atrazine, atrazine-desethyl and IPU in the Riiver Main during the period of 1st April to 30th June over the years 1993–2000; atrazine 1989–2000

rivers, respectively. Metamitron was detected in the Nidda at concentrations of up to 0.32 $\mu g L^{-1}$, whereas it was never found in the Main.

The behavior of atrazine and its metabolite atrazine-desethyl in the Main revealed some interesting trends. A reduction in the transport of atrazine was observed during the period spanning the beginning of April to the end of June over the years 1989 to 1993, from about 200 kg to 50 kg. However, in 1994 a dramatic rise occurred, to almost 250 kg, mainly related to a significant drop in water flow. Following this, an almost continuous reduction took place again, until a value of 20 kg was reached in 2000. Atrazine-desethyl, monitored since 1993, was found at almost identical concentrations to the parent compound atrazine, and it exhibited similar transport rates. After 1995, the overall transport rates obtained were actually slightly higher for the metabolite.

Analysis of the annual concentration patterns of atrazine and its main metabolite in the River Main in 2000 demonstrated that the concentration of the metabolite is always slightly higher except during the main pesticide application time (week 20 to 30), where concentrations of up to 0.08 $\mu g L^{-1}$ were observed.

Since atrazine is only poorly degradable in WWTP, it was detected in almost all of the surface water samples investigated. Also, according to the literature, the half-life of atrazine in soil is about 50 d. Thus the annual trends can be attributed to both WWTP sources from the main application time and to leaching and run-off during the rest of the year.

The occurrence of both the herbicide glyphosate (N-(phosphonomethyl) glycine) and its main metabolite aminomethylphosphonic acid (AMPA) in the River Rhine is also of growing interest. Glyphosate is a broad-spectrum, non-selective, post-emergence, highly polar and widely used herbicide. It exhibits low toxicity to animals and has been accepted worldwide as an environmentally friendly agent for agriculture application.

Even when the concentrations of glyphosate are below 0.05 $\mu g\,L^{-1}$, the concentrations of AMPA reach almost 0.5 $\mu g\,L^{-1}$ [32]. An additional input of AMPA comes from the use of nitrogen-containing organic phosphonates, which are additives in detergents.

The determination of glyphosate and AMPA in the aquatic environment is somewhat problematic. Glyphosate rapidly degrades into AMPA. Both are highly soluble in water, poorly adsorbable and nonfluorescent. These properties mean that existing analytical methodologies always need to include a derivatization step prior to instrumental analysis. Numerous articles on methods for determining glyphosate and AMPA in the aquatic environment are available [37–43].

In Germany there are 39 products containing glyphosate alone and five products that contain combinations with diuron [32]. In 1999, a project was presented that demonstrated their behavior in the different steps of a raw water facility [32]:

Different samples from the soil passage, the potable water treatment, the raw water treatment facility and from surface waters (Main and Nidda) were analyzed with ion chromatography, post-column derivatization and fluorescence detection (LOD = 0.05 $\mu g\,L^{-1}$). Extraction of the analytes from the water phase was carried out by ion exchange.

Occurrence of glyphosate and its metabolite in the Main and Nidda rivers in the Frankfurt area:
Glyphosate is present in the river Main from April to September at a concentration of up to 0.1 $\mu g\,L^{-1}$. In the Nidda it is present over the whole year at a maximum concentration of 0.4 $\mu g\,L^{-1}$, which is due to the higher amount of waste water in the Nidda. The concentration of the metabolite AMPA exceeds the glyphosate concentration by several times.

Their behavior during the soil passage:
The results from bank filtration experiments show that glyphosate is removed after a distance of about 200 m, whereas AMPA needs about 300 to 500 m to be completely eliminated. The experiments were carried out at the waterside of the Main.

Glyphosate and AMPA are not detectable in groundwater, even though they have been applied in massive amounts around rail tracks since 1991.

The efficiencies of different water treatment steps at eliminating glyphosate and AMPA:
Water treatment at this facility included several steps, namely floc filtration, gravel filtration, and activated carbon filtration. In order to evaluate the efficiencies of those steps, samples were taken before and after each step so that the glyphosate and AMPA could be quantified. The first step, flocculation with activated silicic acid and addition of potassium permanganate and aluminum salts, gave an elimination rate of 39 ± 14% for glyphosate and 22 ± 15% for AMPA. Gravel filtration reduced both by less than 10%. Activated carbon filtration also reduced glyphosate by < 10%, and AMPA by 21 ± 9%. Those re-

sults show that glyphosate and its metabolite are not completely removed in a raw water treatment facility.

4.3 Wastewater Treatment Plant Effluents

4.3.1 Long-Term Transport in a Single WWTP

Daily mixed effluent samples from the WWTP in Woelfersheim were investigated for pesticides and metabolites over the years 1994 to 1998 during a three-week period (4th May to 24th May) (Table 3). Calculated transport values for the priority pesticides, obtained by multiplying the measured concentrations by the effluent flow rates, are listed in Table 3. In 1994 and 1995, the mean total transport of the eight selected pesticides was 484 ± 119 g, with the maximum mean values for metamitron, MCPP and IPU being 200 g, 110 g and 110 g, respectively. Atrazine, 2,4-D, 2,4-DP, MCPA and simazine had mean transport rates of between 30 and 4 g. The relatively high variations observed result from single events, such as the cleaning of spraying tools on paved farmyards, as determined by discussion with the farmers responsible.

In 1995 the farmers in Woelfersheim, Germany were instructed in the proper handling of pesticides and supplied with modern spraying machines that result in almost no pesticide loss after application. This resulted in a drastic decrease in the total pesticide transport, from 484 g in 1994/1995 to 65 g in 1996. The calculated transport rates for 1997 and 1998 were also reduced, with values of 81 g and 50 g, respectively. Comparing the values from 1998 with those from 1994/1995, the reduction obtained was around 90% for all of the compounds investigated.

Table 3 Comparison of the *mean* transport rates of priority pesticides in a single WWTP effluent (from Woelfersheim, Hesse, Germany) during the period of 4th May to 24th May over the years 1994–1998

Pesticide	Transport (g) 1994/1995 a)	1996	1997	1998	Reduction (%) 1994/1995–1998
Metamitron	200 ± 120	15	2	19	
MCPP	110 ± 60	19	20	6	
IPU	110 ± 38	5	7	3.4	
2,4-D	30 ± 7	15	18	14	
2,4-DP	25 ± 15	4	19	1.5	
MCPA	8.0 ± 1.3	4.2	14	5.4	
Simazine	4.0 ± 1.2	2.8	2	0.2	
Atrazine	17 ± 10	2.6	4	0.5	
Sum	484 ± 119	65	81	50	90

a) mean value

4.3.2
Analysis of 106 WWTP Effluents in Spring 1999

Approximately 800 WWTP are located in the state of Hesse, Germany. From these, a representative number of 106 WWTP were chosen, all of which treat mainly municipal wastewater and are located in agricultural areas. During the period of 6th April to 17th May 1999, all of the WWTP effluents were sampled twice as three-week mixed samples for 66 selected major pesticides and metabolites. Altogether, 18 pesticides were detected more than twice, six pesticides were detected only once, and the remaining 42 compounds were never detected above the LOD.

Amongst the pesticides investigated, IPU, 2,4-DP and MCPP were detected in more than 70 samples with concentrations higher than the LOD, and these showed the highest mean concentrations (up to 16.5 $\mu g\,L^{-1}$ for IPU, 8.43 $\mu g\,L^{-1}$ for 2,4-DP, 6.39 $\mu g\,L^{-1}$ for MCPP) over the six-week period. The 50-P-values could only be determined for IPU, 2,4-DP, MCPP and MCPA and they ranged between 0.38 $\mu g\,L^{-1}$ and 0.08 $\mu g\,L^{-1}$, respectively. The 90-P-values could be determined for 12 of the compounds, namely IPU, 2,4-DP, MCPP, MCPA, fluoxypyr, bentazone, ethofumesate, metamitron, atrazine, 2,4-D, simazine and metribuzine. Overall, the values obtained were considerably lower than those reported by Seel et al. [5], which was thought to be due to the analysis of three-week mixed samples, which reduces the occurrence of peak values.

The total herbicide diuron, although not included in the statistical calculations since it is not an indicator of agricultural application, was detected 93 times with values $>$LOD.

4.3.3
Analysis of 35 WWTP Effluents in Spring 2000

In 2000, again from 6th April to 17th May, the 35 WWTP effluents in Hesse, Germany that were shown to be the most polluted during the 1999 sampling campaign were again investigated twice as three-week mixed samples. Between the two sampling campaigns, an education program was given to the farmers, similar to that used at Woelfersheim but with less interaction due to the larger area and increased number of WWTPs.

However, comparison of the calculated pesticide transport values for the 35 WWTP between 1999 and 2000 showed no major changes. Some of the investigated WWTPs had somewhat lower emission rates, whereas others had elevated transport rates. The sum of the transport rates for the investigated pesticides, excluding diuron, over the 35 WWTP was nearly constant at 50 kg in 1999 and 47 kg in 2000. The transport rate for both years was determined to be between 0.1 and 1.0 kg for about 50% of the WWTP, leading to the conclusion that the farmers had not modified their pesticide application and equipment treatment methods following the education program.

Analysis in 2000 of the 35 WWTP observed to be most polluted in 1999 revealed more than 24 samples with IPU, MCPA, 2,4-DP, MCPP, bentazone, tebuconazole (not investigated in 1999), metamitron, fluoxypyr and ethofumesate at concentrations higher than the LOD. The maximum concentrations found were 18.2 $\mu g\,L^{-1}$ for IPU, 10.6 $\mu g\,L^{-1}$ for MCPA, 9.88 $\mu g\,L^{-1}$ for bentazone and 7.96 $\mu g\,L^{-1}$ for metamitron. The 50-P-values for the 13 pesticides detected at the highest concentrations were between 0.96 $\mu g\,L^{-1}$ and 0.07 $\mu g\,L^{-1}$ respectively. Altogether in 2000, 28 pesticides, including two metabolites, were detected more than twice in the investigated three-week mixed samples. Additionally, six pesticides were detected only once, and 32 compounds were never detected above the LOD in these WWTPs.

4.4 Conclusions

In order to study the diffuse pollution of pesticides, it is of crucial importance to combine a validated method with a suitable sampling strategy. It is not sufficient to focus only on the parent compounds; metabolites must also be taken into account. For the challenging task of determining pesticides and their metabolites, mass spectrometric detection seems to be a prerequisite for unequivocal identification and quantification. In order to identify possible pollution sources and to investigate the long-term trend in pesticide behavior, a fairly dense sampling program is also required.

Sampling from sewage drains leading to WWTPs showed that farms connected to sewage drains are the most important source of pollution. Analysis of puddles on roads and paths also showed pesticide contamination, which will also be a source of pesticide entry into sewage drains through rainfall wash-off. It can be assumed, however, that the main sources of pollution are the cleaning of spraying tools in farmyards and the pesticide lost from spraying machines traveling by road.

The diffuse pollution problem is a difficult one to tackle. Mitigation measures to circumvent diffuse pollution, even those resulting from many small point pollutions (such as those that were partially successful in this study) depend strongly on the motivation of the pesticide users and the level of interaction with them achieved, since a measurable result will only be obtained through the responsible application and use of pesticides by the farmers.

For those pesticides exhibiting high polarity and low degradability, only restriction of the number of pesticide-users and substitution by more sustainable compounds will solve the problem of transport and persistence.

Acknowledgements This work is supported by the European Commission (6th Framework programme, Project Aqua Terra GOCE-CT-2005-505428, www.eu-aquaterra.de). We thank Stanislava Gabriel (Hessian Agency for the Environment and Geology, Wiesbaden) for providing us with data and helpful contributions in carrying out this work. We also acknowledge proofreading and helpful discussions of Dr. Friedhelm Karrenbrock, Rheinenergie Köln.

References

1. Industrieverbands Agrar e.V. (2004) Jahresbericht 2003/2004. Industrieverbands Agrar e.V., Frankfurt am Main, Germany
2. Hock B, Fedtke C, Schmidt RR (1995) Herbizide – Entwicklung, Anwendung, Wirkungen, Nebenwirkungen, 1st edn. Georg Thieme Verlag, Weinheim
3. BVL (2004) List of authorized plant protection products in the Federal Republic of Germany (Date: 1st July 2004). Bundesamt für Verbraucherschutz und Lebensmittelsicherheit (BVL), Braunschweig
4. BMU (2005) Daten zur Entwicklung der Grundwasserbelastung durch Pflanzenschutzmittel (webpage). Bundesministerium für Umwelt, Naturschutz und Reaktorsicherheit, Berlin (available at http://www.bmu.de/files/gewaesserschutz/downloads/application/pdf/pflanzenschutzmittel_grundwasser.pdf, last accessed 21st October 2005)
5. Seel P, Knepper TP, Gabriel S, Weber A, Haberer K (1996) Vom Wasser 86:247
6. Seel P, Knepper TP, Gabriel S, Weber A, Haberer K (1994) Vom Wasser 76:357
7. Kuhlmann B, Skark C, Zullei-Seibert N (2003) Vom Wasser 101:55
8. Stalikas CD, Konidari CN (2001) J Chromatogr A 907:1
9. Hu JY, Aizawa T, Magara Y (1999) Water Res 33:417
10. Mills GR (1998) J Chromatogr A 813:63
11. Curini R, Gentili A, Marchese S, Marino A, Perret D (2000) J Chromatogr A 874:187
12. Thurman EM, Ferrer I, Parry R (2002) J Chromatogr A 957:3
13. Bonnington LS, Barcelò D, Knepper TP (2003) J Mass Spectrom 38:1054
14. Hennion MC (1999) J Chromatogr A 856:3
15. Houben A, Meulenberg E, Noij T, Gronert C, Stoks P (1999) Anal Chim Acta 399:69
16. Nistor C, Emnéus J (1999) Waste Manage 19:147
17. Dinelli G, Vicari A, Bonetti A (1995) J Chromatogr A 700:195
18. Galletti GC, Bonetti A, Dinelli G (1995) J Chromatogr A 692:27
19. Dinelli G, Vicari A, Brandolini V (1995) J Chromatogr A 700:201
20. Liu J, Chao J, Jiang G, Cai Y, Liu J (2003) J Chromatogr A 995:21
21. Furlong ET, Burkhardt MR, Gates PM, Werner SL, Battaglin WA (2000) Sci Total Environ 248:135
22. Cao J, Liu J, Wen M, Liu J, Cai Y, Jiang G (2002) J Chromatogr A 955:183
23. Baskaran S, Lauren DR, Holland PT (1996) J Chromatogr A 746:25
24. Ayano E, Kanazawa H, Ando M, Nishimura T (2004) Anal Chim Acta 507:215
25. Penmetsa KV, Leidy RB, Shea D (1997) J Chromatogr A 766:225
26. Geschäftsstelle der ARW (1993) Jahresbericht der ARW. Geschäftsstelle der Arbeitsgemeinschaft Rhein-Wasserwerke e.V. (ARW), Köln, ISSN 0343-0391
27. Geschäftsstelle der ARW (1994) Jahresbericht der ARW. Geschäftsstelle der Arbeitsgemeinschaft Rhein-Wasserwerke e.V. (ARW), Köln, ISSN 0343-0391

28. Geschäftsstelle der ARW (1995) Jahresbericht der ARW. Geschäftsstelle der Arbeitsgemeinschaft Rhein-Wasserwerke e.V. (ARW), Köln, ISSN 0343-0391
29. Geschäftsstelle der ARW (1996) Jahresbericht der ARW. Geschäftsstelle der Arbeitsgemeinschaft Rhein-Wasserwerke e.V. (ARW), Köln, ISSN 0343-0391
30. Geschäftsstelle der ARW (1997) Jahresbericht der ARW. Geschäftsstelle der Arbeitsgemeinschaft Rhein-Wasserwerke e.V. (ARW), Köln, ISSN 0343-0391
31. Geschäftsstelle der ARW (1998) Jahresbericht der ARW. Geschäftsstelle der Arbeitsgemeinschaft Rhein-Wasserwerke e.V. (ARW), Köln, ISSN 0343-0391
32. Post B, Allendorf A, Lange AFTh, Landrieux T (1999) Jahresbericht der ARW. Geschäftsstelle der Arbeitsgemeinschaft Rhein-Wasserwerke e.V. (ARW), Köln, ISSN 0343-0391
33. Geschäftsstelle der ARW (2000) Jahresbericht der ARW. Geschäftsstelle der Arbeitsgemeinschaft Rhein-Wasserwerke e.V. (ARW), Köln, ISSN 0343-0391
34. Geschäftsstelle der ARW (2001) Jahresbericht der ARW. Geschäftsstelle der Arbeitsgemeinschaft Rhein-Wasserwerke e.V. (ARW), Köln, ISSN 0343-0391
35. Geschäftsstelle der ARW (2002) Jahresbericht der ARW. Geschäftsstelle der Arbeitsgemeinschaft Rhein-Wasserwerke e.V. (ARW), Köln, ISSN 0343-0391
36. Fleig M, Brauch HJ, Lindner K, Krauss-Kalweit I (2003) Jahresbericht der ARW. Geschäftsstelle der Arbeitsgemeinschaft Rhein-Wasserwerke e.V. (ARW), Köln, ISSN 0343-0391
37. Nedelkoska TV, Low GKC (2004) Anal Chim Acta 511:145
38. Hidalgo C, Rios C, Hidalgo M, Salvadó V, Sancho JV, Hernández F (2004) J Chromatogr A 1035:153
39. Kudzin ZH, Gralak DK, Drabowicz J, Łuczak J (2002) J Chromatogr A 947:129
40. Patsias J, Papadopoulou A, Papadopoulou-Mourkidou E (2001) J Chromatogr A 932:83
41. Mallat E, Barceló D (1998) J Chromatogr A 823:129
42. Vreeken RJ, Speksnijder P, Bobeldijk-Pastorova I, Noij THM (1998) J Chromatogr A 794:187
43. Cikalo MG, Goodall DM, Matthews W (1996) J Chromatogr A 745:189

Hdb Env Chem Vol. 5, Part L (2006): 177–184
DOI 10.1007/698_5_058

Published online: 1 February 2006

Pharmaceuticals in the River Rhine

H. Weil · T. P. Knepper (✉)

Europa Fachhochschule Fresenius, Limburger Str. 2, 65510 Idstein, Germany
knepper@fh-fresenius.de

Abstract Pharmaceuticals are introduced into the River Rhine mainly via wastewater and can be detected almost permanently in mixed samples taken over the entire year. Depending on use and recalcitrance, the concentrations are in general below 0.1 $\mu g L^{-1}$, except for carbamacepine, diclofenac, ibuprofen and several X-ray contrast media. Thus the detected concentrations are quite comparable to the pesticide concentrations found in the River Rhine, but since pharmaceuticals are applied over the entire year and not depending on the seasonal application, the yearly transport rates can even be higher. Due to the high water flow in the River Rhine, it is quite unlikely, other than in small streams, that peak concentrations above 0.5 $\mu g L^{-1}$ can be detected. Thus there is a strong correlation between the water flow and the share of wastewater.

Keywords Endocrine disruptors · Occurrence · Pharmaceuticals

Abbreviations

STP	sewage treatment plant
LOD	limit of detection
Log K_{OW}	octanol water coefficient
SPE	solid phase extraction
GC/MS	gas chromatography mass spectrometry
MS/MS	tandem mass spectrometry
LC/MS/MS	liquid chromatography tandem mass spectrometry
ARW	association of waterworks in the River Rhine catchment area

1 Introduction

Going hand in hand with an achieved increase of analytical sensitivity, the amount of analytes also detected in the River Rhine increased tremendously during the recent decades. Besides agrochemicals, industrial chemicals, personal care products and detergents as well as pharmaceuticals, endocrine disrupters, disinfectants, and contrast media came into focus. Most investigations on pharmaceuticals in the environment, food or drinking water concentrate on highly consumed compounds like anaesthetics [1, 2], analgesics [1], lipid regulators [3, 4], anticonvulsants [3, 5], hormones [6, 7] and antibiotics [8, 9]. First findings of pharmaceuticals in the aquatic environment were reported in the 1970s [1]. Since the 1980s investigations of the River Rhine and its tributaries showed that some pharmaceutical compounds may reach concentrations above the since then permanently decreasing limit of detection (LOD) [10]. These findings triggered many waterworks along the River Rhine to incorporate the monitoring of pharmaceuticals in their routine monitoring program. The large number of pharmaceuticals in use, with several thousands of active ingredients (in Germany more than 3000) which are used in Europe in around 150 000 different products can enter the aquatic environment in several ways (Fig. 1).

The human drugs are introduced primarily after medication via urinal or faecal excretion into sewage treatments plants (STPs), where they are eliminated to great extent [10, 11]. However, those pharmaceuticals which are polar and poorly biodegradable are finally emitted into surface waters. In

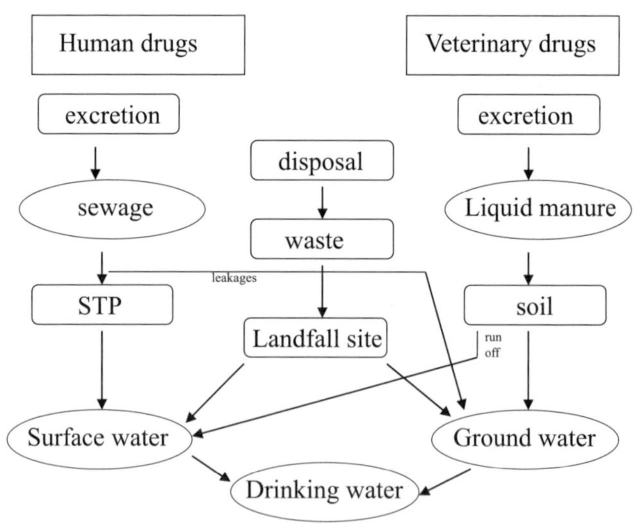

Fig. 1 The fate of pharmaceuticals in the environment [12]

addition, direct disposal of unused medications via wastewater by manufacturers and individuals has been described. Veterinary drugs have further dispersion possibilities, such as the conventional use of liquid manure as fertilisers [12].

2
Groups of Pharmaceuticals

As mentioned above, there are almost countless individual pharmaceuticals and endocrine disruptors, which can be released into the aquatic environment. The most relevant classes of pharmaceuticals can be grouped as follows (see also Table 1): analgesics, antiphlogistics, antipyretics, antirheumatics, lipid-lowering agents, antiepileptica, vasodilators, makrolide and sulfonamide antibiotics, beta-blockers, cytostatics and iodinated X-ray contrast media.

This widespread number of applied pharmaceuticals requires a selection upon those single compounds, which could be relevant for the aquatic environment. The procedure drawn in Fig. 2 summarizes the selection for individual compounds being relevant for the River Rhine and the drinking water produced thereof. Compounds are selected if they have been already detected in drinking water or if their concentrations in raw water exhibit a threshold value going hand in hand with a polarity of a Log $K_{OW} < 3$.

From these, in terms of toxicity, concentration and polarity, the following compound classes seem to play the mayor role in the River Rhine.

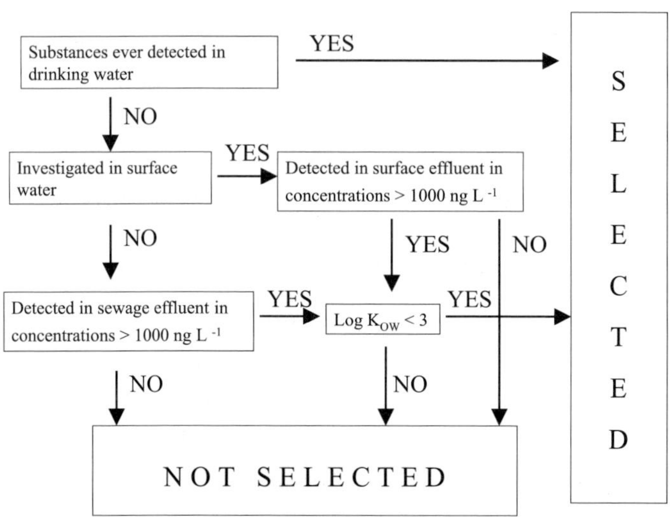

Fig. 2 Selection scheme [12]

Table 1 Routinely investigated pharmaceuticals during different monitoring programs in the River Rhine catchment area; substances in bold were detected at concentrations above the LOD [13, 14]

Analgesics, Antiphlogistics, Antipyretics, Antirheumatics	**Indometacine**
	Diclofenac
	Ibuprofen
	Phenazone
	Propylphenazone
	Phenacetin
	Indometacin
	Fenoprofen
	Ketoprofen
Lipid-lowering agents	**Bezafibrate**
	Fenofibric acid
	Clofibric acid
	Gemfibrozil
	Fenobrifate
Antiepileptica	**Carbamazepine**
Vasodilators	**Pentoxifylline**
Beta-blockers	**Metoprolol**
	Atenolol
	Sotalol
Iodinated X-ray Contrast media	**Iopamidol**
	Iopromide
	Iomeprol
	Amidotrizoic acid
	Iohexol
	Ioxaglic acid
	Ioxitalamic acid
	Iopamide
	Iopan acid
	Iotalamin acid
Makrolide Antibiotics	**Clarithromycin**
	Anhydro-Erythromycin A
	Roxithromycin
	Clindamycin
	Azithromycin
	Oleandomycin
	Spiramycin
	Tylosin
Sulfonamide Antibiotics	**Sulfamethoxazole**
	Trimethoprim
	Sulfadimidin
	Sulfamerazin
	Sulfamethoxol
Others	Chloramphenicol
	Dapson
	Furazolidin
	Metronidazol
	Ronidazol

2.1 Antibiotics

A large body of literature exists upon the occurrence of antibiotics in the environment [15]. Usage in veterinary and animal husbandry, especially aquaculture, plays a major role in the emission of antibiotics into the aquatic environment and results in an eventual human consumption. In addition, antibiotics may be released via hospital effluents.

Nowadays it is discussed that the rampant, widespread (and sometimes indiscriminate) use of antibiotics, coupled with their subsequent release into the environment is the leading proposed cause of spreading resistance among bacterial pathogens.

2.2 Endocrine Disruptors

Estrogenic drugs, primarily synthetic xenoestrogens, are used extensively in estrogen-replacement therapy and in oral contraceptives, in veterinary medicine for growth enhancement, and in athletic performance enhancement. Hormones have a high efficiency and low effect threshold. For example the synthetic oral contraceptive (17α-ethynylestradiol) occurs generally at low concentration ($< 7\ \text{ng L}^{-1}$) in STP effluents, it is still suspected, in combination with the steroidal estrogens 17β-estradiol and estrone [16], of causing vitellogenin production (feminization) in male fish. Xenoestrogens detected in the river Rhine are bisphenol A and nonylphenolic compounds [17].

2.3 Iodinated X-ray Contrast Media, Cytostatics and Disinfectants

Contrast media, such as the iodated X-ray contrast media and disinfectants, are very often not biodegradable and spread widely in the aquatic environment. Especially orally taken pharmaceutics are quite recalcitrant against enzymatic degradation and hydrolysis in the stomach. Contrast media (gadolinium), disinfectants (mercury) and cytostatic agents (platinium) contain heavy metals, which are toxic and not degradable.

Disinfectants are applied in contrast to many other chemical substances because of their specific biological effect, they have a high bacterial toxicity. Cytostatic agents have high carcinogenic, mutagenic or embryotoxic properties.

2.4 Others

In view of the criteria outlined in Fig. 2, additional pharmaceuticals of interest in the River Rhine are:

- Carbamazepine
- Diclofenac

Carbamazepine used for medicinal treatment of schizophrenia in doses up to 1200 mg per day is absorbed slowly and erratically after oral administration. The reported Log K_{OW} is 2.45 [12]. Diclofenac sodium is an anti-inflammatory agent approved for the long-term treatment of rheumatic arthritis, osteoarthritis and ankylosing spondylitis. With a daily dosage of up to 200 mg, it is rapidly and completely adsorbed. Reported Log K_{OW} of diclofenac is 4.4 and of the sodium salt 1.1 [12].

3 Analytical Procedures

Nowadays a series of modern analytical methods exists in order to analyse more than a 100 different compounds in concentrations down to less than 1 ng L^{-1}. And this either in waste, surface water or drinking water. After an enrichment step, which is highly likely a sophisticated solid phase enrichment (SPE), the detection of the individual compounds is either via GC/MS (MS/MS for the endocrine disrupting chemicals) or LC/MS/MS. The tremendous technology evolution in the latter technique nowadays even makes it possible to analyse surface water, such as River Rhine water without any sample treatment [18].

4 Occurrence of Pharmaceuticals in the River Rhine

An overview of various substances, classified according to their pharmaceutical mode of action, which are routinely analysed for during various monitoring programs of waterworks located at the River Rhine is given in Table 1. The first monitoring campaigns started as early as 1996. Going hand in hand with the development of more robust and sensitive analytical procedures, this list is permanently updated. Those compounds which are marked bold have been detected more than once at concentrations above the LOD in the River Rhine. Sampling campaigns can be permanently over the entire year as e.g., at the sampling point in Cologne, Germany, where monthly mixed samples are taken [13] or over a distinct period of time. The latter was the case for a monitoring program in 2002 of the Association of River Waterworks in the Dutch part of the Rhine catchment area [14]. From the 78 pharmaceuticals being looked for, almost each compound could be detected in nearly every River Rhine water sample from the Dutch German border.

Table 2 Concentrations of pharmaceuticals in River Rhine water 2000 (90-percentile in $\mu g\,L^{-1}$) [19]

Substance Rhine-km	Basel, CH 150	Karlsruhe, D 360	Mainz, D 500	Düsseldorf, D 735
Bezafibrate	< 0,010	0,033	0,048	0,063
Carbamazepine	0,023	0,160	0,141	0,220
Dicofenac	0,026	0,037	0,067	0,080
Ibuprofen	0,019	0,016	< 0,005	0,012
Gemfibrozil	< 0,005	< 0,005	< 0,005	0,010

CH = Switzerland; D = Germany

Additional investigations of River Rhine water for estrogens and special antibiotics were also conducted, but did not show any relevant concentrations [data not shown].

Comparing the 90-percentile concentrations of the key pharmaceuticals during the course of the River Rhine from Basel, Switzerland to Düsseldorf, Germany [19], one can observe a strong correlation of the concentrations of the poorly degradable pharmaceuticals and the share of wastewater. There is an concentration increase of all selected pharmaceutical except for ibuprofen, which shows a rapid primary biodegradation during wastewater treatment (Table 2).

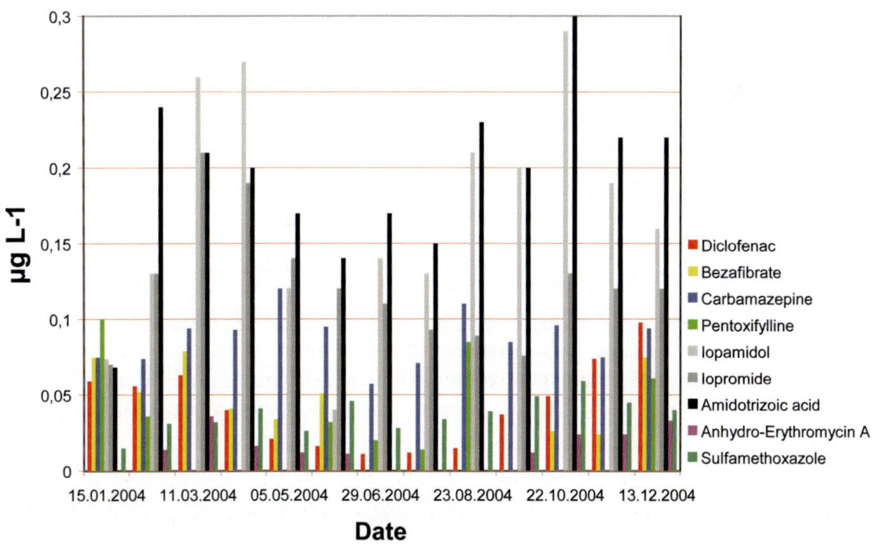

Fig. 3 Pharmaceuticals in Rhine water at Cologne Rhine km 684,5, 15.01.2004–13.12.2004 [13]

Figure 3 gives detailed results of those substances which were regularly detected in the water samples from the River Rhine. From the selected compounds according to the selection procedure mentioned above there are several compounds which can be quantified even in the mixed samples in concentrations above 0.05 $\mu g\,L^{-1}$, e.g., diclofenac, ibuprofen, carbamazepine and several X-ray contrast media. During the last 10 years since the first measurements in the River Rhine started, the concentration pattern of the detected pharmaceuticals is quite conservative and thus in strong correlation with amount of drugs in use. For those compounds, which are not eliminated to great percentage during wastewater treatment [20, 21] the concentrations only vary under the influence of waterflow.

References

1. Kümmerer K (2001) Chemosphere 45:957
2. Hilton MJ, Thomas KV (2003) J Chromatogr A 1015:129
3. Oellers S, Singer HP, Fassler P, Muller SR (2001) J Chromatogr A 911:225
4. Stumpf M, Ternes TA, Wilken RD, Rodrigues SV, Baumann W (1999) Sci Total Environ 225:135
5. Ternes TA, Bonerz M, Schmidt T (2001) J Chromatog A 983:175
6. Richardson SD, Ternes TA (2005) Anal Chem 77:3807
7. Kolpin DW, Furlong ET, Meyer MT, Thurman EM, Zaugg SD, Barber LB, Buxton HT (2000) Environ Sci Technol 36:1202
8. Thiele-Bruhn S (2003) J Plant Nutr Soil Sci 166:145
9. Hirsch RU, Ternes T, Haberer K, Kratz KL (1999) Sci Total Environ 225:109
10. Ternes T (1998) Water Res 32:3245
11. Ternes T et al. (1999) Sci Total Environ 225:81
12. Mons MN, van Genderen J, van Dijk-Looijaard AM (2000) Inventory on the presence of pharmaceuticals in Dutch water. KIWA Nieuwegein, The Netherlands
13. Jahresbericht 2004 der Arbeitsgemeinschaft Rhein-Wasserwerke e.V. ARW 61. Bericht ARW Cologne, Germany 2005, ISSN 0343–0391
14. Sacher F, Stoks P (2003) Pharmaceutical Residues in Waters in the Netherlands. RIWA Amsterdam, The Netherlands
15. Daughton CG, Ternes TA (1999) Environmental Health Perspectives 107:907
16. Desbrow C, Routledge EJ, Brighty GC, Sumpter JP, Waldock M (1998) Environ Sci Techno 32:1549
17. Jahresbericht 2001 der Arbeitsgemeinschaft Rhein-Wasserwerke e.V. ARW 58. Bericht ARW Cologne, Germany 2005, ISSN 0343–0391
18. Seitz W, Weber WH (2005) Oral presentation: Anreicherungsfreie LC-MS-Analytik im Ultra-Spuren-Bereich held on Langenauer Wasserforum 07.11.2005, Langenau, Germany
19. RIWA: Jahresbericht 1999–2000, Teil A Der Rhein (2002) Amsterdam, The Netherlands ISBN 90-6683-096-4
20. Joss A, Keller E, Alder AC, Göbel A, McArdell CS, Ternes T, Siegrist H (2005) Water Res 39:3139
21. Marco B, Müller J, Knepper TP (2006) Water Res (submitted)

Hdb Env Chem Vol. 5, Part L (2006): 185–210
DOI 10.1007/698_5_035

Published online: 10 January 2006

Analysis, Occurrence, and Fate of Aromatic Sulfonates in the Rhine and Its Tributaries

Frank Thomas Lange (✉) · Heinz-Jürgen Brauch

DVGW-Technologiezentrum Wasser, Karlsruher Str. 84, 76139 Karlsruhe, Germany
lange@tzw.de

Abstract An overview is given of the occurrence of aromatic sulfonates in the Rhine river and its tributaries. Since 1990, 25 different monomeric sulfonates and oligomeric sulfonated naphthalene-formaldehyde condensates (SNFC), with up to seven naphthalenesulfonate units in the molecule have been quantitatively analyzed in the aqueous phase by high-performance liquid chromatography coupled with ultraviolet or fluorescence detection. Typical concentrations of permanently occurring sulfonates are in the $\mu g\,L^{-1}$ range. In order to achieve the appropriate low LOQ an extraction is necessary, prior to the chromatographic determination. Usually, ion-pair extraction on reversed-phase C_{18} material is used for this purpose. Due to accidental spills, concentrations of up to several hundred $\mu g\,L^{-1}$ can occur temporarily. SNFC with three or more naphthalenesulfonate building blocks are adsorbed to suspended solids and sediments to more than 99%. The multi-barrier approach applied by the waterworks along the Rhine river including the treatment steps of bankside filtration, flocculation/sedimentation, ozone treatment, and activated carbon filtration is effective in removing most of the sulfonates from the raw waters. However, there are several very persistent compounds—for example, 1,5-naphthalenedisulfonate—which may not be removed through the water purification process in the water treatment plants. Therefore, small residues can also break through into drinking water.

Keywords Analysis · Aromatic sulfonates · Fate · Monitoring data · Partitioning

Abbreviations

1,3,5-NTS	1,3,5-Naphthalenetrisulfonate
1,3,6-NTS	1,3,6-Naphthalenetrisulfonate
1,5-NDS	1,5-Naphthalenedisulfonate
3-NBS	3-Nitrobenzenesulfonate
ARW	Arbeitsgemeinschaft Rheinwasserwerke e.V. (Working group of water treatment plants along the Rhine river)
CAS	Chemical Abstracts Service
DAS	4,4′-Dinitro-2,2′-stilbenedisulfonate (Flavonic Acid)
DOC	Dissolved organic carbon
DNS	4,4′-Dinitro-2,2′-stilbenedisulfonate
ESI	Electrospray ionization
GAC	Granular activated carbon
HPLC	High-performance liquid chromatography
LAS	Linear alkylbenzenesulfonates
LOD	Limit of detection
LOQ	Limit of quantification
MS	Mass spectrometry
SNFC	Sulfonated naphthalene-formaldehyde condensates
SPE	Solid-phase extraction
UV	Ultraviolet

1 Introduction

Aromatic sulfonates are among the oldest industrially produced chemicals. Manufactured since the nineteenth century, they have been used in a variety of applications in aqueous system, e.g., as dyes, as fluorescent whitening agents, as dispersants in textile, leather, and paper manufacturing, as plasticizers for concrete, and in numerous other processes [1–7]. Monomeric benzene-, naphthalene-, anthraquinone-, and stilbenesulfonates can enter the aquatic environment as components or impurities of technical blends and as intermediates or by-products in the synthesis of other sulfonated products. This review comprises the present knowledge of sources and occurrence of these polar monomeric sulfonates and oligomeric naphthalenesulfonate-formaldehyde condensates (SNFC) in the Rhine basin area, excluding sulfonated surfactants like the well-known linear alkylbenzenesulfonates (LAS). For an overview of the occurrence and fate of surfactants, the reader is referred to the contribution of Knepper and Karrenbrock (in this volume).

Important manufacturing sites for a broad range of sulfonated products have existed along the Rhine and its tributaries since before the IG Farbenindustrie AG was established. Therefore, river pollution with polar aromatic sulfonates in this catchment area has been anticipated for a long time.

Possible sources of surface water pollution are accidental spills, industrial wastewaters from the production sites, and municipal wastewaters, which discharge aromatic sulfonates from various applications. However, due to a lack of sensitive, compound-specific methods of analysis, the first report on the occurrence of a stilbenesulfonate in the Rhine river was only published in 1990 [8]. Since that time the number of analytical papers, the number of identified sulfonates (Fig. 1), and the number of reports on quantitative monitoring data (Table 1) have rapidly increased.

Also, the problem of non-availability of reference compounds for quantitative determinations was addressed in a few cases, either by synthesizing such standards on a laboratory scale [9, 10] or isolating them from technical products [11]. Possibilities for trace-analytical determinations of polar aromatic sulfonates are explored in Sect. 2.

Although existing data suggested that the acute toxicity of aromatic sulfonates is low [12], the research in this field was largely driven by the concern that these polar, anthropogenic, and xenobiotic high-production-volume chemicals could reach the drinking-water wells of water treatment plants in the Rhine catchment area. These facilities supply drinking water to about 20 million people.

In laboratory and field investigations polar aromatic sulfonates turned out to comprise a broad spectrum of biodegradabilities [9, 12–32]. Some of these compounds, especially several two- and threefold sulfonated naphthalenes, were found to be persistent under aerobic conditions.

Although most of the treatment plants in the Rhine catchment area possess a high technical standard based on a multi-barrier approach for surface-water purification, some of these compounds can pass natural attenuation processes during riverbank filtration and underground passage and are only incompletely removed by conventional treatment steps like oxidative treatment by ozone and clean-up by activated carbon filtration. This topic is addressed in Sect. 6.

There might be different reasons why, to date, there has been no comprehensive risk assessment of these chemicals. One reason is that a number of products, which contained aromatic sulfonates, were introduced into the market at times when environmental testing was less important than today. A second factor that hampers chemical testing for these compounds is the fact that many of these sulfonates are single components or impurities of technical products or intermediates in production processes. For non-consumer products sometimes pure substances for testing purposes are not available. Furthermore, aromatic sulfonates occur in commercial products either as sulfonic acids or as alkali salts, alkali earth salts, and ammonium salts of the free acids as well as mixtures of these. All of these different forms and technical mixtures have individual CAS registry numbers, which complicates both the collection of existing data and the risk assessment for single compounds.

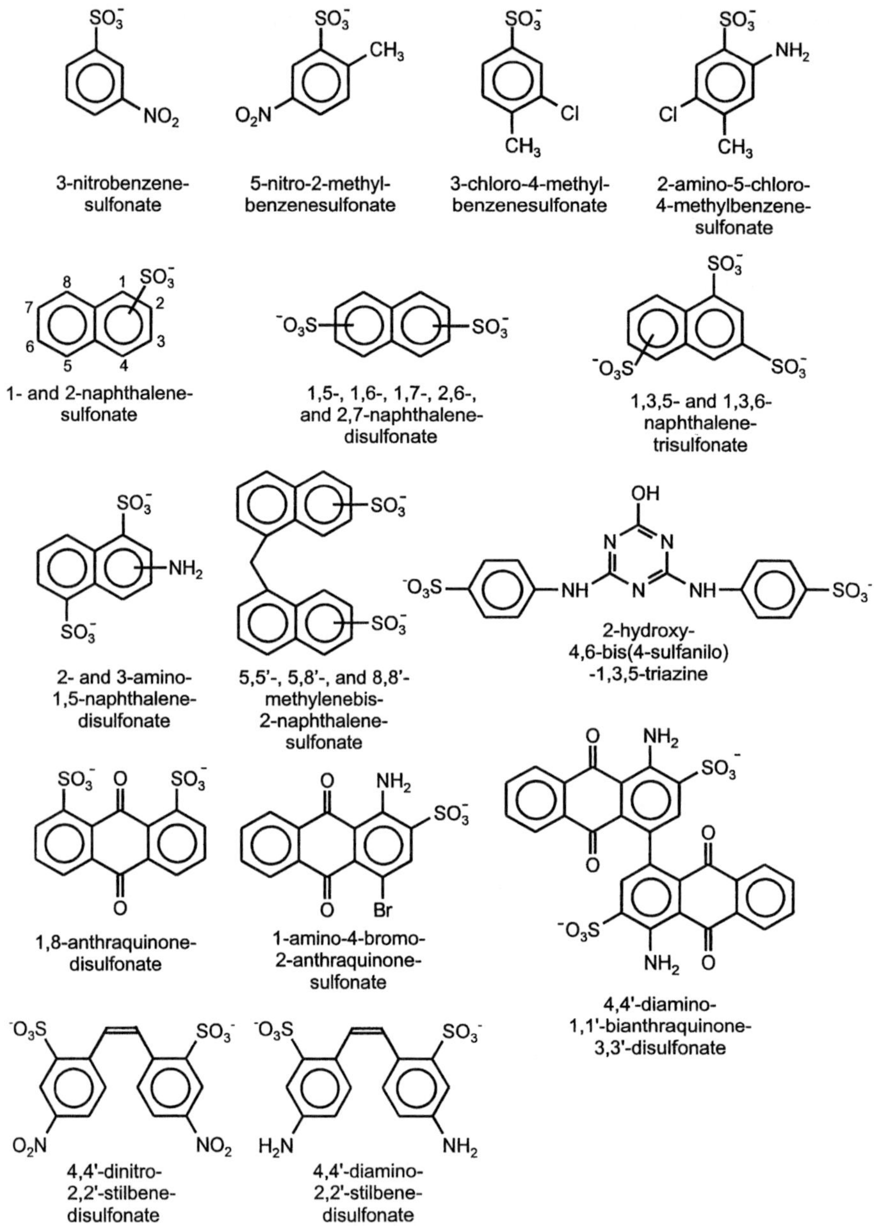

Fig. 1 Structures of aromatic sulfonates identified in the Rhine river and its tributaries

Table 1 Polar aromatic sulfonates identified in the Rhine river or its tributaries

	River	Concentration in $\mu g\,L^{-1}$ [a]	Refs.
Benzenesulfonates			
3-Nitrobenzenesulfonate	LC-trib	27	[64]
	R	0.3–209	[22, 44]
5-Nitro-2-methylbenzene-sulfonate	R	1.8–4.1	[42]
3-Chloro-4-methylbenzene-sulfonate	M	≤ 2.3	[9]
	S/R	≤ 1.1	[48]
2-Amino-5-chloro-4-methylbenzenesulfonate	M	≤ 5.0	[9]
	R	1.0–1.2	[50]
	S/R	≤ 6.2	[48]
Naphthalenesulfonates			
1-Naphthalenesulfonate	LC-trib	≤ 0.2	[63, 64]
	R	≤ 0.25	[44, 54, 55, 62–64, 68, 83]
	L	≤ 0.16	[68]
	N	≤ 0.30	[68]
	K	≤ 0.10	[68]
	Ko	≤ 2.1	[68]
	S/R	≤ 0.45	[48]
2-Naphthalenesulfonate	LC-trib	≤ 0.63	[63, 64]
	R	≤ 2.2	[44, 51–55, 60–64, 68, 83]
	W	≤ 0.03	[68]
	L	≤ 0.91	[68]
	N	≤ 0.55	[68]
	K	≤ 0.25	[68]
	Ko	≤ 10	[68]
	S/R	0.1–1.9	[48]
1,5-Naphthalenedisulfonate	LC-trib	0.09–3.0	[63, 64]
	R	0.1–5.6	[22, 39, 42, 44, 50–55] [59–64, 68, 83–85]
	W	≤ 0.08	[68]
	L	≤ 0.60	[68]
	N	≤ 0.52	[68]
	K	≤ 0.12	[68]
	Ko	≤ 14	[68]
	S/R	0.05–0.53	[48]

Table 1 (continued)

	River	Concentration in $\mu g\,L^{-1}$ [a]	Refs.
1,6-Naphthalenedisulfonate	LC-trib	0.27–4.3	[63, 64]
	R	$\leq$ 0.97	[44, 51–55, 59, 61–64, 68, 83]
	W	$\leq$ 0.04	[68]
	L	$\leq$ 0.81	[68]
	N	$\leq$ 0.63	[68]
	K	$\leq$ 0.62	[68]
	Ko	$\leq$ 7.2	[68]
	S/R	0.06–1.8	[48]
1,7-Naphthalenedisulfonate	LC-trib	0.32–6.4	[63, 64]
	R	$\leq$ 0.94	[44, 52–55, 62–64, 68, 83]
	W	$\leq$ 0.10	[68]
	L	$\leq$ 0.78	[68]
	N	$\leq$ 1.2	[68]
	K	$\leq$ 0.41	[68]
	Ko	$\leq$ 21	[68]
	S/R	0.27–3.1	[48]
2,6-Naphthalenedisulfonate	LC-trib	$\leq$ 0.56	[63, 64]
	R	$\leq$ 0.22	[44, 51–55, 59, 61–64, 68, 83]
	L	$\leq$ 0.17	[68]
	N	$\leq$ 0.20	[68]
	K	$\leq$ 0.10	[68]
	Ko	$\leq$ 1.3	[68]
	S/R	0.2–0.62	[48]
2,7-Naphthalenedisulfonate	LC-trib	0.21–3.1	[63, 64]
	R	$\leq$ 0.48	[44, 50–55, 59–64, 68, 83]
	W	$\leq$ 0.05	[68]
	L	$\leq$ 0.39	[68]
	N	$\leq$ 3.8	[68]
	K	$\leq$ 0.23	[68]
	Ko	$\leq$ 11	[68]
	S/R	0.14–1.5	[48]
1,3,5-Naphthalenetrisulfonate	R	$\leq$ 0.27	[44, 53–55, 63, 64, 68, 83]
	L	$\leq$ 0.06	[68]
	N	$\leq$ 0.23	[68]
	K	$\leq$ 0.03	[68]
	Ko	$\leq$ 0.60	[68]
	S/R	$\leq$ 0.24	[48]

Table 1 (continued)

	River	Concentration in $\mu g\,L^{-1}$ [a]	Refs.
1,3,6-Naphthalenetrisulfonate	LC-trib	≤ 0.85	[63, 64]
	R	≤ 1.1	[44, 52–55, 63, 64, 68, 83]
	L	≤ 0.03	[68]
	N	≤ 0.34	[68]
	K	≤ 0.51	[68]
	Ko	≤ 1.7	[68]
	S/R	0.15–2.1	[48]
1,3,7-Naphthalenetrisulfonate	R	≤ 0.04	[55, 64]
5,5′-Methylenebis-2-naphthalenesulfonate	R	0.02	[10]
5,8′-Methylenebis-2-naphthalenesulfonate	R	0.03	[10]
8,8′-Methylenebis-2-naphthalenesulfonate	LC-trib	0.05–0.48	[10, 63, 64]
	R	≤ 0.69	[44, 53–55, 63, 64, 68, 83]
	W	≤ 0.04	[68]
	L	≤ 0.35	[68]
	N	≤ 0.20	[68]
	K	≤ 0.10	[68]
	Ko	≤ 8.7	[68]
	S/R	0.05–2.5	[48]
2-Amino-1,5-naphthalene-disulfonate	LC-trib	0.11–3.3	[63, 64]
	R	≤ 6.6	[22, 44, 51–55, 61–64, 68, 83]
	W	≤ 0.10	[68]
	L	≤ 0.28	[68]
	N	≤ 0.52	[68]
	K	≤ 0.40	[68]
	Ko	≤ 0.33	[68]
	S/R	0.03–0.49	[48]
3-Amino-1,5-naphthalene-disulfonate	LC-trib	≤ 0.62	[63, 64]
	R	≤ 1.2	[44, 50–55, 60–64, 68, 83]
	W	≤ 0.04	[68]
	L	≤ 0.07	[68]
	N	≤ 0.10	[68]
	K	≤ 0.19	[68]
	Ko	≤ 0.14	[68]
	S/R	0.02–0.25	[48]

Table 1 (continued)

	River	Concentration in $\mu g\,L^{-1}$ [a]	Refs.
Anthraquinonesulfonates			
1,8-Anthraquinone-disulfonate	R	≤ 30	[43]
1-Amino-4-bromo-2-anthraquinonesulfonate	R	≤ 1.5	[52]
4,4′-Diamino-1,1′-bianthraquinone-3,3′-disulfonate	R	≤ 2.0	[55, 63, 64, 68]
Stilbenesulfonates			
cis-/trans-4,4′-Diamino-2,2′-stilbenedisulfonate	R	≤ 3.6	[44, 50, 53, 83]
cis-/trans-4,4′-Dinitro-2,2′-stilbenedisulfonate	R	≤ 9.2	[44, 50–54, 61, 63, 64]
Other sulfonates			
2-Hydroxy-4,6-bis(4-sulfanilo)-1,3,5-triazine	R	≤ 0.93	[50, 51, 54, 61, 63, 64]
	Ko	≤ 1.8	[68]

[a] The concentrations given as "≤ value" indicate that there were values reported above the LOD/LOQ given in the original papers, and the maximum is reported in the table.

K	Körsch river
Ko	Kocher river
L	Leopoldskanal
LC-trib	Lake Constance tributaries
M	Main river
N	Neckar river
R	Rhine river
S/R	Schwarzbach and Rodau rivers (Rhine-Main area)
W	Wiese river

2 Methods of Analysis

As already mentioned in the introduction, quantitative methods for trace-analytical determinations of polar aromatic sulfonates from different kinds of waters were only available since the early 1990s [8]. A good review of the methods for the analysis of benzene-, naphthalene-, anthraquinone-, and stilbenesulfonates covering the analytical progress in this field up to 1995 is given by Reemtsma [33]. Although other methods of analysis have also been reported, solid-phase extraction (SPE) followed by high-performance liquid chromatography (HPLC) coupled with different kinds of detectors is the method of choice. In the last years advances have been made especially by using high-performance liquid chromatography coupled with electrospray-mass spectrometry or with electrospray-tandem mass spectrometry, HPLC-ESI-MS or HPLC-ESI-MS-MS, in the negative ionization mode for identification and quantification of polar aromatic sulfonates [34–38]. However, for monitoring purposes, HPLC with UV diode-array detection or fluorescence detection [8, 39] is still the most widely applied technique. Examples of chromatograms obtained with this methodology are given in Fig. 6 (Sect. 4). The advantages of this technique are the robustness and lower price of optical detectors compared to mass spectrometers, the high sensitivity of fluorescence detection, and the possibility to distinguish isomeric molecules by means of UV and fluorescence spectra.

In order to achieve limits of detection and quantification in the $\mathrm{ng\,L^{-1}}$ to $\mathrm{\mu g\,L^{-1}}$ range, a preconcentration step is required. All methods applied so far in measurements in the Rhine catchment area were based on solid-phase extraction. Often ion-pair extraction using tetrabutylammonium salts as ion-pairing reagents was used. But high-capacity solid-phase materials like styrene-divinylbenzene copolymers [36, 38] and carbon-based material [40] have since become available for this purpose.

When comparing data from different authors it is sometimes not clear which form of the sulfonates was taken as the basis for quantification, i.e., free acid, sodium salts, anhydrous or hydrated salts, etc. To give an example: the difference in concentrations due to the difference in molecular masses between the dianion and the disodium salt of an anhydrous naphthalenedisulfonate is 16% of the smaller mass—that of the dianion. However, for an assessment of the order of magnitude and for monitoring relative changes, e.g., the effectiveness of purification steps during drinking water preparation, these systematic deviations are not relevant.

Besides the analysis of aqueous samples, recently an analytical method for extraction of sulfonated naphthalene-formaldehyde condensates (SNFC) from solid samples like river sediment, suspended solids, and sewage sludge was developed. Details of this method are given elsewhere [41]. Briefly, the solid sample is extracted with an alkaline water–acetonitrile mixture, assisted by ultrasonication, and the extract is analyzed by HPLC with fluorescence detection according to [11]. The LOQ achieved by this procedure are between 0.05 and 0.18 $\mathrm{mg\,kg^{-1}}$ for oligomeric SNFC ranging from two ($n = 2$) to seven ($n = 7$) naphthalenesulfonate units in the molecule. The combined analyses of aqueous samples and suspended solids allowed for the determination of the SNFC partition between these two river compartments (cf. Sect. 5).

3
Occurrence of Aromatic Sulfonates due to Emissions from Production Sites

Short-term emissions, e.g., from shock loads of industrial wastewater treatment plants due to batch processes in the synthesis of certain sulfonated aromatic products, are one possible source of short-term pollution of the Rhine and its tributaries. Another way aromatic sulfonates can enter the rivers are through accidental spills during production processes or during handling of these compounds. Usually after such short-term emissions much higher concentrations up to several hundred $\mu g\,L^{-1}$ were observed in the Rhine or its tributaries than in long term monitoring programs [9, 39, 42–44]. The temporary occurrence of some individual aromatic sulfonates in high concentration can be ascribed to emissions of certain production companies. The following sulfonates from such short-term emissions were detected in the Rhine river or one of its tributaries: 3-nitrobenzenesulfonate, 3-chloro-4-methylbenzenesulfonate, 2-amino-5-chloro-4-methylbenzenesulfonate, *cis*-4,4′-dinitro-2,2′-stilbenedisulfonate (*cis*-DNS), 1,5-naphthalenedisulfonate (1,5-NDS, Armstrong Acid), and 1,8-anthraquinonedisulfonate.

After an accidental spill on 12 January 1991, Brouwer et al. [45] detected anthraquinone-1,8-disulfonate at a maximum concentration of 30 $\mu g\,L^{-1}$ in the Rhine river at Hagenstein (the Netherlands) near the German/Dutch border. Comparable values of 20–44 $\mu g\,L^{-1}$ were reported by the Landesamt für Wasser und Abfall Nordrhein-Westfalen (LWA—today known as Landesumweltamt Nordrhein-Westfalen, or LUA) in Düsseldorf (Germany), and a value of 14 $\mu g\,L^{-1}$ was found in a water sample at Lobith (the Netherlands) on 14 January 1991.

Brouwer et al. also identified 5-nitro-2-methylbenzenesulfonate (4-nitrotoluene-2-sulfonate) after a reported discharge of this compound into the Rhine river in July 1992 [42]. 5-nitro-2-methylbenzenesulfonate readily forms stilbene derivatives, which are used as intermediates in the production of fluorescent whitening agents (see below) and yellow, water-soluble cotton dyes [6]. The maximum concentration of 5-nitro-2-methylbenzenesulfonate measured was 4.1 $\mu g\,L^{-1}$.

Another benzenesulfonate, 3-nitrobenzenesulfonate acid (3-NBS, Ludigol®), was emitted in an accidental spill in 1998 [46] by a manufacturing plant of the BASF AG in Ludwigshafen. 3-NBS is a mild oxidant used in the textile industry. The discharge of 3-NBS was studied in detail by several German and Dutch laboratories. It is currently the best analyzed example of an accidental spill in which an aromatic sulfonate was involved. The emission was reported by the BASF AG to be 3.4 t compared to calculated values of 4.6 and 4.7 t based on the measured concentrations in the Rhine river. The maximum concentration measured was 209 $\mu g\,L^{-1}$ evaluated in a 6 h composite sample taken 4 March 1998, at the monitoring station in Worms, located 10 km downstream of the production plant. Six days after the spill began, the 3-NBS concentration decreased to a maximum concentration of about 7 μg L^{-1} at the German/Dutch border, and the load calculated from the monitoring data decreased to 1.6 t. This decrease of the 3-NBS load was attributed to natural attenuation processes, predominantly by biodegradation. Based on a maximum concentration of about 20 μg L^{-1} on the lower Rhine and an assumed dilution factor of 100, the expected concentration in drinking-water wells of treatment plants using bankside filtered Rhine river water as raw water was estimated to be a value around the LOD of the analytical methods used. There was no report on the detection of 3-NBS in the raw or drinking water of the Rhine treatment plants since that spill, which indicates a further natural attenuation of the 3-NBS concentrations during the bank filtration process.

At the monitoring station Bischofsheim on the Main river, which ends up meeting the Rhine river at Mainz, two more benzenesulfonates, 3-chloro-4-methylbenzenesulfonate and 2-amino-5-chloro-4-methylbenzenesulfonate, were identified in concentrations of up to 2.3 and 4.1 μg L^{-1}, respectively [9]. The latter compound is used as a so-called CLT acid or Lake Red C acid as a diazo component for azo dyes and azo pigments [1]. One important pigment is Pigment Red 53 : 1 [47], patented in 1902 by Farbwerke vorm. Meister Lucius & Brüning in Höchst a. M., the company preceding the Hoechst AG. In Germany Pigment Red 53 : 1 was produced by only one company (Hoechst AG, known today as Clariant), which is situated on the Main river. In 1991

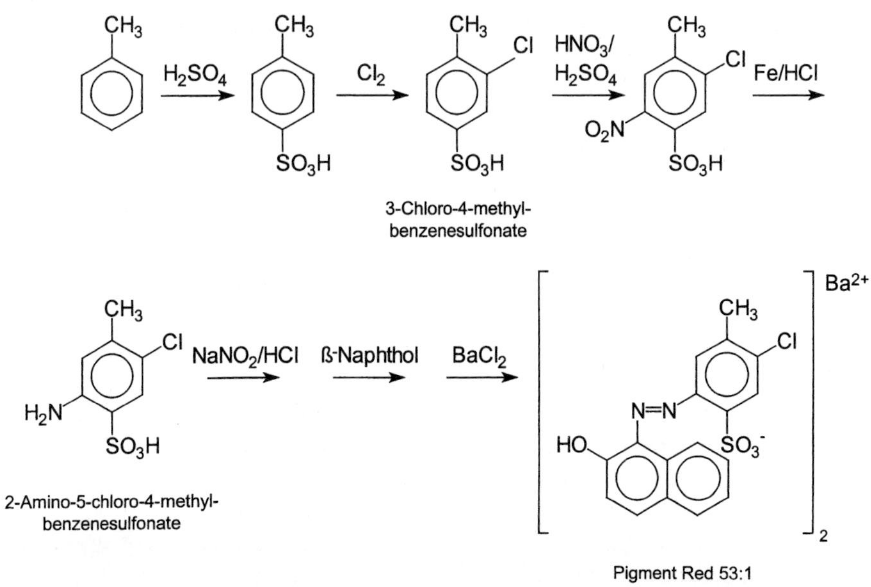

Fig. 2 3-Chloro-4-methylbenzenesulfonate and 2-amino-5-chloro-4-methylbenzenesulfonate (CLT acid) as intermediates in the production of Pigment Red 53 : 1 (adopted from [9])

the production volume was 1500 t. CLT acid is produced from a nitrated precursor, which is synthesized from 3-chloro-4-methylbenzenesulfonate by nitration (Fig. 2).

CLT acid is not completely removed in the sewage treatment plant of the manufacturer. In seven out of 21 samplings of the sewage treatment plant effluent, the concentration of CLT acid ranged between 1 and 2 $mg\,L^{-1}$. In the receiving water, the Main river, both CLT acid and its precursor, 3-chloro-4-methylbenzenesulfonate, were found several times during a sam-

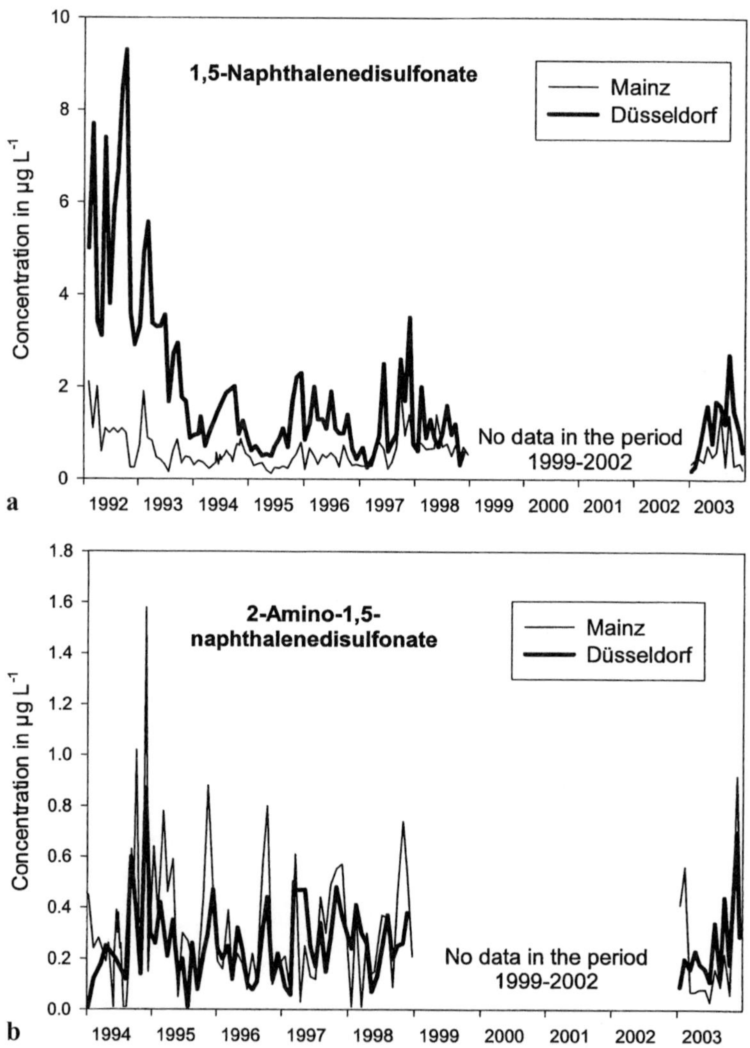

Fig. 3 Concentrations of **a** 1,5-naphthalenedisulfonate (1,5-NDS) and **b** 2-aminonaphthalene-1,5-disulfonate in the Rhine river

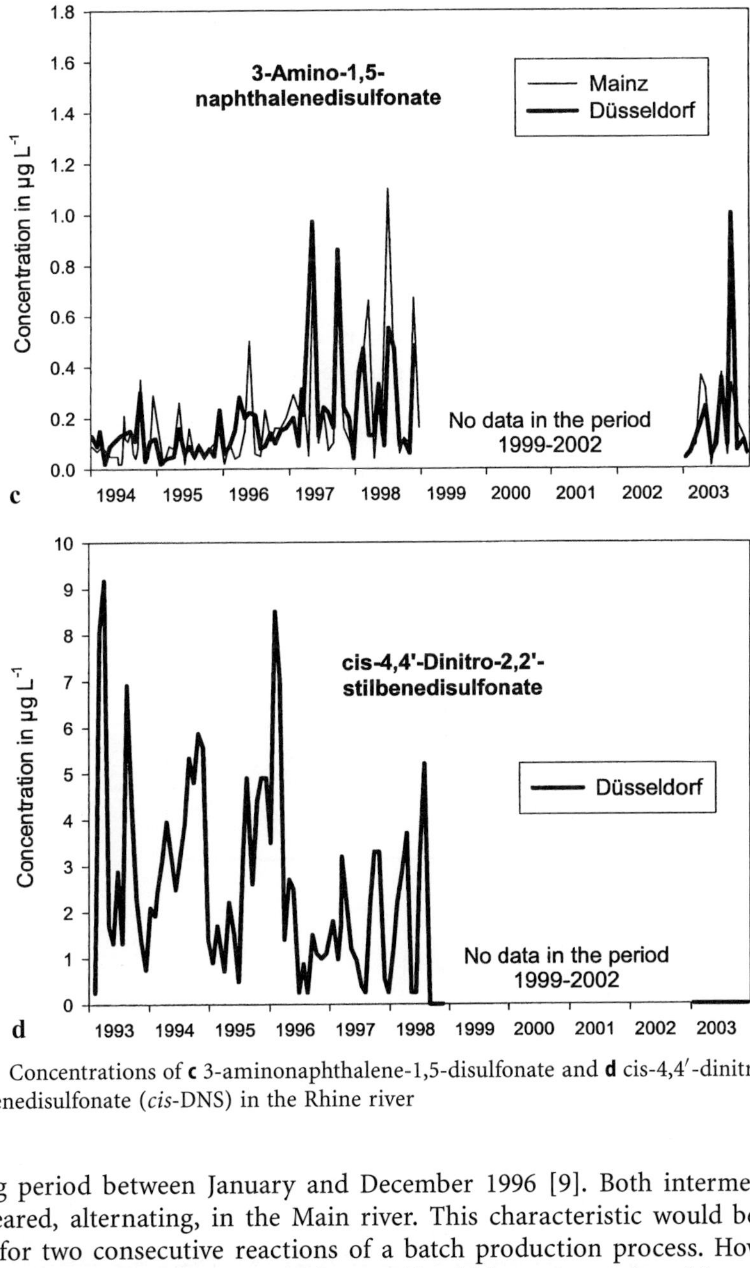

Fig. 3 Concentrations of **c** 3-aminonaphthalene-1,5-disulfonate and **d** cis-4,4′-dinitro-2,2′-stilbenedisulfonate (*cis*-DNS) in the Rhine river

pling period between January and December 1996 [9]. Both intermediates appeared, alternating, in the Main river. This characteristic would be typical for two consecutive reactions of a batch production process. However, at the production site in Frankfurt, CLT acid has not produced been produced since 1992, but it reacted with β-naphthol to give Pigment Red 53 : 1 [9]. Therefore, it is possible that other reactions involving 3-chloro-4-methylbenzenesulfonate were the reason for the appearance of both benzenesulfonates in the Main river. 3-Chloro-4-methylbenzenesulfonate can also be discharged with production wastewater of the synthesis of an intermediate for pesticide production [48].

Examples of a naphthalenesulfonate and two stilbenesulfonates, which were emitted by the Bayer AG at Leverkusen on the lower Rhine, are 1,5-NDS, *cis*-DNS, and *cis-/trans*-4,4′-diamino-2,2′-stilbenedisulfonate (*cis-/trans*-DAS, Flavonic Acid) [9]. 1,5-NDS is an intermediate in the production of building blocks for azo dyes and is also described as a stabilizer for diazo compounds [6]. The concentration of 1,5-NDS in the Rhine river at Düsseldorf, downstream of the industrial wastewater treatment plant of the Bayer AG, is depicted in Fig. 3a.

Since monitoring began, the 1,5-NDS concentration at Düsseldorf decreased from high values of up to 9 $\mu g\,L^{-1}$ to values typically between the LOQ of 0.02 $\mu g\,L^{-1}$ and about 3 $\mu g\,L^{-1}$.

DNS and DAS are important intermediates in the production of fluorescent whitening agents (Blankophor®, Bayer) and water-soluble cotton dyes [6, 49]. DNS is the first aromatic sulfonate, which was identified in the Rhine at Düsseldorf in 1990 by Schullerer et al. [8] and measured in the monitoring programs of the working group of the Rhine treatment plants, ARW (Fig. 3d) [44,

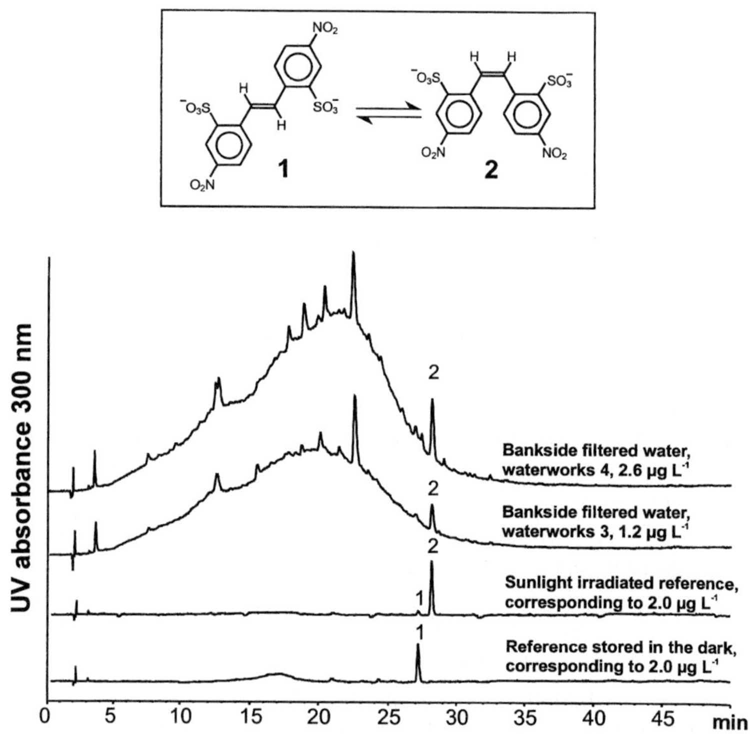

Fig. 4 HPLC study on the *cis/trans* photo-isomerism of DNS (modified from [59])

50–55]. The production of fluorescent whitening agents was estimated to increase from the end of the 1990s until the year 2003 to 230 000 t a^{-1} worldwide. In 1997, with a production of 40 000 t that year, Bayer held a market segment of nearly 25% [46]. In spite of the increasing production volumes of fluorescent whitening agents at the Leverkusen site of the Bayer AG, since 1998, when the Bayer AG put a new production plant for fluorescent whitening agents into operation, *cis*-DNS has not been detected in the Rhine river at Düsseldorf above its LOD of 0.5 μg L^{-1}.

DNS is an interesting compound from an analytical point of view. Like other stilbene derivatives, e.g., DAS and the fluorescent whitening agents produced from DAS [56–58], it undergoes a photo-isomerization reaction in sunlight [9, 59]; see Fig. 4. *Cis*- and *trans*-forms possess different physicochemical properties. Therefore, they can be distinguished by retention times in HPLC analysis and by their UV spectra. In industrial wastewater both *cis*- and *trans*-isomers were identified [9]. Also, the dipotassium salt of DNS, which was used as a reference compound, contains the *trans*-form. However, if a standard solution or an aqueous sample containing DNS traces is exposed to sunlight for several hours, the equilibrium solution contains the *cis*-isomer as the main product.

4
Long-term Pollution of the Rhine River with Aromatic Sulfonates

In contrast to short-term emissions like most of the examples given in Sect. 3 there are a number of sulfonates that were regularly detected in the monitoring programs of the Rhine river water treatment facilities between 1993 and 2003 [44, 50–55, 60–64]. The most frequently found sulfonated compounds were naphthalenesulfonates and aminonaphthalenedisulfonates (Fig. 1, Fig. 3a–c, Fig. 6). Naphthalenemono-, -di-, and -trisulfonates are ubiquitarily distributed in anthropogenic influenced rivers [65]. These compounds originate both from production sites and from their multiple applications. One important source of naphthalenemono- and -disulfonates are technical blends, consisting of condensates of naphthalenesulfonates with formaldehyde (SNFC = sulfonated naphthalene-formaldehyde condensates; see Figs. 5 and 6) [9–11, 38, 66–72].

These SNFC are used as synthetic tanning agents and as dispersants in many applications, e.g., as dye dispersants in textile dyeing or as concrete plasticizers. The simultaneous occurrence of the monomeric naphthalenesulfonates together with low molecular oligomeric condensation products (Fig. 7) supports the expectation that the monomers are emitted into the Rhine mainly due to synthesis and application of such technical products [9, 10, 44, 53, 54, 62–64]. As demonstrated in Fig. 7, the concentration of the SNFC oligomers in the aqueous phase decreases with increasing molecular

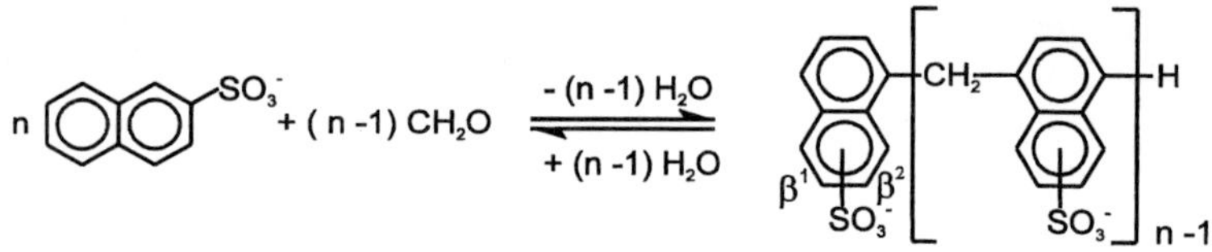

Fig. 5 Formation of sulfonated naphthalene-formaldehyde condesnates (SNFC) by condensation of 2-naphthalenesulfonate with formaldehyde

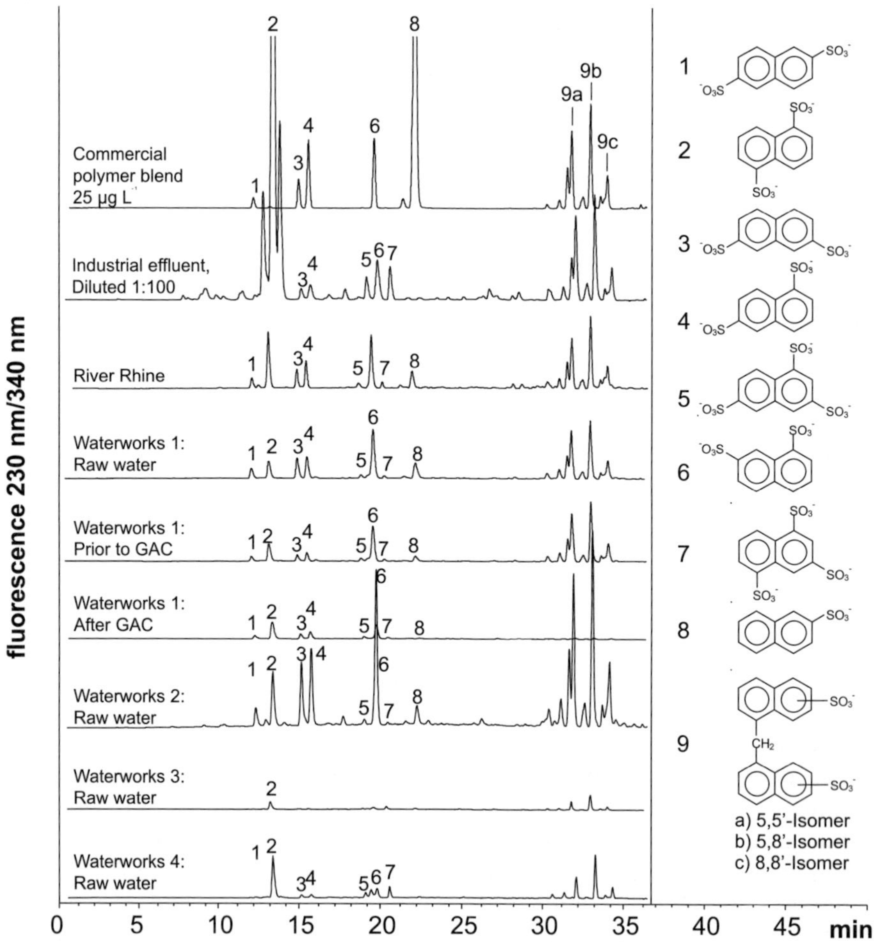

Fig. 6 Examples of chromatograms recorded by ion-pairing HPLC with fluorescence detection at a wavelength combination of $\lambda_{ex} = 230$ nm/$\lambda_{em} = 340$ nm (reconstructed from [9] and [59])

weight. The assumption that the higher molecular oligomers, which cannot be detected in the aqueous phase, are adsorbed to suspended solids and sediments was proved in a survey of the state of Baden-Württemberg in the upper

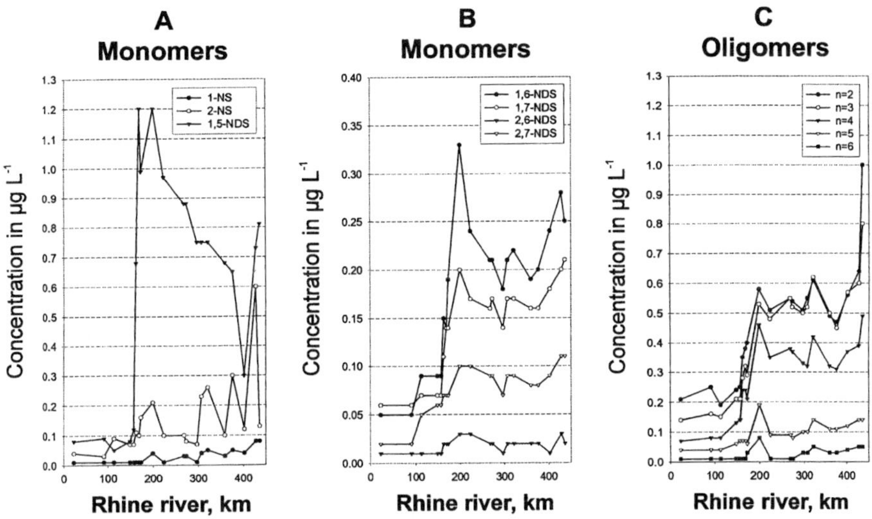

Fig. 7 Concentration profiles of SNFC and their monomers in the aqueous phase along the Rhine river (reproduced from [41])

Rhine valley in the year 2000 [41, 68]. This topic is addressed in the next section.

5
Partition of Sulfonated Naphthalene–Formaldehyde Condensates between Aqueous Phase and Suspended Solids/Sediments

The major goal of a monitoring program in the year 2000 was to evaluate the partition of SNFC and their monomers in the surface waters in the state of Baden-Württemberg, in southwest Germany, including the Rhine in the Upper Rhine Valley and its tributaries. Prerequisites for this kind of quantitative investigation were the isolation of reference compounds from a technical product [11] and the development of an extraction method for suspended solids and sediments [41]. At two sampling sites on the Rhine river (Iffezheim, km 334 and Mannheim, km 426) and two additional sites on the Neckar river and on the Danube river, water samples and suspended solids were simultaneously analyzed monthly from April to December 2000 [41]. As an example, the concentrations of the different SNFC oligomers at the sampling site Iffezheim is shown in Fig. 8. From the concentrations measured in the solids and in the aqueous-phase field-partition coefficients K_d were derived for SNFC from two to seven naphthalenesulfonate units ($n = 2$–7) in the molecule. The resulting K_d values at the Rhine sampling sites increased from $5.8 \times 10^2\ \mathrm{L\,kg^{-1}}$ for the condensates with $n = 3$ to $7.3 \times 10^4\ \mathrm{L\,kg^{-1}}$ for

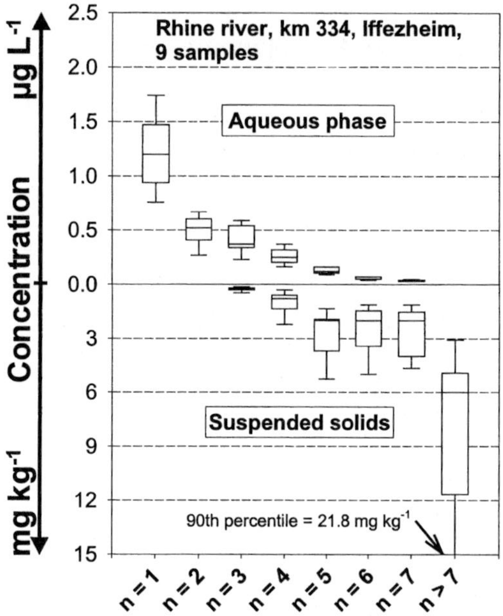

Fig. 8 Box-plot presentation of the partition of SNFC and their monomers between aqueous phase and suspended solids in the Rhine river at Iffezheim in the Upper Rhine valley (km 334). The *solid line* within the box represents the 50th percentile; the *box* delimits the 25th and 75th percentiles; the *bars* indicate the 10th and 90th percentiles (reproduced from [41])

$n = 7$. This means that the oligomers with three or more naphthalenesulfonate units in the molecular chain are adsorbed to more than 99% onto the solids. Monomers were not detected in the extracts of suspended solids, and the concentration of adsorbed oligomers with $n = 2$ was below their LOQ (0.05 mg kg^{-1}). Therefore, higher molecular SNFC are not expected to be a problem for drinking water preparation.

6
Occurrence and Fate of Aromatic Sulfonates in Drinking Water Preparation from Rhine River Water

The water treatment plants along the Rhine river usually apply a multi-barrier approach for purification of the river water. Most often, the first purification step is the natural attenuation by bankside filtration. Oxidation by ozone and adsorption by activated carbon filtration are additional important purification steps in these facilities. In the following sections, only these three steps are discussed.

6.1
Removal of Aromatic Sulfonates by Bankside Filtration

Nine polar aromatic sulfonates were identified in bankside filtered waters of different water treatment plants along the Rhine river [59]. These sulfonates, which occurred in the wells of the facilities, were also frequently measured in the Rhine river water (see Sects. 3 and 4). Except for 2-naphthalenesulfonate, all of these compounds are two- or threefold sulfonated naphthalene derivatives or disulfonated intermediates from the production of fluorescent whitening agents (*cis*-DNS and 2-hydroxy-4,6-bis(4-sulfanilo)-1,3,5-triazine. The concentrations measured in bank filtrates are compiled in Table 2. These findings reflect the high production volumes, the mobility, and the recalcitrant nature of these compounds. The exceptional persistence of certain multiply sulfonated/substituted aromatic sulfonates are in good agreement with degradation studies on naphthalenesulfonates [9, 22–32] and stilbenesulfonates [9, 12] and with studies on their behavior in sewage treatment [9, 10, 22, 27, 73–75]. In particular, multiple substitution and sulfonation in the α-positions of the naphthalene unit enhance the persistence of naphthalenesulfonates [22, 29, 59].

Table 2 Polar aromatic sulfonates identified in bank filtrates of the Rhine river

	Concentration in µg L	Refs.
2-Naphthalenesulfonate	0.08–0.16	[59]
1,5-Naphthalenedisulfonate	0.1–1.32	[39, 59]
1,6-Naphthalenedisulfonate	0.02–0.33	[59]
2,6-Naphthalenedisulfonate	0.03–0.12	[59]
2,7-Naphthalenedisulfonate	0.02–0.20	[59]
2-Amino-1,5-naphthalenedisulfonate	0.03–0.56	[59]
3-Amino-1,5-naphthalenedisulfonate	0.02–0.37	[59]
cis-/trans-4,4′-Dinitro-2,2′-stilbenedisulfonate	0.57–5.82	[39, 59]
2-Hydroxy-4,6-bis(4-sulfanilo)-1,3,5-triazine	0.53–0.71	[59]

6.2
Removal of Aromatic Sulfonates by Ozone Treatment

Brauch and Lange [76], Lange et al. [59], and Kühn and Baldauf reported field studies on the elimination of 1,5-NDS and *cis*-DNS by ozone treatment in Rhine river facilities.

In a two-stage ozone treatment for purifying surface water with an ozone dose of 0.7 mg L^{-1} in the first stage and 1.0 mg L^{-1} in the second stage, about

40–50% and 10–20% of 1,5-NDS remains unchanged after the first and second stages, respectively (Fig. 9) [59]. However, between the first and second stages, there was also a flocculation/sedimentation to remove the more polar oxidation products formed during primary ozone treatment. Therefore, the concentration decrease due to flocculation/sedimentation and that due to ozone treatment in the second stage could not be distinguished.

In another facility using Rhine river bank filtrate as raw water, about 80% of the *cis*-DNS was removed by a one-step ozone treatment [59]. In addition to these field investigations, the primary reaction of ozone with 1,5-NDS, 2-amino-5-chloro-4-methylbenzenesulfonate, anthraquinone-2,6-disulfonate, *cis*-DNS, and 3-amino-1,5-naphthalenedisulfonate was studied in bench-scale experiments with sulfonate concentrations between 5 and 100 $\mu g\,L^{-1}$ and realistic ozone doses of about 1 $mg\,L^{-1}$ [77]. In these experiments the influence of ozone dose, initial sulfonate concentration, pH, hydrogen carbonate content, and DOC on the reaction of ozone with 1,5-NDS was studied. Even in deionized water, where there is very little concurrence of natural organic matter (DOC $\approx$ 0.2 $mg\,L^{-1}$) 1,5-NDS was oxidized incompletely. Using an initial ozone concentration of 0.7 $mg\,L^{-1}$, 15% of the 1,5-NDS was recovered after a reaction time of 45 min. In most of the water treatment plants the reaction times are shorter than 45 min and the DOC values higher than 1 $mg\,L^{-1}$. Thus, the trace concentrations of 1,5-NDS after ozone treatment are in general 40–60% of the raw water value [59].

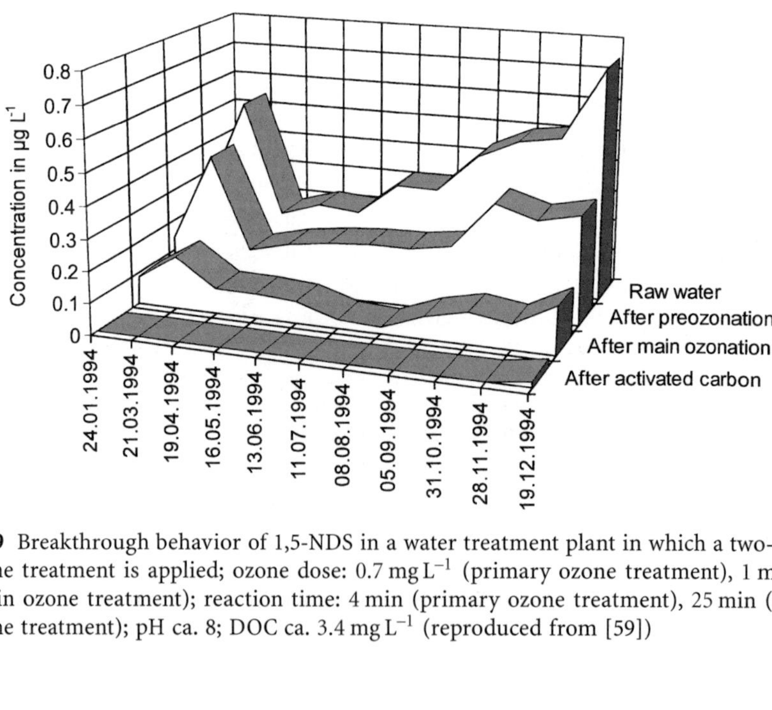

Fig. 9 Breakthrough behavior of 1,5-NDS in a water treatment plant in which a two-stage ozone treatment is applied; ozone dose: 0.7 $mg\,L^{-1}$ (primary ozone treatment), 1 $mg\,L^{-1}$ (main ozone treatment); reaction time: 4 min (primary ozone treatment), 25 min (main ozone treatment); pH ca. 8; DOC ca. 3.4 $mg\,L^{-1}$ (reproduced from [59])

With an initial concentration of *cis*-DNS of 5 μg L^{-1} and an ozone dose of about 1 mg L^{-1} after 10 min, 10% of this compound remained unchanged. The elimination was complete after 30 min. The concentration of the other tested compounds dropped below the LOQ after only 2 min [77]. By-products formed during oxidation by ozone were not identified in these investigations.

6.3 Removal of Aromatic Sulfonates by Activated Carbon Filtration

Few papers report on field studies of the removal of aromatic sulfonates by activated carbon filtration [59, 76, 78]. It turns out that the elimination efficiency strongly depends on the process conditions in the different water treatment plants. However, the removal of aromatic sulfonates from bankside-filtered Rhine river raw water is a side effect of the general water purification process. To date, the water treatment facilities are not specially optimized for the elimination of polar aromatic sulfonates.

In a treatment plant, where the treatment consists of three main steps, i.e., iron removal, manganese removal, and a final filtration step by two parallel granular activated carbon (GAC) adsorbers, the polar naphthalenesulfonates could not always be efficiently removed (Fig. 10).

The naphthalenesulfonates that remained in the water after iron and manganese removal showed an early breakthrough on GAC [59]. At loads larger than 35 m^3 of water per kg of GAC the breakthrough of 1,5-NDS at a concentration of 0.6 μg L^{-1} was complete [78].

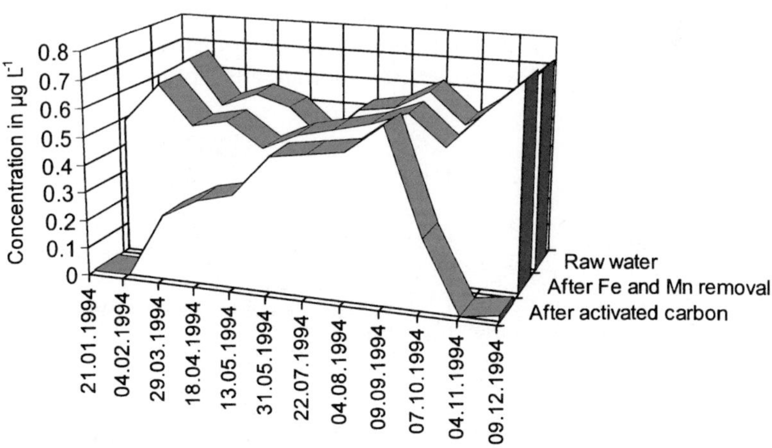

Fig. 10 Breakthrough behavior of 1,5-NDS in a water treatment plant. Two parallel adsorbers, each containing 50 m^3 of activated carbon; operation time and load of the activated carbon filters: filter 1: 15.12.1993–27.09.1994 (1 240 220 m^3), filter 2: 15.06.1993–11.10.1994 (1 648 160 m^3) (reproduced from [59])

In another treatment facility, where a greater fraction of the sulfonates was eliminated by ozone treatment and flocculation/sedimentation before entering into the charcoal filters, a nearly complete removal was quantitatively observed for 1,5-NDS; see Fig. 9.

Data from a facility with ozone treatment followed by biologically active charcoal filtration indicate almost complete removal of polar aromatic sulfonates with low concentrations ($\leq 0.1\ \mu g\,L^{-1}$) of 1,5-NDS and 1,3,5-/1,3,6-NTS in the drinking water (Smeenk H, 1999, private communication).

In addition to these field measurements, laboratory investigations of the adsorption of polar aromatic sulfonates were carried out by several researchers [28, 29, 74, 79–82]. The results of this more basic research is not explored in detail here, because it beyond the scope of this review.

Briefly, thermodynamic and kinetic aspects were studied. While the results of the investigations on the kinetics [81, 82] are not consistent and, therefore are not discussed further here, the results of thermodynamic considerations [79, 80, 82] give a more concise picture. The findings of the studies on the adsorption thermodynamics suggest that the adsorption of singly-charged monomeric aromatic sulfonates is mainly dominated by Van der Waals interaction between sorptive and sorbens, while the adsorption of multiply charged sulfonates is controlled by anion exchange. This conclusion is supported by several authors who reported on a concentration-dependence of the adsorption/desorption equilibria, especially for two- and threefold sulfonated compounds, like 1,5-NDS and 1,3,5-NTS. For these compounds the adsorption ability strongly decreased with increasing initial sulfonate concentrations. Furthermore, the pH of the sulfonate solution in contact with charcoal increased during equilibration. The concentration dependence could be suppressed if the pH of the sulfonate solution was kept constant until equilibrium was reached. These observations are strong indications of an ion-exchange mechanism of the di- and trisulfonate adsorption at cationic sites on the charcoal surface.

7
Concluding Remarks

From an environmental point of view, aromatic sulfonates belonged to the class of "forgotten compounds" for a long time, although they have been industrially produced in the Rhine catchment area since before 1900. In the 1990s they were one of the "emerging compound classes" due to the new analytical methods, which allowed them to be analyzed down to a few tens of $ng\,L^{-1}$. In the monitoring programs of the waterworks along the Rhine, aromatic sulfonates were regularly measured in the river for about one decade. In other large German rivers, like the Elbe river or the Danube, they have been analyzed, too. Nowadays, there is a large amount of data on the occurrence of aromatic sulfonates in the Rhine, i.e., enough data from the immisions. Although it is obvious that individual persistent compounds cannot be removed by natural attenuation processes alone, e.g., through bankside filtration, and, furthermore can reach drinking water, legislation has not yet responded in establishing parametric values for sulfonates as a precaution. This might be due to the fact that most of the single compounds discussed here are not consumer products, but intermediates in the production of other chemicals or by-products/impurities of technical blends. From the point of view of the observed concentrations in river water, aromatic sulfonates are more relevant in the Rhine catchment area than pesticides, for example. Nevertheless, from the monitoring data of some single compounds, e.g., DNS, it is obvious that the chemical industry has undertaken efforts to reduce emissions of chemicals from the production site when taking new production plants into operation.

References

1. Lindner O (1985) Benzenesulfonic acids and their derivatives. In: Gerhartz W (ed) Ullmann's encyclopedia of industrial chemistry, 5th edn, Vol A3. Wiley, New York, p 507
2. Dresler H (1995) Naphthalene derivatives. In: Grayson M (ed) Kirk-Othmer, Encyclopedia of chemical technology, 4th edn, Vol 16. Wiley, New York, p 980
3. Bien HS, Wunderlich K (1974) Anthrachinon-Farbstoffe und Zwischenprodukte. In: Bartholomé E (ed) Ullmanns Encyklopädie der technischen Chemie, 4. Auflage, Vol 7. Verlag Chemie, Weinheim, p 585
4. Booth G (1991) Naphthalene derivatives. In: Gerhartz W (ed) Ullmann's encyclopedia of industrial chemistry, 5th edn, Vol A17. Wiley, New York, p 9
5. Heusch R, Reizlein K (1987) Disperse systems and dispersants. In: Gerhartz W (ed) Ullmann's encyclopedia of industrial chemistry, 5th edn, Vol A8. Wiley, New York, p 586
6. Siegrist AE, Eckhardt C, Kaschig J, Schmidt E (1991) Optical brighteners. In: Gerhartz W (ed) Ullmann's encyclopedia of industrial chemistry, 5th edn, Vol A18. Wiley, New York, p 153
7. Rixom MR, Mailvaganam NP (1986) Chemical admixtures for concrete, 2nd edn. Spon, London
8. Schullerer S, Brauch HJ, Frimmel FH (1990) Vom Wasser 75:83
9. Lange FT, Redín C, Brauch HJ, Eberle SH (1998) Vom Wasser 90:121
10. Redín C, Lange FT, Brauch HJ, Eberle SH (1999) Acta Hydrochim Hydrobiol 27:136
11. Wolf C, Storm T, Lange FT, Reemtsma T, Brauch HJ, Eberle SH, Jekel M (2000) Anal Chem 72:5466
12. Greim H, Ahlers J, Bias R, Broecker B, Hollander H, Gelbke HP, Klimisch HJ, Mangelsdorf I, Paetz A, Schön N, Stropp G, Vogel R, Weber C, Ziegler-Skylakakis K, Bayer E (1994) Chemosphere 28:2236
13. Kölbener P, Baumann U, Cook AM, Leisinger T (1994) Wat Res 28:1855
14. Wellens H (1990) gwf-Wasser/Abwasser 23:85
15. Zürrer D, Cook AM, Leisinger T (1987) Appl Environ Microbiol 53:1459
16. Thurnheer T, Locher HH, Cook AM (1988) Appl Microbiol Biotechnol 29:605
17. Gerike P, Fischer WK (1979) Ecotoxicol Environ Saf 3:159

18. Gerike P (1984) Chemosphere 13:121
19. Baumann U (1991) Z Wasser Abwasser Forsch 24:237
20. Deshpande SD, Chakrabarti T, Subrahmanyam PVR, Sundaresan BB (1985) Water Res 19:293
21. Beratergremium für umweltrelevante alte Stoffe (1992) BUA-Stoffbericht 63. Wiley, New York
22. Altenbach B (1996) Determination of Substituted Benzene- and Naphthalenesulfonates in Waste Water and Their Behavior in Sewage Treatment, Dissertation, ETH No. 11437, ETH Zurich
23. Brilon C, Beckmann W, Hellwig M, Knackmuss HJ (1981) Appl Environ Microbiol 42:39
24. Brilon C, Beckmann W, Hellwig M, Knackmuss HJ (1981) Appl Environ Microbiol 42:44
25. Nörtemann B, Knackmuss HJ (1988) gwf-Wasser/Abwasser 129:75
26. Wittich RM, Rast HG, Knackmuss HJ (1988) Appl Environ Microbiol 54:1842
27. Reemtsma T, Jochimsen J, Jekel M (1993) Vom Wasser 81:353
28. Fichtner S, Lange FT, Schmidt W, Brauch HJ (1995) Fresenius J Anal Chem 353:57
29. Neitzel PL, Abel A, Grischek T, Nestler W, Walther W (1998) Vom Wasser 90:245
30. Soeder CJ, Luther M, Kneifel H (1988) gwf-Wasser/Abwasser 129:82
31. Luther M, Soeder CJ (1991) Water Res 25:299
32. Luther M, Shaaban MM, Soeder CJ, Shafiy YH, El-Fouly MM (1991) gwf-Wasser/Abwasser 132:411
33. Reemtsma T (1996) J Chromatogr A 773:473
34. Suter MJ, Riediker S, Giger W (1999) Anal Chem 71:897
35. Storm T, Reemtsma T, Jekel M (1999) J Chromatogr A 854:175
36. Loos R, Alonso MC, Barceló D (2000) J Chromatogr A 890:225
37. Loos R, Riu J, Alonso MC, Barceló D (2000) J Mass Spectrom 35:1197
38. Crescenzi C, Di Corcia A, Marcomini A, Pojana G, Samperi R (2001) J Chromatogr A 923:97
39. Lange FT, Wenz M, Brauch HJ (1995) J High Resol Chromatogr 18:243
40. Altenbach B, Giger W (1995) Anal Chem 67:2325
41. Lange FT, Merklinger M, Wenz M, Brauch H, Lehmann M, Pinter I (2005) Environ Sci Technol 39:1523
42. Brouwer ER, Tol TM, Lingeman H, Brinkman UAT (1993) Quím Anal 12:88
43. Brouwer ER, Slobodnik J, Lingeman H, Brinkman UAT (1992) Analysis 20:121
44. Arbeitsgemeinschaft Rhein-Wasserwerke e.V. (ARW) (1998) 55. Jahresbericht der Arbeitsgemeinschaft Rhein-Wasserwerke e.V. (ARW). Köln
45. Debets AJJ, Hupe KP, Kok W, Brinkman UAT (1992) J Chromatogr A 600:163
46. Hauthal HG (1998) Nachrichten aus der Chemie 46:987
47. OECD SIDS (1999) SIDS Initial Assessment Report for 9th SIAM. Paris, 29 June–1 July 1999
48. Fooken C, Gihr R, Häckl M, Seel P (1997) Umweltplanung, Arbeits- und Umweltschutz. Heft 233. Orientierende Messungen gefährlicher Stoffe – Landesweite Untersuchungen auf organische Spurenverunreinigungen in hessischen Fließgewässern, Abwässern und Klärschlämmen 1991–1996. Wiesbaden
49. L. C. C. Bayer Chemicals AG (2004) Blankophor® fluorescent whitening agents – constantly updated, always state-of-the-art. Available via http://www.solutionsforpaper.com/sfp/global/products/technology/blankophor. Cited 28 April 2004
50. Arbeitsgemeinschaft Rhein-Wasserwerke e.V. (ARW) (1993) 50. Jahresbericht der Arbeitsgemeinschaft Rhein-Wasserwerke e.V. (ARW). Köln
51. Arbeitsgemeinschaft Rhein-Wasserwerke e.V. (ARW) (1994) 51. Jahresbericht der Arbeitsgemeinschaft Rhein-Wasserwerke e.V. (ARW). Köln

52. Arbeitsgemeinschaft Rhein-Wasserwerke e.V. (ARW) (1995) 52. Jahresbericht der Arbeitsgemeinschaft Rhein-Wasserwerke e.V. (ARW). Köln
53. Arbeitsgemeinschaft Rhein-Wasserwerke e.V. (ARW) (1996) 53. Jahresbericht der Arbeitsgemeinschaft Rhein-Wasserwerke e.V. (ARW). Köln
54. Arbeitsgemeinschaft Rhein-Wasserwerke e.V. (ARW) (1997) 54. Jahresbericht der Arbeitsgemeinschaft Rhein-Wasserwerke e.V. (ARW). Köln
55. Arbeitsgemeinschaft Rhein-Wasserwerke e.V. (ARW) (2003) 60. Jahresbericht der Arbeitsgemeinschaft Rhein-Wasserwerke e.V. (ARW). Köln
56. Husain S, Narsimha R, Alvi SN, Nageswara R (1992) J Chromatogr 596:127
57. Poiger T, Field JA, Field TM, Giger W (1993) Anal Meth Instr 1:104
58. Stoll JMA, Giger W (1997) Anal Chem 69:2594
59. Lange FT, Wenz M, Brauch HJ (1995) Anal Meth Instr 2:277
60. Arbeitsgemeinschaft Wasserwerke Bodensee-Rhein (AWBR) (1994) 25. Jahresbericht der Arbeitsgemeinschaft Wasserwerke Bodensee-Rhein, St. Gallen
61. Arbeitsgemeinschaft Wasserwerke Bodensee-Rhein (AWBR) (1995) 26. Jahresbericht der Arbeitsgemeinschaft Wasserwerke Bodensee-Rhein, St. Gallen
62. Arbeitsgemeinschaft Wasserwerke Bodensee-Rhein (AWBR) (1996) 27. Jahresbericht der Arbeitsgemeinschaft Wasserwerke Bodensee-Rhein, St. Gallen
63. Arbeitsgemeinschaft Wasserwerke Bodensee-Rhein (AWBR) (1997) 28. Jahresbericht der Arbeitsgemeinschaft Wasserwerke Bodensee-Rhein, St. Gallen
64. Arbeitsgemeinschaft Wasserwerke Bodensee-Rhein (AWBR) (1998) 29. Jahresbericht der Arbeitsgemeinschaft Wasserwerke Bodensee-Rhein, St. Gallen
65. Lange FT, Furrer R, Brauch HJ (2000) Polar aromatic sulfonates and their relevance to waterworks. Association of River Water Companies (RIWA), Amsterdam
66. Knepper TP, Sacher F, Lange FT, Brauch HJ, Karrenbrock F, Roerden O, Lindner K (1999) Waste Management 19:77
67. Menzel CM, Lange FT, Käss W, Hötzl H (2002) Environ Geol 41:731
68. Landesanstalt für Umweltschutz Baden-Württemberg (2002) Aromatische Sulfonate in Oberflächengewässern, Schwebstoffen und Sedimenten Baden-Württembergs. Oberirdische Gewässer, Gewässerökologie, Karlsruhe
69. Ruckstuhl S (2001) Environmental Exposure Assessment of Sulfonated Naphthalene Formaldehyde Condensates and Sulfonated Naphthalenes Applied as Concrete Superplasticizers, Dissertation, ETH No. 14477, ETH, Zurich
70. Ruckstuhl S, Suter MJF, Giger W (2001) Analyst 126:2072
71. Ruckstuhl S, Suter MJF, Kohler HPE, Giger W (2002) Environ Sci Technol 36:3284
72. Ruckstuhl S, Suter MJF, Giger W (2003) J Contam Hydrol 67:1
73. Lindner K, Knepper TP, Karrenbrock F, Rörden O, Brauch HJ, Lange FT, Sacher F (1996) Abschlußbericht zum ARW/VCI-Forschungsvorhaben. Internationale Arbeitsgemeinschaft der Wasserwerke im Rheineinzugsgebiet (IAWR). Rhein-Themen 1. ISBN 90-6683-080-8, DVGW-Technologiezentrum Wasser (TZW), Karlsruhe, pp 54, 94, 128
74. Bastian B, Haberer K, Knepper TP (1995) Vom Wasser 84:369
75. Reemtsma T, Jekel M (1997) Water Res 31:1035
76. Brauch HJ, Lange FT (1995) Gewässerschutz Wasser Abwasser 152:45/1
77. Jähnke S (1994), Diploma thesis, University of Karlsruhe
78. Kühn W, Baldauf G (1996) gwf-Wasser/Abwasser 137:110
79. Schullerer S (1992) Summarische Bestimmung und chromatographische Identifizierung organischer Schwefelverbindungen im Wasser auf der Basis der Ionenpaar-Extraktion, Dissertation, University of Karlsruhe
80. Johannsen K, Bodemann I, Groß HJ, Sontheimer H (1994) Vom Wasser 83:179
81. Johannsen K, Groß HJ, Gaukel V, Sontheimer H, Frimmel FH (1994) Vom Wasser 83:169

82. Ulmer M (1998) PhD thesis, University of Karlsruhe
83. Arbeitsgemeinschaft Wasserwerke Bodensee-Rhein (AWBR) (1999) 30. Jahresbericht der Arbeitsgemeinschaft Wasserwerke Bodensee-Rhein, St. Gallen
84. Schullerer S, Koschenz G, Brauch HJ, Frimmel FH (1992) Vom Wasser 78:229
85. Schullerer S, Frimmel FH (1993) Anal Chim Acta 283:251

Hdb Env Chem Vol. 5, Part L (2006): 211–234
DOI 10.1007/698_5_019

Published online: 23 November 2005

Occurrence, Fate and Relevance of Aminopolycarboxylate Chelating Agents in the Rhine Basin, Germany

Carsten K. Schmidt (✉) · Heinz-Jürgen Brauch

DVGW-Water Technology Center (TZW), Chemical Analysis Department,
Karlsruher Straße 84, 76139 Karlsruhe, Germany
c.schmidt@tzw.de

Abstract Aminopolycarboxylates are used in many products and industrial applications because of their ability to form stable complexes with metal ions. The most commonly used aminopolycarboxylates are ethylenediaminetetraacetic acid (EDTA), nitrilotriacetic acid (NTA) and diethylenetriaminepentaacetic acid (DTPA). More recently, 1,3-propylenediaminetetraacetic acid (1,3-PDTA), β-alaninediacetic acid (β-ADA) and methylglycinediacetic acid (MGDA) were introduced. Due to their high polarity and extensive usage, aminopolycarboxylates found their way into the aquatic environment. In particular the Rhine has been subject of major aminopolycarboxylate pollution in the past. However, joint efforts of drinking water suppliers, industries, and German authorities resulted in an impressive reduction of the general pollution. Besides presenting long-term and actual occurrence data, this report also focuses on aminopolycarboxylate use volumes, analysis, their speciation and fate in the environment, as well as on their toxicological and ecotoxicological properties in order to discuss the detected levels of contamination.

Keywords Aquatic environment · Chelating agents · EDTA · Occurrence · Speciation

Abbreviations

β-ADA	β-Alaninediacetic acid
[*S*,*S*]-EDDS	[*S*,*S*]-Ethylenediaminedisuccinate
EDTA	Ethylenediaminetetraacetic acid
DIN	Deutsches Institut für Normung e.V.
DTPA	Diethylenetriaminepentaacetic acid

ICP	Inductively coupled plasma
IDS	Iminodisuccinic acid sodium salt
MGDA	Methylglycinediacetic acid
MS	Mass spectrometry
NTA	Nitrilotriacetic acid
1,3-PDTA	1,3-Propylenediaminetetraacetic acid

1 Introduction

Aminopolycarboxylates such as ethylenediaminetetraacetic acid (EDTA), nitrilotriacetic acid (NTA), and diethylenetriaminepentaacetic acid (DTPA) have been used as chelating agents in many industrial processes and domestic products for years (Fig. 1) [1–3]. With their ability to solubilize and inactivate metal ions by complex formation, these compounds are used for a variety of purposes. Recently, some alternative aminopolycarboxylates such as 1,3-propylenediaminetetraacetic acid (1,3-PDTA), β-alaninediacetic acid (β-ADA), and methylglycinediacetic acid (MGDA) were introduced [1, 2, 4, 5].

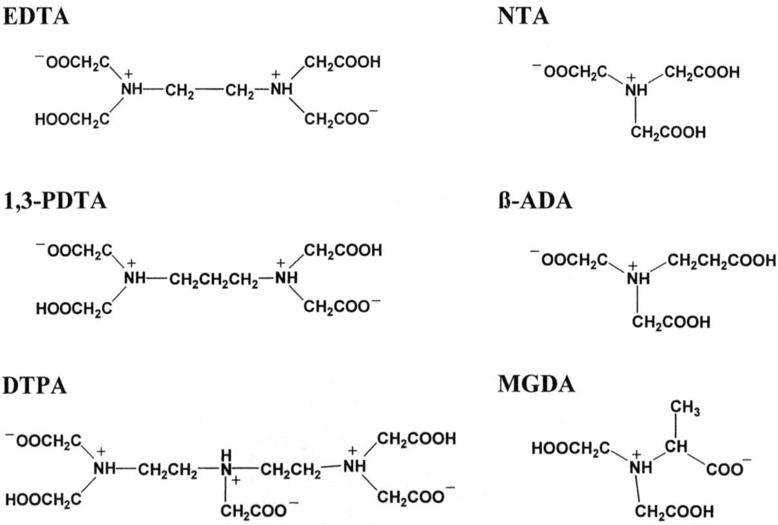

Fig. 1 Chemical structures of important aminopolycarboxylates

2 Applications and Use Volumes

Important fields of application comprise water softening and treatment, pulp and paper industry, agro chemicals, photochemicals, detergents in industry and private households, food processing, galvano industry, textiles, cosmetics, pharmaceuticals, and medical detoxification [1, 2, 6, 7]. Due to their broad range of fields of industrial and domestic application, EDTA, NTA, and DTPA are used in huge quantities worldwide. The annual global consumption is estimated as approximately 200 000 t (calculated as 100% acid), about half is EDTA, and a quarter each are NTA and DTPA [4]. The segmentation in Western Europe and Germany is very similar [8]. Europe consumes approximately 50% of the worldwide aminopolycarboxylic acids production volume [9]. Almost half of the amount sold in Europe is allocated to Germany, Sweden, and the UK [8]. Outside Europe, aminopolycarboxylates are mostly used in North America, and to a smaller extent in Japan and Southeast Asia [4, 9].

More than 50% of the total aminopolycarboxylic acid consumption takes place in washing and cleaning and the pulp and paper industries [9]. In 1998, more than 100 000 t of aminopolycarboxylic acids (EDTA, NTA, and DTPA) were used in these two fields. On a quantity basis, further important buyer branches are photo industries (approx. 15 000 t) and agriculture (approx. 13 000 t) [9]. Although only limited data are available with respect to the consumption of MGDA, β-ADA and 1,3-PDTA, the production of these compounds seems to be relatively low compared to EDTA, NTA and DTPA. In Germany, approx. 130 t/a MGDA and 0.15 t/a β-ADA (data from 2001) are used in detergents and cleaning agents [10, 11], approx. 28 t/a 1,3-PDTA (data from 2000) are used in the German photo industry [8].

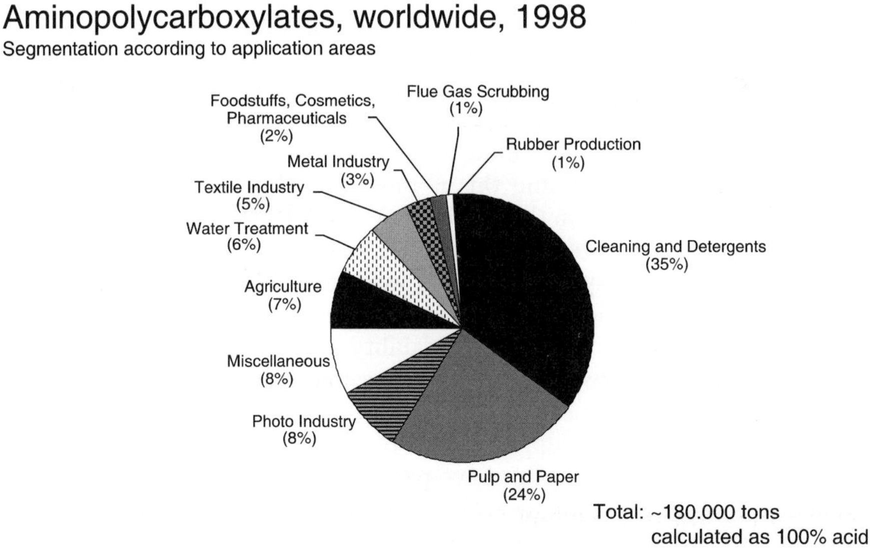

Fig. 2 Segmentation of aminopolycarboxylates (EDTA, NTA, and DTPA) according to application areas worldwide (1998) [9]

Within the last 10 years, the sales volumes of EDTA, DTPA and NTA have steadily increased in Europe [12]. While the German situation is comparable with respect to DTPA and NTA, it is different regarding EDTA. Mean EDTA sales in Germany show a downward trend, despite a slight uptrend from 1998 onwards. From 1990 to 2001, EDTA and NTA sales in Europe increased approx. 31 and 19%, respectively. DTPA sales increased approx. 123% between 1993 and 2001. In Germany, EDTA sales decreased 38% from 1991 to 2001, while the percentage of NTA and DTPA sales increased comparable to Europe (+12% and +84%).

In contrast to the European development between 1990 and 2000, in particular EDTA sales in the area of commercial detergents and cleaning agents have significantly decreased in Germany [12]. Meanwhile, the use of EDTA in commercial detergents has been discontinued [13]. At the same time, the sales volume in Western Europe has increased, in particular in the areas of paper and agriculture [12]. EDTA (and also DTPA) consumption in paper industries increased dramatically when the manufacturers switched over their production to chlorine-free bleaching processes [9]. Somewhat later, this trend was even intensified by the increased recycling of paper. However, no comparable growth rates are expected in this branch for the future. It should be mentioned that the use of EDTA in paper industries is actually an ecological advance. The renouncement of chlorine bleaching of pulp prevents environmental pollution with organic chlorine compounds. The alternative peroxide or dithionite bleaching process, however, requires strong complexing agents like EDTA or DTPA [4, 8]. A major area of development is the use of EDTA in agriculture to compensate for trace nutrient deficiencies in the soil. Micronutrient fertilizers are particularly used in wine and citrus fruit growing. Due to pronouncedly larger cultivated areas, increased EDTA sales for agricultural applications are observed, especially in southern Europe [13]. In western Europe, EDTA sales show a clear downward trend as far as the textile industry, electroplating industry, photo industry and household detergents and cleaning agents are concerned, and this is most probably due to the increased application of substitutes in these fields (see discussion below) [12].

The decrease in the overall German EDTA sales volume demonstrates that in the last years Germany has intensively worked on a reduction of EDTA consumption and its substitution by alternative products [4, 14]. Environmental issues, in particular its low biodegradability and the potential remobilization of heavy metal ions have been reasons for an impressive reduction of EDTA consumption in Germany. In 1991, representatives of the chemical industry, drinking water suppliers, and the three federal ministries of environment, research and health have passed a voluntary agreement on the reduction of water pollution by EDTA, with the target to reduce the pollution of surface water by EDTA in Germany by 50% within 5 years. The promising approaches to reach this target were wastewater avoidance, improved wastewater treatment, and in particular the reduction of EDTA use, combined with

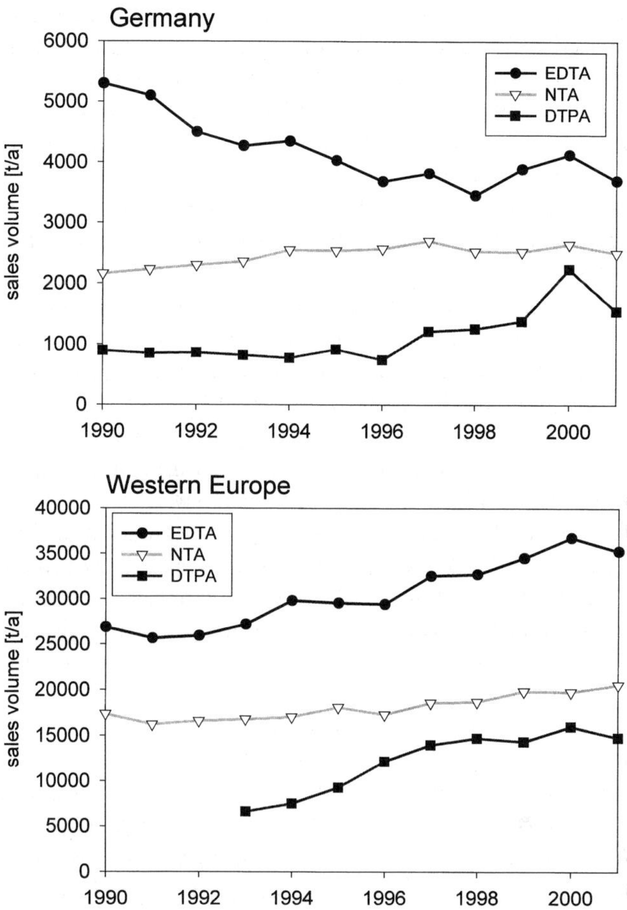

Fig. 3 Sales volume development of the aminopolycarboxylates EDTA, NTA, and DTPA in Germany and Western Europe [12]

its substitution by suitable alternative products that appear more favorable from the drinking water suppliers viewpoint. Documentation and assessment of the progress is performed in annual meetings of all involved parties directed by the federal environmental office. In the following years, the original circle has been extended by more associations (e.g., food and cosmetic industries) to speed up the reduction process [15, 16].

Although this voluntary agreement in Germany represents a specific German program, it is also a project with pilot characteristics. On the one hand, it helps to assess to which extent such an approach will result in an effective reduction of an unwanted compound in the environment; on the other hand, the development in Germany may act as a signal for other countries. EDTA is included in the first priority list of the EU program on existing chemicals and currently undergoes a risk assessment according to EU regulations [4, 13, 15, 17–19]. The use of alternative products is a major request of the voluntary agreement on the reduction of water pollution by EDTA in Germany. While some alternative products are already in use for a long time in parallel to EDTA, others have been developed only in recent years as a result of the EDTA discussion. In principle, EDTA is optimally suited for all listed applications, but its use should be restricted to applications where (due to technical reasons) no replacement is possible. The search for EDTA replacement compounds focuses on complexing agents that in contrast to EDTA are readily biodegradable. However, no all-round EDTA substitute has been developed so far [4]. Alternative products designed and used so far are restricted to few applications and are often more expensive than EDTA [4, 5, 15, 20]. NTA, MGDA, [*S*,*S*]-ethylenediaminedisuccinate ([*S*,*S*]-EDDS) and iminodisuccinic acid sodium salt (IDS) are readily biodegradable substitutes for the detergent and cleaning-agents industry; tartrates are used in semiconductor production; the photo industry utilizes β-ADA [2, 4, 5, 18, 21–23]. In the long term, this development will result in a shift of complexones in major applications. However, no complete substitution of EDTA in all relevant application areas is to be expected for the near future, since a substitution is always evaluated in consideration of technological, economic and ecological aspects [18].

3 Analysis

A number of assays exist for the analytical determination of aminopolycarboxylates [24, 25]. The most reasonable approach for sensitive aminopolycarboxylate determination is gas chromatography. However, aminopolycarboxylic acids have to be converted to readily volatile alkyl ester derivatives during sample preparation. Various metal complex species of an individual aminopolycarboxylate are decomposed in this step. Therefore, gas chromatographic methods are unfortunately not suitable for the determination of individual aminopolycarboxylate-metal species and determine only the integral content of all existing species of a specific aminopolycarboxylic acid [24]. For the analysis of aqueous samples, a DIN standard has been published that allows the simultaneous determination of EDTA, DTPA, NTA, β-ADA, 1,3-PDTA and MGDA [26]. Initially, the water sample is evaporated to dryness, first without additive, then after addition of hydrochloric or formic acid for conversion of metal species into the free acid form. A subsequent esterification step converts the complexing agents into their propyl or butyl esters. Finally, water is added to the reaction mixture, and the formed esters are extracted from the mixture with *n*-hexane. Finally, they are separated by gas chromatography and quantified with a nitrogen-sensitive or preferably mass-spectrometric detector. This method can be applied to the analysis of different water types in a concentration range of 0.5 to 200 μg/L, utilizing sample volumes from 50 to 100 mL [26].

Individual aminopolycarboxylate-metal species at environmentally relevant concentrations (i.e., low μg/L range) can be determined by ion chromatography coupled to ICP/MS [27, 28] or by capillary electrophoresis coupled to ion spray tandem mass spectrometry [24, 29].

4 Occurrence

Due to their high polarity and partly low degradability (EDTA, 1,3-PDTA, DTPA and β-ADA), aminopolycarboxylates enter our creeks, rivers, and lakes mainly via industrial and domestic wastewaters. On the basis of a recent comprehensive data evaluation, concentrations that can be typically expected in runoffs of industrial sewage treatment plants are in the low to middle mg/L-range [30]. Pollution levels of municipal wastewaters are usually in the middle μg/L-range [30].

Aminopolycarboxylate concentrations in Lake Constance and in the upper Rhine upstream to Basel are relatively low. Long-term observations at different sampling points at Lake Constance revealed mean EDTA concentrations about 1.5 μg/L, whereas levels of NTA were below the detection limit of 0.5 μg/L [31]. DTPA could be regularly detected in this lake at mean concentrations of 0.7 μg/L, whereby the DTPA input results predominantly from the tributary Schussen, that is influenced by the wastewaters of two paper industry plants [31, 32]. According to data from 2002, EDTA concentrations roughly double at the upper Rhine between Basel (mean: 1.2 μg/L) and Karlsruhe (mean: 3.1 μg/L) [33].

Figure 4 gives an overview of the mean EDTA concentrations of the Rhine and its tributaries Neckar, Main, Mosel, and Ruhr between Karlsruhe and Lobith since 1991. It is obvious that mean EDTA concentrations generally show a decreasing tendency since the enactment of the voluntary agreement with its arrangements in the year 1991. While the Rhine at Karlsruhe shows comparably low EDTA concentrations nowadays with around 5 μg/L, influents by the major Rhine tributaries and by industrial wastewaters result downstream in an increase of EDTA concentrations alongside the Rhine. A major increase in EDTA concentrations is observed between Karlsruhe and Mainz. Between these two sampling points, near Worms, the plant of the largest EDTA producer in Germany is located (BASF AG). As can be seen from the sampling point of Worms, which is influenced by the EDTA wastewater trail of this company, the pollution is caused by high EDTA loads of the company's wastewater. However, the yearly measured EDTA concentrations also reflect the continuous efforts of BASF AG to reduce the EDTA amount in its wastewater since 1991. Furthermore, Fig. 4 demonstrates the major pol-

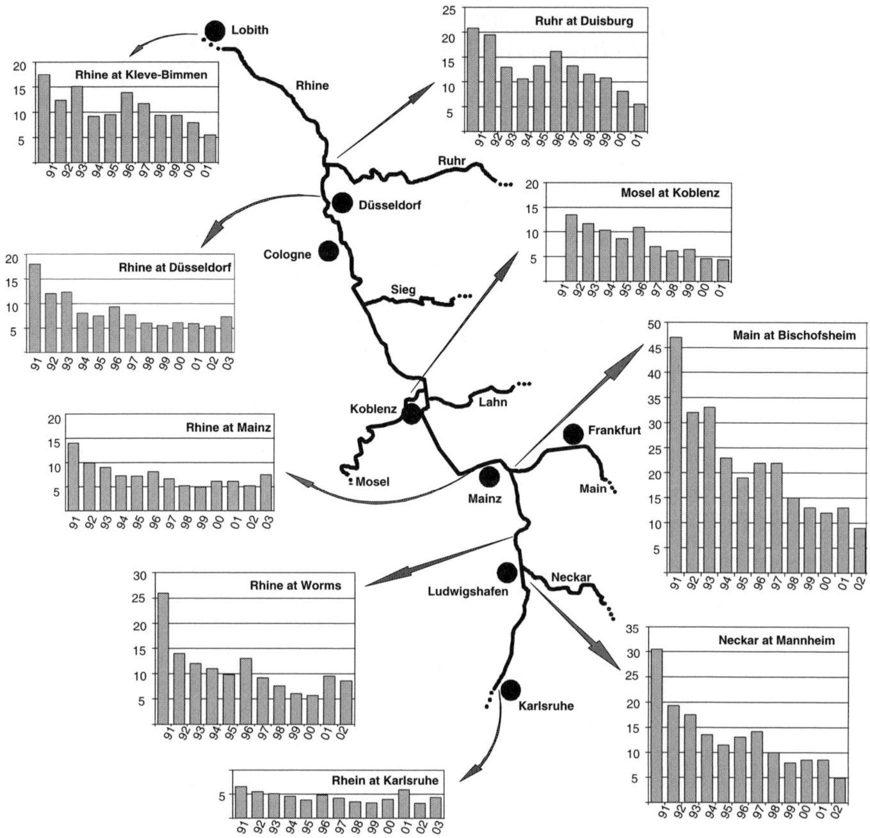

Fig. 4 Long-term development of EDTA concentrations (annual mean values in μg/L) in the Rhine between Karlsruhe and Kleve-Bimmen and in the Rhine tributaries Neckar, Main, Mosel, and Ruhr since 1991 (modified and extended from [30]). Sampling procedure: Rhine at Kleve-Bimmen, composite samples (C), number of samples $n = 26$ per annum; Rhine at Düsseldorf, Mainz, Worms, and Karlsruhe, C, $n = 13/a$; Ruhr at Duisburg, single samples (S), $n = 13/a$, Mosel at Koblenz, C, $n = 26/a$; Main at Bischofsheim, C, $n = 52/a$; Neckar at Mannheim, S, $n = 13/a$ [31, 34–42]

lution of the Rhine tributaries Neckar, Mosel, Ruhr, and particularly Main with EDTA in the early 1990s. Since then, EDTA concentrations in these rivers decreased with the consequence that EDTA concentrations also became lower in the Rhine at Mainz and Düsseldorf. However, annual mean values of EDTA concentrations in the lower reaches of Main and Neckar are still relatively high (up to 15 μg/L). Along the Rhine, DTPA levels increase starting with values below the quantitation limit of 1 μg/L at Basel to mean values of 1.3 μg/L at Karlsruhe (data from 2002) [33]. This value continues to increase until the sampling point Mainz to 1.5 μg/L and then fairly remains constant in the Rhine's onward journey [33]. Detailed measurements in the

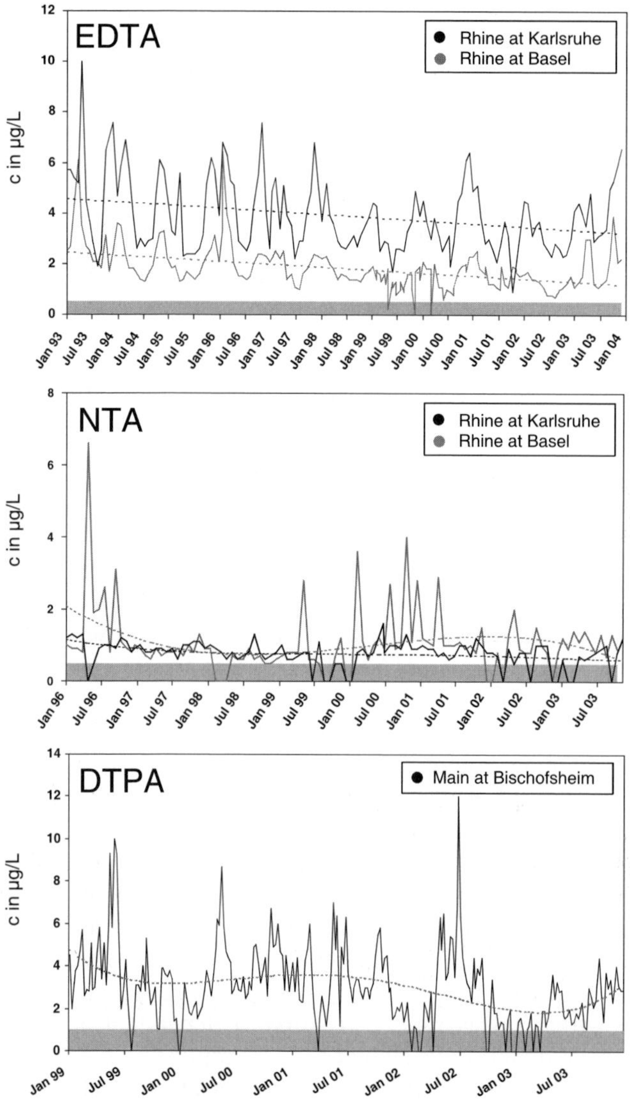

Fig. 5 Concentration curves of EDTA and NTA in the Rhine at Basel and Karlsruhe (composite samples, $n = 13$ per annum) and of DTPA in the Main at Bischofsheim (composite samples, $n = 52/a$) (modified and extended from [30, 31, 45])

longitudinal and cross profile of the Rhine could trace the DTPA input back to certain pulp and paper industries in the regions of Straßbourg, Karlsruhe and Mannheim [31]. No increase in NTA concentrations can be observed along the Rhine, which may be drawn back to NTA's good microbial degradability. Typical NTA concentrations in the Rhine are in the range of 0.5–2.1 µg/L

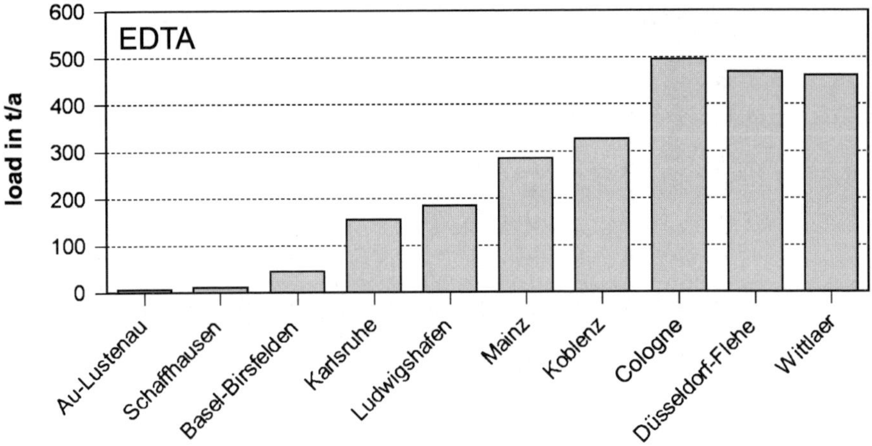

Fig. 6 EDTA loads along the Rhine in the year 2002 [33, 43]

nowadays [33, 43]. The aminopolycarboxylates β-ADA and 1,3-PDTA were only occasionally found in the Rhine with concentrations close to the quantitation limit of 1 μg/L [31, 44]. This is expected due to the low consumption volumes compared to EDTA, NTA and DTPA. For MGDA, no positive findings were reported.

Figure 5 shows the annual concentration curves of EDTA, NTA, and DTPA at different sampling points along the Rhine and Main. It is obvious that EDTA concentrations have decreased during the last years, whereas no clear tendency can be established for NTA and DTPA. On closer examination of the EDTA concentration curves of the Rhine at Karlsruhe and Basel, one notices that there are generally lower EDTA concentrations during the summer than during the winter months. This is probably due to an intensified photodegradation of Fe(III)-EDTA species in the brighter summer months (see discussion in Sect. 6).

Since measured concentrations are directly dependent on the runoff conditions of a river, quantitative changes of anthropogenic pollution in running waters can only be assessed when loads are considered instead of concentrations [46]. Thereby, the load characterizes the amount of substance that is transported within 1 year through the river's cross section.

Figure 6 gives an overview of the 2002 EDTA load estimates for the Rhine. EDTA loads of the Rhine are clearly increasing in the longitudinal section of the river, showing an elevation from the sampling point Au-Lustenau to the sampling point Wittlaer roughly by a factor of 100. At the lower Rhine, estimates of EDTA loads were between 460 and 500 t in the year 2002. Of this load, 45 t can be assigned to the Rhine part upstream from Basel and further 100 t originate from the upper Rhine between Basel and Karlsruhe. Further inputs particularly stem from the tributaries Neckar (2002: $\sim$ 30 t), Main

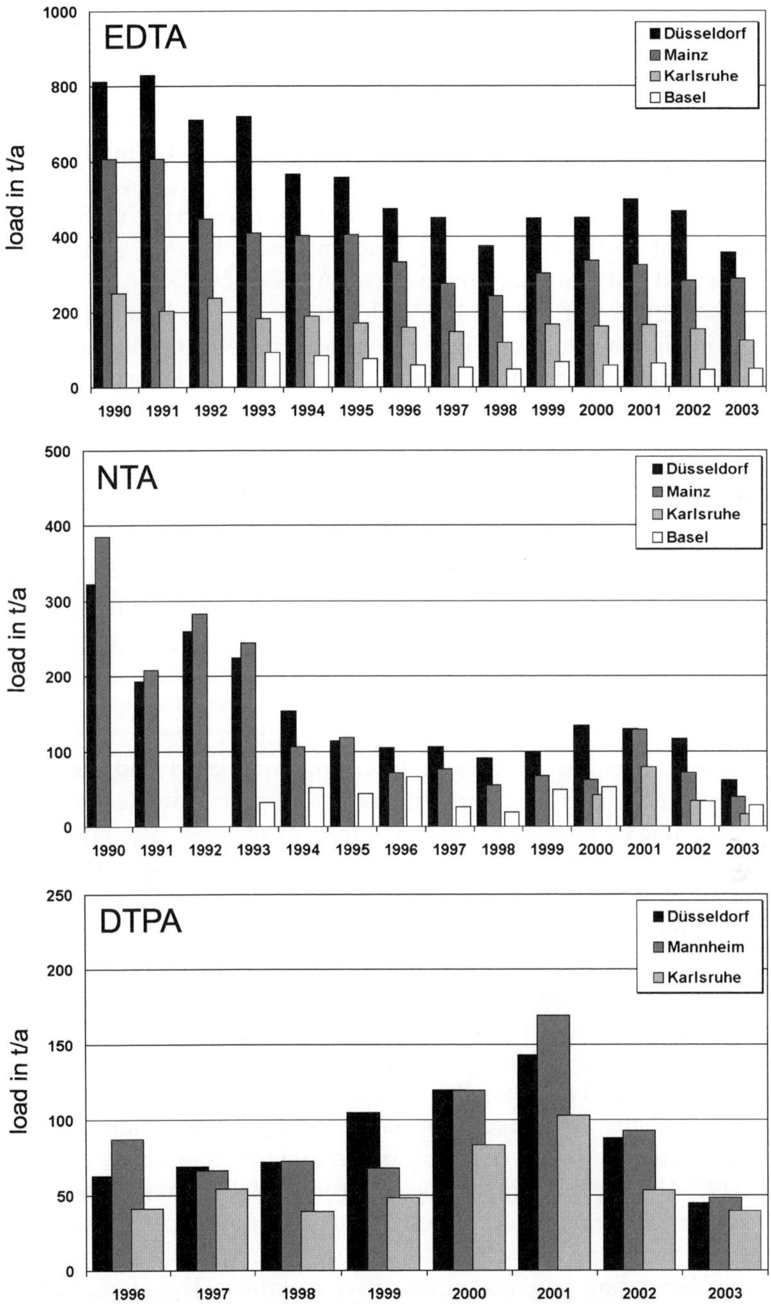

Fig. 7 Long-term development of EDTA, NTA (1990–2003) and DTPA (1996–2003) loads of the Rhine at the monitoring sites Basel (EDTA, NTA), Karlsruhe (EDTA, NTA, DTPA), Mannheim (DTPA), Mainz (EDTA, NTA), and Düsseldorf (EDTA, NTA, DTPA) (modified and extended from [30], data included from [31, 34–39, 41, 42])

(2002: ∼ 70 t), and Mosel (2001: ∼ 50 t) as well as from direct industrial inputs by the main producer BASF AG near Ludwigshafen (2001: ∼ 50 t) [33, 43, 47].

Figure 7 shows EDTA, NTA and DTPA loads of the Rhine at different sampling points along its longitudinal profile in the long-term. EDTA loads declined along the whole course of the Rhine until 1998, slightly increased until 2001, and then decreased notably again. Despite the obvious reduction of EDTA loads in the Rhine, the decline of 50% (referring to the loads in 1991) aimed at by the voluntary declaration, has only been achieved in the years 1998 and 2003. In the years in between (1999–2002) the load reduction was significantly lower (about 40%). The future has to show if EDTA loads of the Rhine can be stabilized at the level aimed at or if pollution can even be further reduced in the longer term. Since the early 1990s, NTA loads in the Rhine were clearly reduced and then remained almost unaltered during recent years. For DTPA loads along the Rhine, an increasing tendency could be established between 1996 and 2001; however, recent estimates have been again significantly lower.

5 Speciation

Basically, the input of aminopolycarboxylic acids into the aquatic environment may result in a number of direct and indirect effects on the ecosystem. Thereby, reaction patterns of free, non-complexed and metal-bound aminopolycarboxylates are clearly different, and even individual aminopolycarboxylate metal-complexes differ in their properties, depending on the bound central atom [1, 48–51]. As a consequence, knowledge on the chemical speciation, i.e., the identity of the respective complexed metal ion (central atom) of aminopolycarboxylic acids, is essential in many ways. Central atoms, which are of particular interest in this respect, are metal ions that already exist in nature due to geochemical reasons or that are major constituents of industrial wastes. Besides the hardness constituents calcium and magnesium, these are in particular copper, zinc, chromium, nickel, lead, cobalt, cadmium, iron, manganese, and mercury [52, 53].

At the currently measured concentrations in the aquatic environment, aminopolycarboxylic acids most probably exclusively exist as metal complexes [17, 54]. The pattern of the environmental aminopolycarboxylate-metal speciation depends on a number of different factors. Stability and tendency of formation of various aminopolycarboxylate metal-complexes in particular depend on potential competitive reactions. Therefore, the chemical composition of the aqueous medium is of vital importance for the species pattern being formed. The speciation of aminopolycarboxylates in waters is a function of the following properties: availability of heavy metal ions, pH-value, alkali and earth alkali ion contents (overall hardness or alkalinity), concentrations of inorganic anions (carbonates, hydroxides, chloride), natural inorganic ligands (exudates and decomposition products of algae, humic compounds) and other synthetic ligands (e.g., phosphonates) [55]. Calculations suggest that even at high aminopolycarboxylate concentrations of > 200 μM, EDTA and DTPA will not exist in their uncomplexed forms, because earth alkali metal ions are usually present in excess [54].

Heavy metal concentrations in German rivers are around 0.5 μmol/L (total of cadmium, copper, mercury, nickel, lead and zinc) [56]; when also iron and manganese are taken into account, the levels even come up to 10–20 μmol/L [17]. The overall concentration of the hardness constituents calcium and magnesium together is around 3000 μmol/L. The stoichiometric EDTA equivalent for the binding of 1 μmol/L metal ions is 300 μg/L. In most surface waters, typical EDTA levels range from 0 to 50 μg/L [1, 53, 57]. As a consequence, under natural conditions the metal ion availability is typically significantly higher than aminopolycarboxylate contents. Compared to natural compounds with chelating properties, aminopolycarboxylate concentrations are relatively low in surface waters that contain for example humate in concentrations of 1–10 mg/L [55]. Humate is composed of naturally occurring high-molecular compounds of a varying, complex structure. Its formation is mostly due to the microbial degradation of plant and animal material. Humic substances can act as natural ligands for heavy metals and can complex considerable metal amounts. Rashid and Leonhard demonstrated in laboratory studies that 1 mg humic acid (from a sediment) can complex up to 11.5 μmol metal (the total of cobalt, copper, manganese, nickel, and zinc) [58]. The metal ions are particularly bound to carboxyl and phenolic hydroxyl groups. While lead, copper and iron have a high affinity towards humate, the affinity of zinc, nickel, cadmium, manganese and cobalt is lower [55, 59]. Even though complex stabilities of heavy metal complexes with humic substances are generally lower than the stabilities of the corresponding aminopolycarboxylate metal-complexes, even in anthropogenically polluted waters, metal ions are preferentially bound by naturally occurring high-molecular organic substance like humic and fulvic acids or various amino acids. Kowalik and Einax demonstrated for highly polluted natural water samples with EDTA concentrations of 0.7–2.5 mg/L that the EDTA portion of the total zinc complexation was as low as 0.15–1.2%, with respect to cadmium it was 0.75–7.8%, and Cu-EDTA complexation was even only 0.03–0.15% (copper has a higher affinity to humic substances) [60].

The application of aminopolycarboxylic acids in certain applications typically aims at the complexation of various heavy metals or hardness constituents. Therefore, in agreement with the intended application, aminopolycarboxylates are released to wastewaters predominantly as metal complexes. During mixture of the initially emitted metal complexes with other wastewaters or their release to receiving waters, they are exposed to changing chemical conditions of the aqueous medium. As the initially emitted metal complexes are no fixed structures, they undergo, depending on current media conditions, continuous metal exchange processes that are directed by their respective complex stabilities at the thermodynamic equilibrium and also by kinetics.

Batchelder et al. reported that an increase in the hardness of surface water also resulted in an elevated Zn-EDTA toxicity, whereas no impact was found on the toxicity of free EDTA acid or its sodium salt [61]. This result indicates that zinc ions can be partially displaced from EDTA complexes by an excess of the hardness constituents calcium and magnesium, even though this is rather improbable according to the corresponding complex stabilities [50]. It has also been demonstrated that an increase of aminopolycarboxylate levels has an impact on speciation. When the complexing agent is present in a molar excess compared to the trace metals, the portion of EDTA or DTPA earth alkali metal species increases [54]. Hering and Morel [62] and Raspor et al. [63] reported that the copper, lead, zinc or cadmium portion complexed by EDTA is considerably reduced in the presence of calcium (in the sea water concentration range) and magnesium. Xue et al. demonstrated that the iron in Fe(III)-EDTA is gradually replaced by other cations in surface waters, in particular by zinc and calcium [64].

The speed of the metal-exchange reactions critically depends on the identity of the complexed metal ion [64, 65]. For example, in Ca-EDTA, the central atom is for orders of magnitude faster exchanged against zinc than in Cu-EDTA or Fe(III)-EDTA [64–66]. It has been demonstrated by a couple of authors that due to the slow exchange kinetics under natural conditions the thermodynamic equilibrium of EDTA speciation is only achieved after days or even weeks [65, 67–69]. The exchange of the iron in Fe(III)-EDTA by calcium in river water has for example a halflife of 20 days [64], and the dissociation of Fe(III)-EDTA under marine conditions was described as a "slow" process [70]. When calcium and magnesium levels exceed those of heavy metals (a standard condition for natural waters) exchange processes concerning the central atom of EDTA-metal complexes are generally relatively slow [1, 54, 62, 63, 65, 68, 71]. Due to the slow kinetics of some metal exchange reactions, the thermodynamic speciation equilibrium is often not achieved between emission points and river mouth [17]. For example, the water of the Rhine needs only one week for the distance from Basel to its mouth. Therefore, the initial speciation of aminopolycarboxylates at their emission points is an important factor determining their actual speciation in this river.

So far, only very few studies have investigated the speciation of aminopolycarboxylic acids in the environment, and more detailed knowledge is restricted to EDTA and NTA [1]. Only a few papers describe reliable methods for the differentiated analysis of aminopolycarboxylate species at environmentally relevant levels and no large measurement programs have been established in natural waters so far [1, 27, 28, 57]. Due to the lack of analytical methods, in most cases, model calculations assuming a thermodynamic equilibrium were performed [72–79]. However, the real speciation occurring in natural waters sometimes differs clearly from the assumed thermodynamic equilibrium state. Due to the complexity of the particular system, an estimate of the speciation by simple model calculations is extraordinary difficult. In addition to kinetics, photodegradation processes [e.g., Fe(III)-EDTA] and sorption of metal complexes also have to be considered [50, 51]. Consequently, aminopolycarboxylic acid speciation studies should not be based only on equilibrium calculations but should also include results of direct analysis of individual metal complexes [1]. More accurate approaches on speciation determination have been recently published by Nowack et al. [57, 80, 81] and Sillanpää et al. [54].

For phosphorus removal in wastewater treatment plants, phosphate is often precipitated by iron(III) salts. In this process, the high iron concentration, the high stability of the Fe(III)-EDTA complex and the relatively low pH-value (locally, after iron salt addition) favor the formation of the Fe(III)-EDTA complex from other metal-EDTA species (e.g. Cu-EDTA) by metal exchange reactions [64, 82]. According to Nirel et al., EDTA metal-complexes (calcium, chrome, copper, magnesium and zinc) can be quickly (within 10 min) converted into Fe(III)-EDTA complexes under the acidic conditions of phosphate precipitation (pH 4) [83]. Bedsworth and Sedlak detected in the primary effluent of a municipal wastewater treatment plant, i.e., prior to phosphate precipitation, 300 nmol/L Cu-EDTA and 310 nmol/L Zn-EDTA [84]. Following phosphate precipitation, Fe(III)-EDTA and also Fe(III)-DTPA were repeatedly detected in the corresponding effluents of wastewater treatment plants [82, 85, 86]. For example, the Fe(III)-EDTA portion of the total EDTA in effluents of wastewater treatment plants along the Glatt (Switzerland) was around 60–70% [57]. More recent studies demonstrated that in addition to Fe(III)-EDTA, wastewater treatment plants mainly discharge Ni-EDTA, Mn-EDTA, Al-EDTA and Ca-EDTA with the effluent stream [54, 67, 69, 83, 87].

In receiving waters, the speciation pattern discharged by wastewater treatment plants changes according to the local chemical environment. Based on a combination of direct analytical measurements and equilibrium calculations, Nowack evaluated the speciation of EDTA in river water [81]. According to his results, 51% of the total EDTA is present as Zn-EDTA and 32% as Fe(III)-EDTA; minor species are Ca-EDTA (7%), Mn(II)-EDTA (5%), Mg-EDTA (2%), Pb-EDTA (2%) and Ni-EDTA (0.1%). Despite its high complex stability, the Cu-EDTA portion is very low (< 1%), since copper is preferentially bound by natural ligands. Two nmol/L Cu-EDTA or less was measured in river water polluted to different degrees [28]. In another study, Nowack et al. detected 24 nmol/L Fe(III)-EDTA and 1.5 nmol/L Ni-EDTA in river water [57]. Considering a number of boundary conditions, also model calculations by Sillanpää et al. established iron and zinc complexes as the dominating EDTA and DTPA species in natural waters [54]. According to Nowack [81], NTA is in rivers predominantly associated with calcium (62% of the total NTA), followed by magnesium (29%) and zinc (3%). The portion of free and uncomplexed NTA is about 5%. Only a minor part of the total NTA is complexed with other heavy metal ions.

Metal exchange reactions may also occur during infiltration of surface water into aquifers, because metal-EDTA complexes can interact with the sediments. These interactions decisively depend on the pH-value present and the metal ion complexed by the aminopolycarboxylate [80, 88, 89]. In principle, the complexed metal ion can be released by exchange with adsorbed or dissolved metal ions (e.g., copper, zinc, lead and cadmium). Also the dissolution of iron, aluminum or manganese oxides under the formation of Fe-EDTA, Al-EDTA or Mn-EDTA may occur [66]. At alkaline pH-values, quite common in calcareous aquifers, the remobilization of adsorbed metals by replacement of the initial EDTA central atom (e.g., Fe(III)) predominates. These reactions affect presumably in particular the mobility of zinc, lead and cadmium. In infiltrated groundwater close to the Glatt (Switzerland) with a calciferous aquifer, EDTA speciation is predominantly characterized by Zn-EDTA, Ca-EDTA and Fe(III)-EDTA, in some groundwaters also elevated Mn(II)-EDTA levels occurred. Furthermore, it could be demonstrated that the Fe(III)-EDTA portion is reduced during the passage of the water through the aquifer. In all investigated groundwaters, Zn-EDTA was the most relevant EDTA species, with a portion of 40–60% of the total EDTA. The Pb-EDTA portion was around 3%, while those of Cu-EDTA and Mg-EDTA always were below 1% [80]. In infiltration pathways with a low pH-value, the dissolution of iron oxides by various metal-EDTA species takes place. Fe(III)-EDTA is formed in an exchange reaction, and the initially complexed metal ion is released and adsorbed by the solid phase [80].

The speciation of aminopolycarboxylates in groundwater therefore considerably depends on the individual geological and chemical properties of the infiltration pathway under investigation and is therefore extremely site-specific. In a recent analysis of a groundwater close to the Glatt (Switzerland), 19 nmol/L Fe(III)-EDTA, 16 nmol/L Cu-EDTA and 6 nmol/L Ni-EDTA were found [28]. Also close to the Glatt, Nowack et al. determined 5–22 nmol/L Fe(III)-EDTA and 0.4–0.8 nmol/L Ni-EDTA in selected groundwaters [57].

6
Behavior During Drinking Water Treatment

For drinking water preparation along the Rhine, surface water is often subjected to an underground passage via riverbank filtration. In recent investigations, the behavior of aminopolycarboxylates during this treatment step was studied at a waterworks station at the lower Rhine River over several months. As sampling points, three observation wells of different depths were chosen. These observation wells are located between the Rhine River and the well gallery used for drinking water production. The aquifer was characterized at the time of sampling by suboxic redox conditions and residence times of the infiltrated water in the aquifer vary depending on the depth of the observation well between 15 and 60 days (well A: ~ 15 days; well B: ~ 30 days; well C: ~ 60 days). Samples from these wells consist only of riverbank-filtrated water that is not influenced by natural landward groundwater. In this study, significant differences were observed for EDTA, NTA, and DTPA. While EDTA was found to be only slightly affected by riverbank filtration, NTA was detected only in the surface water from the Rhine and not in the corresponding bank filtrate

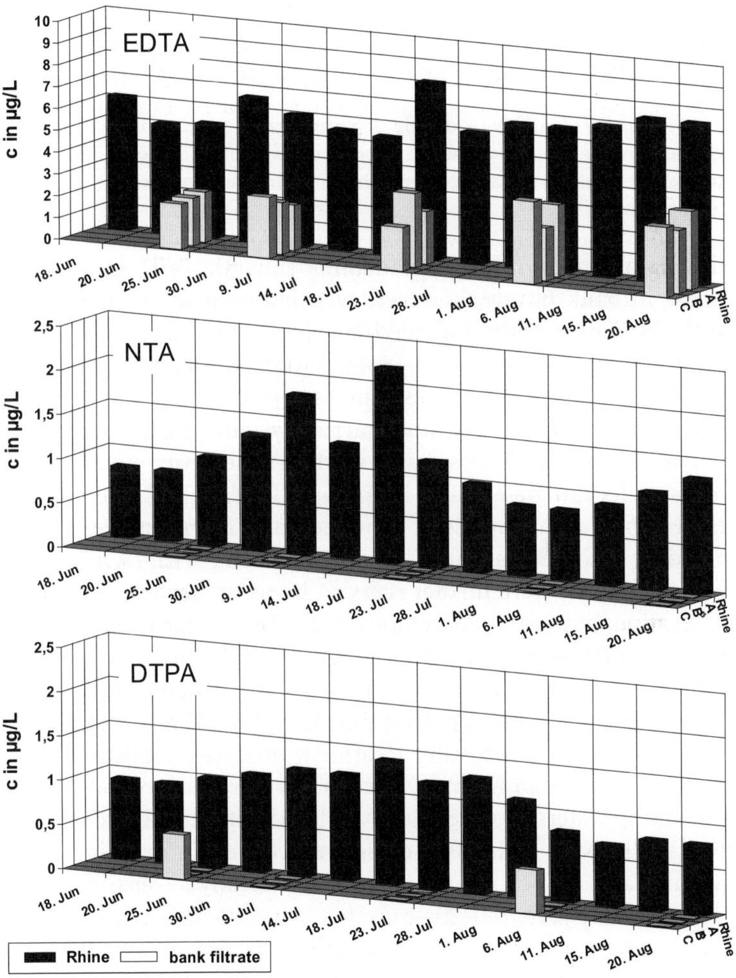

Fig. 8 Behavior of EDTA, NTA and DTPA during suboxic riverbank filtration in a water works at the lower Rhine (data from summer 2003). Measurements were performed based on single samples in Rhine surface water and bank filtrate obtained from different observation wells (A, B, C) located between the Rhine River and the main well gallery used for drinking water production. Limits of quantitation were 0.5 µg/L for EDTA, NTA, and DTPA

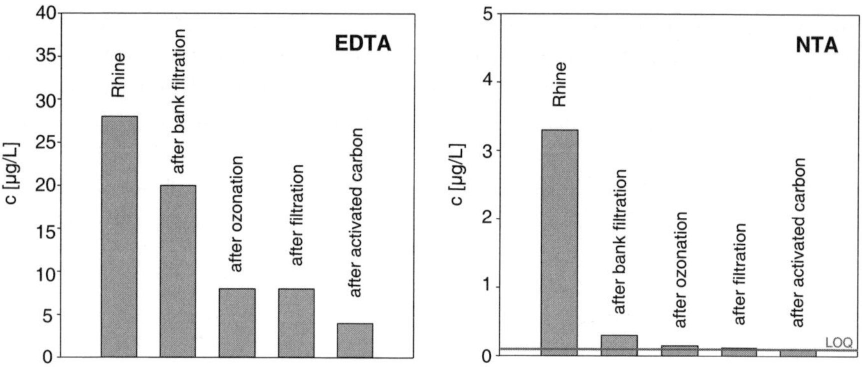

Fig. 9 Behavior of EDTA and NTA during drinking water production at the lower Rhine. Presented data are mean values of weekly composite samples that were obtained by mixture of five single samples per week over a time period of 2 months (data from [90])

indicating its better microbial degradability (Fig. 8). DTPA could only rarely be detected in bank filtrate, however, even levels in the surface water were close to the limit of quantitation making an evaluation generally difficult.

However, riverbank filtration is usually only the first step of the whole drinking water preparation process and is typically followed up by further treatment steps to build up multiple barrier systems. Figure 9 shows the behavior of EDTA and NTA in a water works at the lower Rhine. This water works obtains its raw water from Rhine bank filtrate [90]. Further drinking water treatment steps comprise of ozonation and filtration over a multilayer filter including activated carbon. The concentrations of EDTA and NTA in the raw riverbank filtrate were further reduced by using ozonation and activated carbon filtration; however, total removal of EDTA was not achieved indicating its relevance for drinking water. It is obvious that the ozonation step was the most effective treatment step in decreasing EDTA concentrations [90].

In general, levels of aminopolycarboxylates found in drinking water are decisively dependent on the water treatment process established. Evaluating data from several cities in Germany over the last few years revealed that investigated drinking waters were regularly polluted with EDTA at concentrations of up to 7 µg/L. NTA and DTPA could be detected in drinking water only in individual cases and concentrations were generally lower than 3 µg/L and 4 µg/L, respectively [30].

7
Environmental, Toxicological and Ecotoxicological Properties

In principle, aminopolycarboxylates may be degraded in the environment by biotic and abiotic processes [1, 81]. NTA and MGDA are readily degradable by microbiological processes, while the biotic reduction of EDTA and 1,3-PDTA is relatively low [1, 91–93]. β-ADA and DTPA are in between [94–97]. Photochemical processes are assumed to have the largest impact on the natural reduction of EDTA [81]. Basically, this pathway is restricted to the decomposition of the Fe(III)-EDTA complex. However, its portion of the total EDTA speciation is considerably high in river waters ($\sim$ 30%) and metal exchange reactions of Fe(III)-EDTA are notably slow [81]. In principle, EDTA can be also degraded by microorganisms [1, 98–100]. This has been demonstrated in studies with specially enriched bacterial cultures. The extent of EDTA reduction by this process, however, significantly depends on its metal speciation. In particular, microbial degradation of Zn-EDTA and Fe(III)-EDTA complexes is only poor, although they predominate in the environment. This aspect is the basic difference to microbial NTA decomposition that very efficiently affects the environmentally predominating NTA species Ca-NTA [101].

Aminopolycarboxylates that are not or only slowly eliminated by degradation, may (because of its chelating properties) affect the mobility of metals in the environment. Aminopolycarboxylates that are in contact with sediments can also be adsorbed by them by electrostatic interactions [48, 49, 81, 102, 103]. Furthermore, they can dissolve sediment components (in particular iron and aluminum (hydr)oxides) under formation of the iron and aluminum complexes (in particular in an acidic environment) [81, 102, 104–110] or remobilize already adsorbed or precipitated metal ions (preferentially zinc) from the sediment (in particular in an alkaline environment) [66, 76, 81, 111]. According to the current scientific opinion, effects of aminopolycarboxylic acids on the remobilization of heavy metals are assumed to be rather low in waters with levels in the low to middle µg/L range [1, 17, 53, 77, 112, 113]. However, local metal mobilization cannot be excluded when shock loadings occur.

All aminopolycarboxylic acids do not demonstrate high toxicities against mammals and aquatic organisms, and they do not differ much with respect to their basic toxicological characteristics [114–117]. Toxic effects at increased doses are mainly caused by masking trace elements essential for the studied organism, like zinc or calcium. Consequently, the extent of the observed effects is typically higher for the free aminopolycarboxylic acid than for individual aminopolycarboxylate-metal species. NOEL (no observed effect level) values established for mammals are 250 mg EDTA [117] and 10 mg NTA [116] per kg body weight and day. Therefore, the aminopolycarboxylate levels found presently in drinking water are too low to constitute a risk to human health [116, 117]. Furthermore, recent aminopolycarboxylate levels in surface waters provide most probably no hazard to aquatic life. Typically, acute toxicities towards aquatic organisms start to take place only at concentrations above 40 mg/L aminopolycarboxylate and harmful long-term impacts will most probably not occur at concentrations below 1 mg/L [115]. However, near industrial plants with highly concentrated wastewater discharges, a potential risk to aquatic organisms cannot be completely excluded, since aminopolycarboxylates in receiving waters can locally reach concentrations in the low mg/L-range [17].

Aminopolycarboxylic acids might support algae growth (eutrophication) [50, 115]. In this respect, two mechanisms are of particular interest. On the one hand, complex formation may result in an increased availability of limiting trace elements for the algal cells, on the other hand, the availability of certain toxic heavy metal ions might be reduced by complexation, resulting in the growth of algae species that normally are inhibited in development by the toxic effect. Under natural conditions and at concentrations typically found in surface waters, the eutrophication potential of aminopolycarboxylic acids appears to be comparably low, even though it cannot be completely excluded that it might become relevant under certain conditions (in particular in cases of trace element deficiencies) [17, 115].

Due to the positive findings of aminopolycarboxylates in drinking water, their potential hazard for aquatic organisms close to industrial sewage discharge points and their possible influence on eutrophication, efforts should be made to limit the risks and to prevent the release of highly concentrated aminopolycarboxylate effluents into the aquatic environment. This should be feasible, since several efficient treatment technologies were tested and established in the past few years [118, 119]. The development in Germany demonstrates that adequate regulatory networks and framework conditions are useful in this respect.

References

1. Bucheli-Witschel M, Egli T (2001) FEMS Microbiol Rev 25:69
2. Knepper TP (2003) Trends Anal Chem 22:708
3. Wolf K, Gilbert P (1992) EDTA – Ethylenediaminetetraacetic acid. In: Hutzinger O (ed) The handbook of environmental chemistry, vol 3, Part F. Springer, Berlin Heidelberg New York, p 243
4. Jäger H-U, Schul W (2001) Muench Beitr Abwasser Fisch Flussbiol 54:207
5. Potthoff-Karl B (1994) SÖFW J 120:104
6. Brüggenthies A (1996) PhD Thesis, TU Braunschweig
7. Means JL, Alexander CA (1981) Nucl Chem Waste Manage 2:183
8. UBA (2001) Unterlagen zum 15. EDTA-Fachgespräch "Verringerung der Gewässerbelastung durch EDTA" am 8.11.2001. Umweltbundesamt, Berlin, Germany
9. Davenport RE, Dubois F, DeBoo A, Kishi A (2000) Chelating agents – CEH Product Review. In: Chemical Economics Handbook. SRI International
10. Knepper TP, Driemler J, Maes A, Müller J, Soßdorf D (2001) Einträge synthetischer Komplexbildner in die Gewässer. Final Report FKZ 299 24 284. Umweltbundesamt, Berlin, Germany
11. Knepper TP, Weil H (2001) Vom Wasser 97:193
12. CEFIC (Conseil Européen des Fédérations de l'industrie chimique) (2001) CEFIC-Statistics "Aminopolycarboxylic acids"

13. UBA (1999) Unterlagen zum 13. EDTA-Fachgespräch "Verringerung der Gewässerbelastung durch EDTA" am 4.11.1999. Umweltbundesamt, Berlin, Germany
14. Könen I (1997) PhD Thesis, TH Aachen
15. UBA (1997) Jahresbericht 1997. Umweltbundesamt, Berlin, Germany, p 105
16. Kraus K (2001) Muench Beitr Abwasser Fisch Flussbiol 54:187
17. EU (2001) EU-Risk Assessment: Tetrasodium ethylenediaminetetraacetate (Na_4EDTA)
18. Conrad J (2000) Environmental policy regulation by voluntary aggreements: technical innovations for reducing use and emission of EDTA. FFU-Report 00-04. Forschungsstelle für Umweltpolitik, Freie Universität Berlin
19. Hamm A (1996) Muench Beitr Abwasser Fisch Flussbiol 49:91
20. Bias R (1994) EDTA – Umweltverträglichkeit aus der Sicht des Herstellers. In: Umweltministerium Baden-Württemberg (ed) EDTA im Bodenseeraum. Conference Proceeding Friedrichshafen, Germany
21. Swedish Society for Nature Conservation (2000) Report: Complexing agents
22. Schowanek D, McAvoy D, Versteeg D, Hanstveit A (1996) Aquat Toxicol 36:253
23. Jaworska JS, Schowanek D, Feijtel TCJ (1999) Chemosphere 38:3597
24. Schmidt CK, Brauch H-J (2005) Analysis of aminopolycarboxylates and organophosphonates. In: VanBriesen JM, Nowack B (eds) Biogeochemistry of chelating agents. ACS, Washington, DC, p 76
25. Sillanpää M, Sihvonen M-L (1997) Talanta 44:1487
26. Deutsches Institut für Normung eV (2004) DIN EN ISO 16588
27. Ammann AA (2002) J Chromatogr A 947:205
28. Ammann AA (2002) Anal Bioanal Chem 372:448
29. Sheppard RL, Henion J (1997) Electrophoresis 18:287
30. Schmidt CK, Fleig M, Sacher F, Brauch H-J (2004) Environ Pollut 131:107
31. AWBR (2002) AWBR-Jahresbericht 2001. 33. Bericht. Arbeitsgemeinschaft Wasserwerke Bodensee-Rhein, Stuttgart, Germany
32. Streichfuss D, Brauch H-J, Fleig M, Hoheisel K (2000) Komplexbildner in der Schussen. Final Report UBR-Project II U 14. DVGW-Water Technology Center (TZW), Karlsruhe, Germany
33. AWBR (2003) AWBR-Jahresbericht 2002. 34. Bericht. Arbeitsgemeinschaft Wasserwerke Bodensee-Rhein, Stuttgart, Germany
34. AWBR (2001) AWBR-Jahresbericht 2000. 32. Bericht. Arbeitsgemeinschaft Wasserwerke Bodensee-Rhein, Stuttgart, Germany
35. AWBR (2000) AWBR-Jahresbericht 1999. 31. Bericht. Arbeitsgemeinschaft Wasserwerke Bodensee-Rhein, Stuttgart, Germany
36. ARW (2002) ARW-Jahresbericht 2001. 58. Bericht. Arbeitsgemeinschaft Rhein-Wasserwerke eV, Cologne, Germany
37. ARW (2001) ARW-Jahresbericht 2000. 57. Bericht. Arbeitsgemeinschaft Rhein-Wasserwerke eV, Cologne, Germany
38. ARW (2000) ARW-Jahresbericht 1999. 56. Bericht. Arbeitsgemeinschaft Rhein-Wasserwerke eV, Cologne, Germany
39. ARW (1999) ARW-Jahresbericht 1998. 55. Bericht. Arbeitsgemeinschaft Rhein-Wasserwerke eV, Cologne, Germany
40. UBA (2002) Daten zu organischen Umweltchemikalien der Länderarbeitsgemeinschaft Wasser (LAWA). Internet communication: www.umweltbundesamt.de /hid/index.htm Umweltbundesamt, Berlin, Germany
41. ARW (2004) ARW-Jahresbericht 2003. 60. Bericht. Arbeitsgemeinschaft Rhein-Wasserwerke eV, Cologne, Germany

42. AWBR (2004) AWBR-Jahresbericht 2003. 35. Bericht. Arbeitsgemeinschaft Wasserwerke Bodensee-Rhein, Stuttgart, Germany
43. ARW (2003) ARW-Jahresbericht 2002. 59. Bericht. Arbeitsgemeinschaft Rhein-Wasserwerke eV, Cologne, Germany
44. Sacher F, Lochow E, Brauch H-J (1998) Vom Wasser 90:31
45. Fleig M (2003) Vorkommen von organischen Mikroverunreinigungen in Rhein und Main. Untersuchungsbericht für das Jahr 2002. DVGW-Technologiezentrum Wasser (TZW) Karlsruhe
46. Fleig M, Brauch H-J (2002) EDTA-Frachten 2001 – Vergleich von Schätzmethoden und verwandten Datenreihen. In: ARW (ed) ARW-Jahresbericht 2001. 58. Bericht. Cologne, Germany, p 79
47. UBA (2002) Unterlagen zum 16. EDTA-Fachgespräch "Verringerung der Gewässerbelastung durch EDTA" am 14.11.2002. Umweltbundesamt, Berlin, Germany
48. Nowack B, Sigg L (1996) J Colloid Interface Sci 177:106
49. Nowack B, Lützenkirchen J, Behra P, Sigg L (1996) Environ Sci Technol 30:2397
50. Sillanpää M (1997) Rev Environ Contam Toxicol 152:85
51. Sillanpää M, Rämö J (2001) Chemosphere 45:881
52. Koppe P (1976) Z Wasser Abwasser Forsch 5:153
53. Walterson E, Landner L (1993) EDTA used for detergent applications: an environmental safety assessment for Sweden. Swedish Environmental Research Group (MFG) Stockholm, Sweden
54. Sillanpää M, Orama M, Rämö J, Oikari A (2001) Sci Total Environ 267:23
55. Fent K (1998) Ökotoxikologie. Umweltchemie, Toxikologie, Ökologie. Georg Thieme Verlag, Stuttgart, New York
56. Danish Environmental Protection Agency (2001) Environmental and health assessment of substances in household detergents and cosmetic detergent products. Chapter 7: Compelexing agents
57. Nowack B, Kari FG, Hilger SU, Sigg L (1996) Anal Chem 68:561
58. Rashid MA, Leonard JD (1973) Chem Geol 11:89
59. Nriagu JO, Coker RD (1980) Environ Sci Technol 4:443
60. Kowalik C, Einax JW (2000) Vom Wasser 94:229
61. Batchelder TL, Alexander HC, McCarty WM (1980) Bull Environ Contam Toxicol 24:543
62. Hering JG, Morel FMM (1988) Environ Sci Technol 22:1469
63. Raspor B, Nürnberg HW, Valenta P, Branica M (1980) J Electroanal Chem 115:293
64. Xue H, Sigg L, Kari FG (1995) Environ Sci Technol 29:59
65. Hering JG, Morel FMM (1989) Geochim Cosmochim Acta 53:611
66. Nowack B, Kari FG, Krüger HG (2001) Water Air Soil Pollut 125:243
67. Bedsworth WW, Sedlak DL (1999) Environ Sci Technol 33:926
68. Hering JG, Morel FMM (1990) Environ Sci Technol 24:242
69. Kari FG, Giger W (1996) Water Res 30:122
70. Hudson RJM, Covault DT, Morel FMM (1992) Mar Chem 38:209
71. Xue H, Sigg L (1993) Limnol Oceanogr 38:1200
72. Hennes EC (1990) Working paper to the meeting of the AIS-EDTA-Task Force from 14.2.1989
73. van Dijk-Looyard AM, de Groot AC, Janssen PJCM, Wondergem EA (1990) EDTA in drink- en oppervklaktewater. Rapport-Nr. 718629006. Rijkinstituut voor Volksgenzondheid en Milieuhygiene, Bilthoven, The Netherlands
74. Vuceta J, Morgan JJ (1978) Environ Sci Technol 12:1302
75. Gardiner J (1976) Water Res 10:507

76. Lorenz J (1997) Remobilisierung von Schwermetallen aus ruhenden Gewässersedimenten durch EDTA und NTA bei aerober und anaerober Wasserphase. Report FZKA 5977. Forschungszentrum Karlsruhe, Germany
77. Bernhardt H (1984) NTA – Studie über die Umweltverträglichkeit von Nitrilotriacetat (NTA). Richarz, Sankt Augustin
78. Rubin M, Martell AE (1980) Biol Trace Elem Res 2:1
79. Eberle SH, Hennes EC, Dehnad F (1982) Z Wasser Abwasser Forsch 15:217
80. Nowack B, Xue H, Sigg L (1997) Environ Sci Technol 31:866
81. Nowack B (2002) Environ Sci Technol 36:4009
82. Kari FG (1994) PhD Thesis No. 10698, ETH Zürich
83. Nirel PM, Pardo P-E, Landry J-C, Revaclier R (1998) Water Res 32:3615
84. Bedsworth WW, Sedlak DL (2001) J Chromatogr A 905:157
85. Buchberger W, Mülleder S (1995) Mikrochim Acta 119:103
86. Richardson DE, Ash GH, Harden PE (1994) J Chromatogr A 688:47
87. Schneider JK, Gloor R, Giger W, Schwarzenbach RP (1984) Water Res 18:1515
88. Zachara JM, Smith SC, Kuzel LS (1995) Geochim Cosmochim Acta 59:4825
89. Friedly JC, Kent DB, Davis JA (2002) Environ Sci Technol 36:355
90. Brauch H-J, Schullerer S (1987) Vom Wasser 69:155
91. Pitter P, Sýkora V (2001) Chemosphere 44:823
92. Potthoff-Karl B, Greindl T, Oftring A (1996) SÖFW J 122:392
93. Sýkora V, Pitter P, Bittnerová I, Lederer T (2001) Water Res 35:2010
94. Nitschke L, Wilk A, Cammerer C, Lind G, Metzner G (1997) Chemosphere 34:807
95. Bolton H, Li S, Workman D, Girvin D (1993) J Environ Qual 22:125
96. Stumpf M, Ternes TA, Schuppert B, Haberer K, Hoffmann P, Ortner HM (1996) Vom Wasser 86:157
97. UBA (1991) Unterlagen zum 5. EDTA-Fachgespräch "Verringerung der Gewässerbelastung durch EDTA" am 25.09.1991. Bundesministerium für Umwelt, Naturschutz und Reaktorsicherheit, Bonn-Bad Godesberg, Germany
98. Egli T (2001) J Biosci Bioeng 92:89
99. Klüner T (1996) PhD Thesis, Cuvillier Verlag, Göttingen
100. Nörtemann B (1999) Appl Microbiol Biotechnol 51:751
101. VanBriesen JM, Rittmann B, Xun L, Girvin D, Bolton H (2000) Environ Sci Technol 34:3346
102. Bowers AR, Huang CP (1986) J Colloid Interface Sci 110:575
103. Byegard J, Skarnemark G, Skalberg M (1999) J Radioanal Nucl Chem 241:281
104. Girvin DC, Gassman PL, Bolton H (1993) Soil Sci Soc Am J 57:1
105. Jardine PM, Jacobs GK, O'Dell JD (1993) Soil Sci Soc Am J 57:954
106. Davis JA, Kent DB, Coston JA, Hess KM, Joye JL (2000) Water Resour Res 36:119
107. Jardine PM, Taylor DL (1995) Geoderma 67:125
108. Nowack B, Sigg L (1997) Geochim Cosmochim Acta 61:951
109. Szecscody JE, Zachara JM, Bruckhart PL (1994) Environ Sci Technol 28:1706
110. Kent DB, Davis JA, Anderson L, Rea BA, Coston JA (2002) Geochim Cosmochim Acta 66:3017
111. Twachtmann U, Petrick S, Merz W, Metzger JW (1998) Vom Wasser 91:101
112. AISE (1987) An assessment of the implications of the use of EDTA in detergent products. AISE and AIS Environmental Safety Working Group, Brussels
113. Kernforschungszentrum Karlsruhe (1991) Aquatische Umweltverträglichkeit von Nitrilotriessigsäure (NTA). Eigenverlag, Karlsruhe, Germany
114. Lanigan R, Yamarik T, Andersen F (2002) Int J Toxicol 21:95
115. Schmidt CK, Brauch H-J (2004) Environ Toxicol 19:620

116. World Health Organization (1996) Nitrilotriacetic acid. In: World Health Organization (ed) Guidelines for drinking-water quality. vol 2. Health criteria and other supporting information. Geneva
117. World Health Organization (1998) Edetic acid (EDTA). In: World Health Organization (ed) Guidelines for drinking-water quality. Addendum to vol 2. Health criteria and other supporting information. Geneva
118. Sillanpää M, Pirkanniemi K (2001) Environ Technol 22:791
119. van Ginkel CG, Kester H, Stroo CA, van Haperen A (1999) Water Sci Tech 40:259

Hdb Env Chem Vol. 5, Part L (2006): 235–254
DOI 10.1007/698_5_033

Published online: 9 December 2005

Pollutants as Byproducts and Degradation Products of Chemical Syntheses

T. P. Knepper[1] (✉) · F. Karrenbrock[2]

[1]Europa University of Applied Sciences Fresenius,
Limburger Str. 2, 65510 Idstein, Germany
knepper@fh-fresenius.de

[2]GEW RheinEnergie AG, Parkgürtel 24, 50823 Köln, Germany

Abstract Byproducts of industrial organic syntheses are emitted with the wastewater into the River Rhine; in many cases they are polar and poorly biodegradable. Applying a well-planned strategy as well as combining various modern spectroscopic methods, it was possible to identify a series of relevant byproducts. Among the identified compounds there were halogenated aromatic acids, lactone isomers, sulfinates, sorbose derivatives, as well as substituted phosphorous oxides. All of them could, after identification, also be detected in the River Rhine. In most cases a cooperative discussion with the producers led to a measurable reduction of emissions into the Rhine.

Keywords Analysis · Byproducts · Industrial wastewater · Polar persistent pollutants · Structure elucidation

Abbreviations

WW	Wastewater
FBBR	Fixed bed bioreactor
TPPO	Triphenylphosphinoxid
DPPO	Diphenylphosphinoxid
DPPS	Diphenylphosphinic acid
SPE	Solid phase extraction

HPLC-DAD	High pressure liquid chromatography–diode array detector
WWTP	Wastewater treatment plant
TOC	Total organic carbon
TOX	Total organic halogens
BIOL	Bis-ethyl-*iso*-octanol lactone
GC-MS	Gas chromatography–mass spectrometry
SIM	Single ion monitoring
DAS	Di-isopropylidene-sorbofuranose
DAG	Di-isopropylidene-α-L-xylo-2-hexulofuranosonic acid
LC-ESI-MS-MS	Liquid chromatography electrospray ionization–tandem mass spectrometry
TCPA	Tetrachlorophthalic acid
MS-MS	Tandem mass spectrometry
LC-ESI-TOF	Liquid chromatography–electrospray ionisation–time-of-flight
$M-H^-$	Deprotonated molecular ion
CMST	2-Chloro toluene methylsulfon
R_t	Retention time
SPS	Sarkosin-*N*-(phenylsulfonyl)
HPS	6-[Methyl(phenylsulfonyl)-amino]hexanoic acid
DOC	Dissolved organic carbon

1 Introduction

The River Rhine is a recipient of not only municipal wastewater (WW), but also of the industrial WW of almost all the important European chemical industries. When the river quality decreased soon after World War II, waterworks along the Rhine faced odor and taste problems in drinking water. Bank filtration was no longer sufficient and activated carbon filtration was found to be an appropriate treatment step to solve these problems. Accordingly, activated carbon was introduced in waterworks along the Rhine. In addition, waterworks initiated research projects on the pollution of the Rhine and started monitoring the river quality [1–3].

There is a continuous need for identifying the yet-unknown substances, especially stemming from the chemical industry. However, it is impossible to monitor every substance that might be in a surface water. According to their needs, waterworks along the Rhine set up monitoring programs for selected organic micropollutants that might be relevant for drinking water production, e.g., pesticides, aromatic sulfonates, pharmaceutical and chelating agents (see separate chapters of this book). Also, upgraded analytical capabilities of specialized laboratories of the waterworks continuously enhance the chance to detect further residues of organic compounds in water. The results of all the above mentioned measurements, together with regularly monitored parameters, give fairly good information on the sources of river pollution.

According to the treatment process of the water used for drinking water production one can distinguish two types of substances:

- Substances that are poorly degradable and therefore will not or not completely be removed during bank filtration and hence might arrive at the wells of the waterworks; such substances are classified as "relevant for the waterworks".
- Substances that are not microbially degradable and not adsorbable on activated carbon will neither be removed during bank filtration nor by activated carbon filters in waterworks; such substances are classified as "relevant for drinking water".

The classification of industrial WW important for waterworks was the result of a research project conducted from 1981 through 1985 [4]. Later research projects proposed similar classifications of the river water and bank-filtered water. In those studies only sum and group parameters were taken into account. Knowing the single compounds of the drinking water, relevant fractions of the sum and group parameters characterizing a wastewater (e.g., UV at 254 nm, total organic carbon, total organic halogens), would enable definition of priorities for further needs in the treatment of wastewater. Thus, discharges into natural water bodies could be protected from substances that otherwise would eventually represent a problem for drinking water supply. Based on the concept of Sontheimer [4], a testing device has been developed using a fixed bed bioreactor (FBBR) (Fig. 1) in which industrial WW effluents or spiked surface water can be treated biologically to the highest possible extent, thus simulating degradation under aerobic conditions [5].

Analysis of the mainly polar substances is quite challenging because they are extremely water soluble and therefore difficult to enrich and detect by common methods such as gas chromatography coupled with mass spectrometry (GC-MS). This problem is enhanced by taking into consideration the generally even more polar metabolites. In this case the analysis and assessment of such compounds is more difficult due to the following facts:

- In general the structures of the byproducts and the metabolites are unknown and no or few reference compounds are available
- No analysis procedures are available for the expected concentrations in the aquatic environment
- The amount emitted into the aqueous environment is not known
- No data exist on the toxicity
- Nothing is known about the relevance of these substances for waterworks

The analysis of such compounds may be performed either:

- Directly with the WW effluent, where the analyst has to deal with higher concentrations, but at the same time with samples that are more contaminated

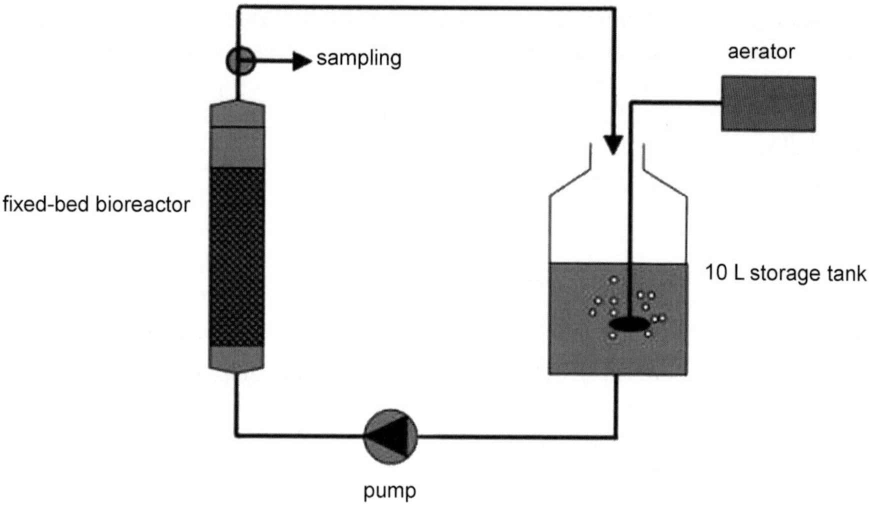

Fig. 1 Fixed bed bioreactor set-up [11]

- With the drinking water, where the matrix problems are not so severe, but due to the lower concentration a more effective enrichment step has to be added

Due to cooperation with the chemical industry it was possible to investigate WW effluents from several major companies directly.

The main compounds identified in the wastewater effluents are:

- Intermediates in chemical synthesis: lactones, TPPO, dikegulac, halogenated benzoic acids, sulfinates
- Metabolites formed during wastewater treatment: e.g., the metabolites of a corrosion inhibitor

2
Formation of Byproducts During Chemical Synthesis

Wastewater treatment plants (WWTP) receiving industrial WW release a complex (and ill-defined) mixture of mainly synthetic chemicals into the aquatic environment, due to their partial or sometimes even complete resistance to biodegradation during the treatment process. Only a minority of the compounds passing through the industrial WWTP, accounting for less than 5% of the total organic carbon (TOC), are characterized.

Industrial organic synthesis has limited selectivity. Besides the target compound, byproducts may be formed. A well-known example is the formation of bis(chloralkyl)ethers during the synthesis of ethylene oxide and propylene oxide via the chlorhydrin process. These haloethers have been found in river water and river sediment [6].

The identification and determination of organic byproducts in waste water and surface water are rather difficult tasks. As these compounds have no use and are of no economic value, little effort has been made to gather information about their physical and chemical properties, their toxicity, and their environmental fate. Their structures are often unknown or only tentatively known and reference compounds are seldom available.

2.1 Bis-ethyl-*iso*-octanol Lactone Isomers (BIOL-isomers)

In the GC-MS chromatograms of an industrial WW effluent, bis-ethyl-*iso*-octanol lactone isomers (BIOL-isomers) were detected regularly as a characteristic peak cluster [7]. The mass spectra of these compounds were quite similar, showing significant fragments at $m/z = 127$ and 170. The mass spectra were unknown and could not be found in the mass spectral databases or in chemical literature. The molecular ion was confirmed to be $m/z = 198$ by applying chemical ionization methods.

According to the manufacturer, these compounds are byproducts from the synthesis of butyraldehyde, formed by the aldol reaction. As shown in Fig. 2, the first step in the reaction is the formation of 2-ethyl-2-hexenal. As described in previous literature [8, 9] the base-catalyzed self-condensation of 2-ethyl-2-hexenal with butyraldehyde leads to the formation of 2,4-diethyl-3-*n*-propyl-δ-valerolactone. As can be seen in Fig. 2, theoretically, there are four possible configuration isomers of 2,4-diethyl-3-*n*-propyl-δ-valerolactone, differing in the axial and equatorial positions of the alkyl-groups.

However, the structural assignment and the stereochemistry of these BIOL-isomers is only tentative. Since there was no reference material available, quantitative determinations were not possible. Using single ion monitoring (SIM) of the characteristic fragments, the BIOL-isomers were also detected in the River Rhine, which is the receiving stream of the industrial WW effluent of the manufacturing company. Due to their presumably poor biodegradability, the BIOL-isomers were also detected in bank-filtered waters [10].

To study the detailed behavior and fate of such compounds and to quantify the BIOL-isomers in a monitoring program, reference compounds are necessary. Since those compounds were not commercially available, they were isolated from industrial WW applying various purification methods with subsequent semipreparative high performance liquid chromatography. Through interpretation of mass spectra after gas chromatographic separation (Fig. 3) the individual BIOL-isomers were identified as stereoisomers of 2,4-diethyl-3-*n*-propyl-δ-valerolactone. Thus, it was possible for the first time to quantify the BIOL-isomers in the Rhine, with a mean sum concentration of 1.6 $\mu g\,L^{-1}$.

Regular analyses performed over a period of almost 2 years always gave the same ratio between the individual isomers [10] .

Drinking water production out of such water was studied, revealing that activated carbon filtration led to a 95% reduction of the BIOL concentration. Additional subsoil passage and a subsequent slow sand filtration led to a total

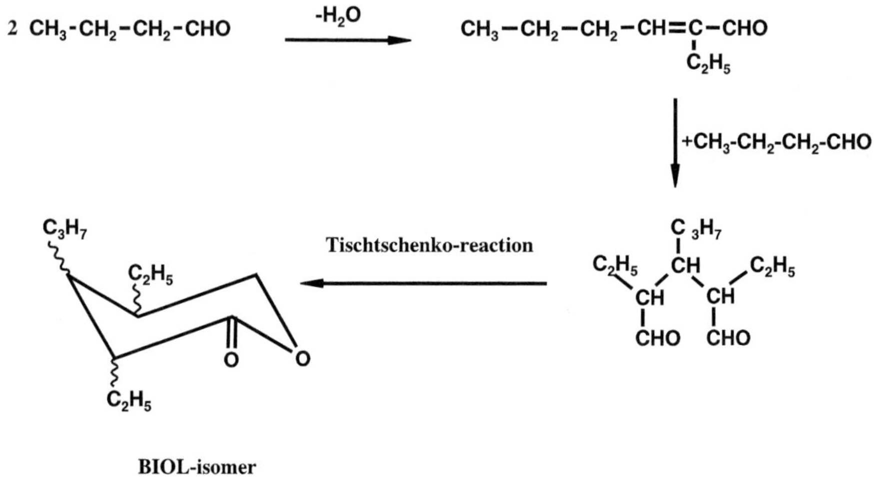

Fig. 2 Potential formation of the stereo isomers of 2,4-diethyl-3-*n*-propyl-δ-valerolactone (BIOL-isomers) via self-condensation of butyraldehyde [10]

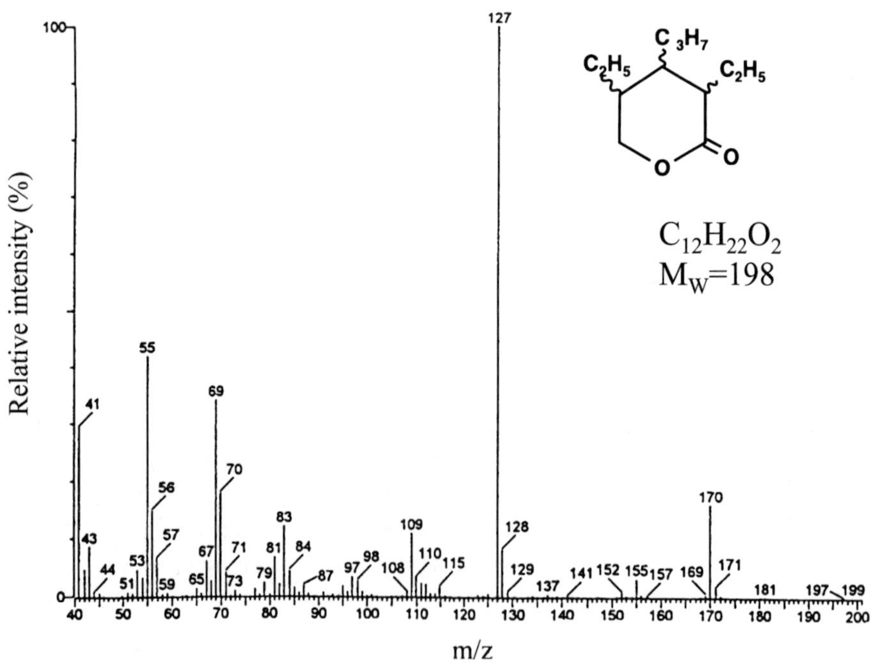

Fig. 3 EI-mass spectrum of the BIOL-isomer A [10]

elimination due to microbial degradation. Even if the BIOL-isomers proved not to be drinking-water relevant, their behavior in the aquatic environment needs to be more thoroughly investigated since these compounds have been discharged for many years in high amounts into the River Rhine.

2.2
Triphenylphosphinoxid

Triphenylphosphinoxid (TPPO), a phosphorous-containing organic compound is a byproduct of the Wittig synthesis, which is used on an industrial scale for the production of vitamin A. Most of the TPPO is extracted from the WW and converted back to triphenylphosphine, the catalyst of the Wittig reaction. Despite this recycling process, TPPO has been found in the Rhine river water for years in concentrations between 0.1 and 1 $\mu g\,L^{-1}$ because of its poor biodegradability [11] (Fig. 4).

In 2004, the mean concentration of TPPO in water samples taken at cologne (Rhine-km 684.5) was 0.11 $\mu g\,L^{-1}$, corresponding to a load of 16 $kg\,day^{-1}$. During the last 10 years, the mean concentration of TPPO dropped from 0,45 $\mu g\,L^{-1}$ in 1994 to 0.11 $\mu g\,L^{-1}$ in 2004. For short periods of time, e.g. 2–3 days, much higher concentrations up to 4 $\mu g\,L^{-1}$ have been observed, which could be traced back to accidental spills in the production plants or temporarily diminished performance of the WWTPs.

TPPO has also been found in concentrations up to 0.1 $\mu g\,L^{-1}$ in the wells of waterworks, which draw bank-filtrated water from the Rhine. However, it

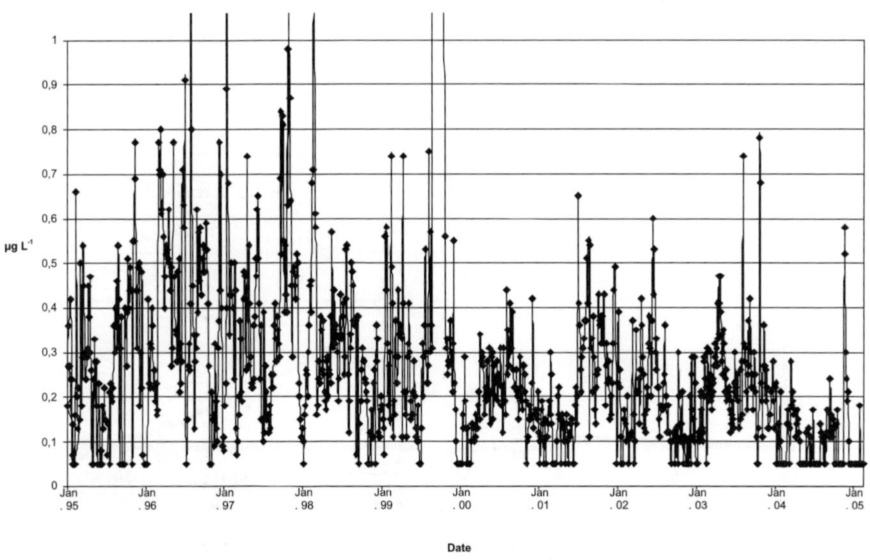

Fig. 4 TPPO in the River Rhine at Cologne, Germany (own data)

can be removed by activated carbon filters during drinking water treatment. As TPPO was supposed to be a substance "relevant to waterworks" further investigations concerning its biodegradability were made.

An FBBR-device according to Fig. 1 was filled with 10 L of surface water from the Rhine and spiked with 10 μg L^{-1} of TPPO [12]. Under aerobic conditions, the concentration of TPPO began to decrease rapidly after 14 days and dropped below the determination limit after 21 days (Fig. 5). At the same time the formation of diphenylphosphinoxid (DPPO) and, to a minor amount, of diphenylphosphinic acid (DPPS) was observed. Both compounds undergo further degradation and were not detectable after 30 days [13].

Whereas DPPO has been already observed, along with TPPO, in treated industrial WW and River Rhine water, the occurrence of DPPS as a degradation product of TPPO was up to now unknown (Fig. 6).

A proven analytical method for the determination of TPPO and DPPO is available, using solid phase extraction (SPE) and HPLC-DAD. However, it was necessary to develop a new method for determination of the more polar and water-soluble DPPS. Solid phase extraction of the acidified water sample with a styren-divinylbenzene-resin, methylation of the eluate with diazomethane, and GC-MS-analysis of the DPPS-methylether served this purpose.

Applying this new method, samples drawn from the Rhine at Worms (Rhine-km 443) and Cologne were analyzed (Table 1). DPPS was found in all

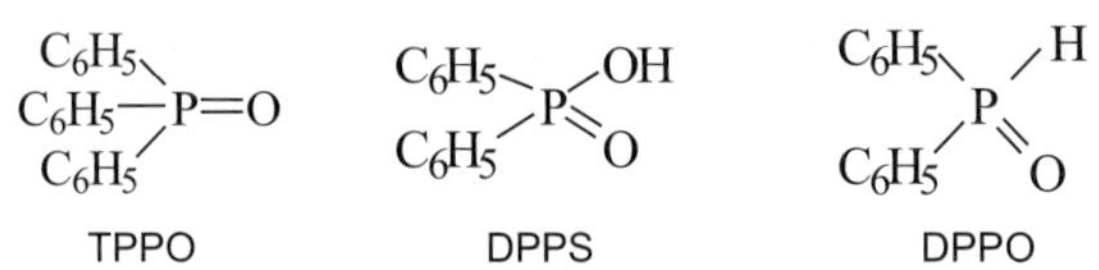

Fig. 5 TPPO and the metabolites identified so far [13]

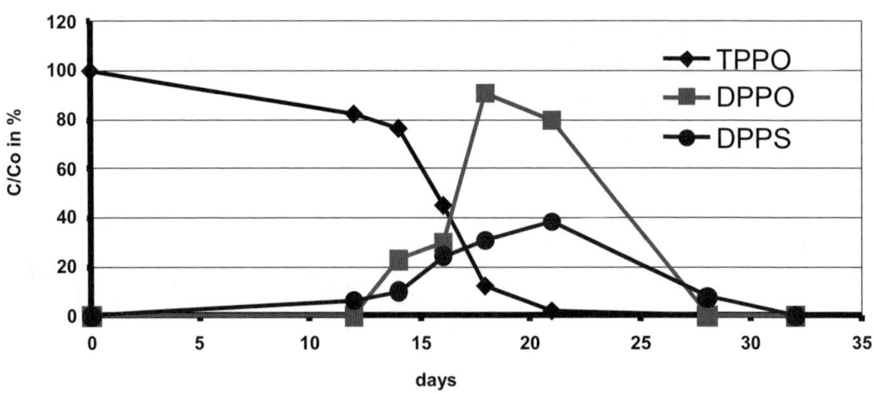

Fig. 6 Degradation test with TPPO in a fixed bed bioreactor filled with Rhine water with concentrations of 10 μg TPPO L^{-1} [13]

Table 1 Contamination of TPPO and DPPS in Rhine water samples

Place	Rhine-km	Date	TPPO [μg L^{-1}]	DPPS [μg L^{-1}]
Cologne	684.5	22.11.2004	< 0.1	0.16
Cologne	684.5	29.11.2004	0.58	0.06
Cologne	684.5	02.12.2004	0.52	0.07
Rhine Water Control Station, left bank	443.1	13.12.2004	0.58	0.17
Rhine Water Control Station, left bank + 100 m	443.1	13.12.2004	0.13	0.06
Rhine Water Control Station, right bank + 100 m	443.1	13.12.2004	< 0.1	0.06

water samples, however, in lower concentrations than TPPO. The highest concentrations of DPPS and TPPO were observed in Worms, Germany at the left river bank. This was due to a major industrial waste water discharge situated at the left bank about 30 km upstream from the sampling point.

2.3 Di-isopropylidene-sorbofuranose and Di-isopropylidene-α-L-xylo-2-hexulofuranosonic acid (DAS/DAG)

In 1993 Haberer and Knepper [14] investigated the behavior of the polar organic drinking water relevant compounds di-isopropylidene-sorbofuranose (DAS) and di-isopropylidene-α-L-xylo-2-hexulofuranosonic acid (DAG) (Fig. 7) during water retrieval in a waterworks area. Both substances are extremely water-soluble and of poor biodegradability and result from the synthesis of vitamin C, which was produced and transported into the Rhine at Grenzach (km 160) tributary of the River Rhine at km 475. Concentrations of DAS up to 2 μg L^{-1} and 0.8 μg L^{-1} of DAG could be measured in the Rhine.

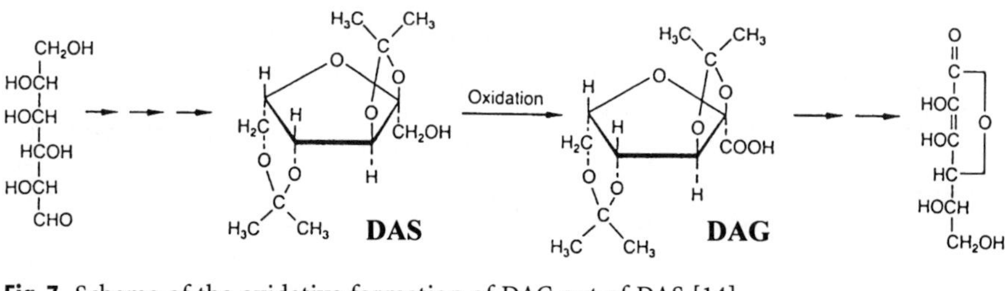

Fig. 7 Scheme of the oxidative formation of DAG out of DAS [14]

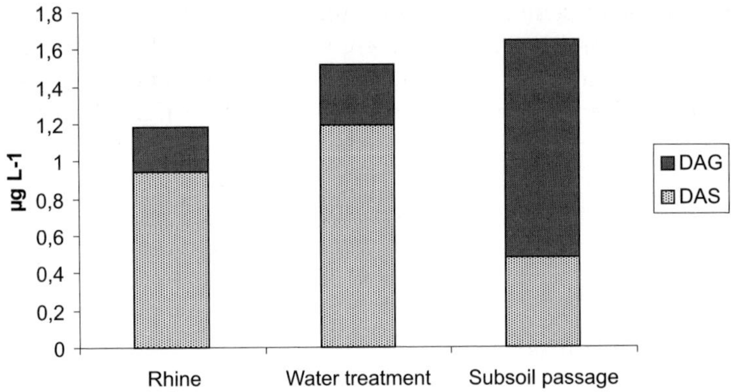

Fig. 8 Change of the ratio of DAG to DAS during physico-chemical drinking water treatment and the subsequent aerobic soil passage

DAS and DAG are not removable during water treatment steps such as coagulation, sand filtration, or activated carbon filtration. Only subsoil passage changed the ratio DAS/DAG through a decrease of the DAS concentration, which was explained by oxidation of DAS to DAG (Fig. 8). The non-toxic DAS and DAG can exemplify the behavior of other polar compounds, which led to the attempt to avoid high loads of WWs containing such poorly degradable organic pollutants from the responsible industrial companies. Since DAG, also known as dikegulac, is also registered as a pesticide, its oxidative formation during the drinking water treatment process was more than unwanted. Making this problem aware to the producing chemical industries led to the implementation of an additional oxidative WW treatment step in the plants and thus to a drastic reduction of emissions into the Rhine.

2.4 Halogenated Aromatic Acids

There are a variety of various halogenated organic compounds produced every year. Some of them are regulated and monitored in the WW effluent, whereas others are only detected as part of the sum parameter "absorbable organic halogens" (AOX). Among the halogenated organic byproducts that are formed during various syntheses, most of them are still unknowns, since specific analytical procedures are missing. For the analysis of halogenated aromatic acids, one can use a UV-detector or a mass spectrometric detector. But only the latter gives enough information to come up with structure elucidation.

The analysis of such complex mixtures as occur in the WW of chemical industries requires a systematic strategy in order to get as much information as possible on the organic compounds present. One strategy is to apply different mass spectrometric detection methods following sequential liquid–liquid extraction procedures at different pH values [15, 16]. For example, the samples enriched at pH 7 can be analyzed by GC-MS and those obtained at pH 0.5 can be either methylated and investigated by GC-MS or introduced without further derivatization into the LC-ESI-MS, with analysis in the negative ionization mode. Additionally, samples extracted at pH 10 can be either derivatized with an excess of trifluoro acetanhydride and analyzed via GC-MS, or analyzed in LC-ESI-MS in the positive ionization mode.

2.5 Halogenated Benzoic Acids

In Table 2 theoretically possible chlorinated and/or brominated benzoic acids are listed and the positive mark (+) indicates that these compounds were detected in industrial WW samples [7]. Since there are almost no reference substances available, identification was mainly done by interpretation of the mass spectra.

As shown in Table 2 halogenated benzoic acids, mainly di- and trichloro benzoic acid, were detected. They appeared not to be further degraded. Other halogenated benzoic acids, mainly tetrabromo-, tribromochloro-, and tetrabromochlorobenzoic acid, were present in the WW effluent of one company.

The evaluation of the chemical structures of the substituted halogenated benzoic acids was obtained by different methods. For example, the methyl ester of 2,3,5,6-tetrachloro-4-trifluoromethylbenzoic acid could only be identified via a collaboration with the industry and thus, even the origin of this compound was determined. It resulted from the hydrolysis of the corresponding acid chloride in the sewage plant and was derivatized to the methyl ester during the sample preparation. The compound was only detected in some of the investigated samples, which resulted from a temporary production of this compound in a pilot plant. This demonstrated the usefulness of such a screening method, allowing detection of the unknown presence of organic acids at low concentrations.

The identity of another halogenated acid could only be proven by the combination of GC-MS analysis after derivatization to the methyl ester and LC-ESI-MS-MS analysis. GC-MS-analysis led to some irritation, because there was an overlap of two halogenated acids, which could not be separated on the GC column and the resulting EI mass spectrum could not be interpreted. On the other hand, the acquisition of a LC-ESI-MS-MS spectrum gave only one signal with an $[M\text{-}H]^-$ ion at $m/z = 268/270$. This polar compound could be identified as 2,4-dichloro-5-sulfamoylbenzoic acid through an MS-MS experiment and the acquisition of a daughter ion mass spectrum. These results were confirmed through the use of a reference compound. 2,4-Dichloro-5-sulfamoylbenzoic acid is a technically important intermediate in the synthesis of different pharmaceuticals. Because of its relevance for waterworks the pro-

Table 2 Occurrence of halogenated benzoic carboxylic acids in industrial WW effluents. The characteristic isotopic patterns of the molecular ions were used for identification [15]

–Benzoic acid	Molecular ions [a]	WW effluent
Chloro-	**170**, 172	(+)
Dibromo-	292, **294**, 296	+
Bromchloro-	248, **250**, 252	n.d.
Dichloro- (2,4- and 2,5-)	204, 206, 208	++
Tribromo-	370, **372, 374**, 376	n.d.
Dibromchloro-	326, **328**, 330, 332	+
Bromdichloro-	282, **284**, 286, 288	n.d.
Trichloro-	**238, 240**, 242, 244	++
Tetrabromo-	448, 450, **452**, 454, 456	++
Tribromchloro-	404, 406, **408**, 410, 412	++
Dibromdichloro-	360, **362**, 364, 366, 368	n.d.
Bromtrichloro-	316, **318**, 320, 322, 324	n.d.
Tetrachloro-	272, **274**, 276, 278, 280	n.d.
Pentabromo-	526, 528, **530, 532**, 534, 536	n.d.
Tetrabromchloro-	482, 484, **486**, 488, 490, 492	+
Tribromdichloro-	438, 440, **442**, 444, 446, 448	n.d.
Dibromtrichloro-	394, 396, **398**, 400, 402, 404	n.d.
Bromtetrachloro-	350, **352**, 354, 356, 358, 360	n.d.
Pentachloro-	306, **308**, 310, 312, 314, 316	n.d.
With additional substituents:		
Aminotrichloro-	253, **255**, 257	++
Dichloromethyl-	**218**, 220, 222	+
Tetrachloro-4-trifluormethyl-	340, **42**, 344, 346	++
Chloromethoxy-	**186**, 188	n.d.
Aminochloro-	**185**, 187	+
2,4-Dichloro-5-sulfamoyl-	281, **283**, 285	+++

[a] Derivatized as methylester; the ions with the highest relative intensity are **bold**. The number of "+"-marks gives the relative signal intensity and/or abundance of the individual halogenated acids; those shown in parenthesis did not occur more than twice
n.d. not detected.

duction of this compound was transferred to another plant utilizing granular activated carbon for sewage treatment and this resulted in a stop of the emission of this compound into the surface water.

2.6 Halogenated Phthalic Acids

From the manifold of the theoretically possible chlorinated and brominated phthalic acids, the tetrasubstituted phthalic acids were occurring regularly and even quite intensely in the WW effluent of one company (Table 3). Tetrachlorophthalic acid (TCPA) always turned out to be the dominating compound of the tetrasubstituted phthalic acids. TCPA was also quantified by using the standard addition procedure. The concentrations detected in the test-filter outlets were almost the same as those measured in the corresponding inlets, adding TCPA to the group of waterworks-relevant compounds. Therefore, TCPA can be routinely detected in all investigated samples of the Rhine. The concentrations measured were within the limit of quantitation ($0.03\ \mu g\ L^{-1}$) and in single samples went up to $0.9\ \mu g\ L^{-1}$ [7].

Table 3 List of the theoretically possible and the detected chlorinated and brominated phthalic acids in industrial WW effluents. The characteristic isotopic patterns of the molecular ions were used for identification [15]

–Phthalic acid	Molecular ions [a]	WW effluent
Bromo-	**272, 274**	n.d.
Chloro-	**228**, 230	n.d.
Dibromo-	350, **352**, 354	n.d.
Bromochloro-	306, **308**, 310	n.d.
Dichloro-	**262**, 264, 266	++
Tribromo-	428, **430, 432**, 434	n.d.
Dibromochloro-	384, **386**, 388, 390	n.d.
Bromodichloro-	340, **342**, 344, 346	n.d.
Trichloro-	**296, 298**, 300, 302	++
Tetrabromo-	506, 508, **510**, 512, 514	++
Tribromochloro-	462, 464, **466**, 468, 470	++
Dibromodichloro-	438, **440**, 442, 444, 446	++
Bromotrichloro-	374, **376**, 378, 380, 382	(+)
Tetrachloro-	330, **332**, 334, 336, 338	+++

[a] Derivatized as methylester; the ions with the highest relative intensity are **bold**. The number of "+"-marks gives the relative signal intensity and the abundance of the individual halogenated acids; those shown in parenthesis did not occur more than twice
n.d. not detected

2.7 Sulfinates

Applying the above developed method for the analysis of unknowns in an industrial chemical plant, various GC-MS-chromatograms were obtained for the neutral and acidic fractions. Thus a series of halogenated intermediates used in various syntheses could be detected. Consequently, it became clear that the information obtained by dual application of GC-MS and LC-ESI-MS was insufficient for the unequivocal assignment of these contaminants. In order to obtain more diagnostic information an LC-ESI-TOF analysis was conducted [17], leading to additional information regarding the elemental composition, which can be calculated based on the determined accurate masses [18]. The assignment of the fragments of all postulated compounds is given in Table 4.

Quite interesting for many of the detected compounds is the stability of the aromatic sulfinate group, which is normally known from literature to be quite instable. A request to a 2-chloro toluene methylsulfon (CMST) producing company did not rule out the assignment of these compounds produced during CMST synthesis as byproducts. But in order to be totally sure of the structural composition and also for quantitation of these compounds, reference substance must be either synthesized or extracted out of waste water. Since this is quite a demanding and time-consuming issue, as for example shown for the identification and quantification of bis-ethyl-iso-octanol lactone isomers [10], this will be a task of its own. Anyhow, since the halogenated toluenesulfinates as well as the corresponding benzoic and sulfonic acids were never detected in river water, this might not be of high priority. Our purpose was to demonstrate the usefulness of such a battery of screening methods, which also allow detection of the yet-unknown presence of organic polar compounds at low concentrations.

3 Metabolites Formed During Wastewater Treatment: Phenylsulfonamides

A further example of a strategy for the detection of a unknown substance was the identification of a compound that appeared regularly in the GC-MS-chromatograms as an intense peak during screening of the River Main in Germany . The substance was identified as Sarkosin-*N*-(phenylsulfonyl) (SPS), an *N*-methylated sulfonamide (Fig. 9) [16]. Further investigations were made, and SPS was also detected in the Rhine River, constantly in concentrations between 0.5 and 1 $\mu g\,L^{-1}$ (Table 5).

In Germany there is no application of SPS known, which could explain the ubiquitous presence in the surface water. Metabolism studies of Knepper et al. [16] showed that the presence of SPS was initially hypothesized to be due

Table 4 Results obtained after comparison of GC-MS after derivatization with diazomethane, LC-ESI-MS, and LC-ESI-TOF-MS of CMST and identified byproducts [17]

Compound [e] Name (No)	Structure	MW (mono-iso-topic)	Rt [min]	m/z (Assignment) GC-MS	LC-ESI-MS LC-ESI-TOF
2-Chloro toluene methylsulfon	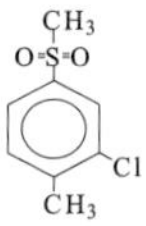	204	23.37 [a]	204/206 ($M^{+\cdot}$) [a] 189/191 ($M-CH_3^{+\cdot}$) 141/143 ($M-SOCH_3^{+\cdot}$) 125/127 ($M-SO_2CH_3^{+\cdot}$) 89 ($M-SO_2CH_3^{+\cdot}$); $-HCl$	Not detected
Di-chloro toluene methylsulfon	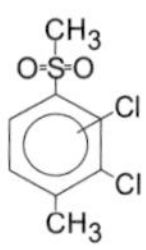	238	27.2 [a]	238/240/242 ($M^{+\cdot}$) [a] 223/225/227 ($M-CH_3^{+\cdot}$) 175/177/179 ($M-SOCH_3^{+\cdot}$) 159/161/163 ($M-SO_2CH_3^{+\cdot}$) 123/125 ($M-SO_2CH_3^{+\cdot}$); $-HCl$	Not detected
2-Chloro methylsulfon benzoate	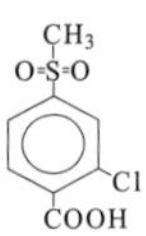	234	30.0 [b] 13.0 [c]	248/250 ($M^{+\cdot}$) [b] 217/219 ($M-OCH_3^{+\cdot}$) 169/171 ($M-SO_2CH_3^{+\cdot}$)	233/235 ($[M-H]^-$) [c] 189/191 ($[M-CO_2H]^-$)
Di-chloro methylsulfon benzoate	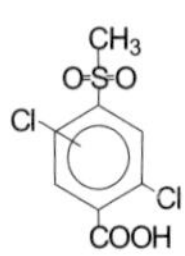	268		Not detected	Not detected
2-Chloro toluene sulfinate	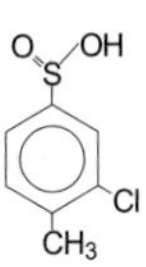	190	19.1 [b] 5.5 [c]	204/206 ($M^{+\cdot}$) [b] 173/175 ($M-OCH_3^{+\cdot}$) 125/127 ($M-SO_2CH_3^{+\cdot}$)	189/191 ($[M-H]^-$) [c] 188.9510/190.9487 ($[M-H]^-$) [d]

[a] GC-MS without derivatization
[b] GC-MS after derivatization with diazomethane; mass assignment of the methylester
[c] LC-ESI-MS
[d] ESI-TOF-MS
[e] position of halogen atoms only tentative

Table 4 Continued.

Compound [e]			Rt [min]	m/z (Assignment)	
Name (No)	Structure	MW (mono-iso-topic)		GC-MS	LC-ESI-MS LC-ESI-TOF
Di-chloro toluene sulfinate	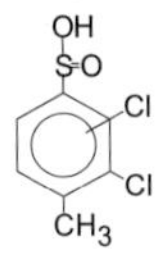	224	23.8 [b] 13.5 [c]	238/240/242 ($M^{+\cdot}$) [b] 207/209/211 ($M - OCH_3^{+\cdot}$) 175/177/179 ($M - SOCH_3^{+\cdot}$) 159/161/163 ($M - SO_2CH_3^{+\cdot}$) 123/125 ($M - SO_2CH_3^{+\cdot}$); – Cl	223/225/227 ($[M-H]^-$) [c] 159/161/163 ($[M - SO_2H]^-$) 222.9470/224.9448/226.9402 ($[M-H]^-$) [d]
2-Chloro toluene sulfonate	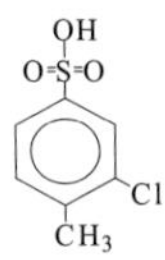	206	5.5 [c]	Not detected	205/207 ($[M-H]^-$) [c] 189/191 ($[M-OH]^-$) 205.946/207.943 ($[M-H]^-$) [d]
Di-chloro toluene sulfonate	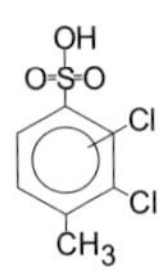	240	7.7 [c]	Not detected	239/241/243 ($[M-H]^-$) [c] 238.9321/240.9291/242.9234 ($[M-H]^-$) [d]

[a] GC-MS without derivatization
[b] GC-MS after derivatization with diazomethane; mass assignment of the methylester
[c] LC-ESI-MS
[d] ESI-TOF-MS
[e] position of halogen atoms only tentative

to the microbial degradation in sewage plants of 6-[methyl(phenylsulfonyl)-amino]hexanoic acid (HPS) (Fig. 9), which is extensively used in Germany as a corrosion inhibiting agent.

Degradation studies performed by spiking the FBBR run with River Rhine water with HPS gave clear evidence. HPS was degraded via several intermediate metabolites to SPS (Fig. 10). The comprehensive degradation assays performed in combination with the results obtained from the measurements in sewage, surface, and drinking waters, showed that SPS did not come from usage of the compound per se. This verified the assumption that it was a metabolite arising from the degradation of HPS. After a more thorough analysis, it was shown that

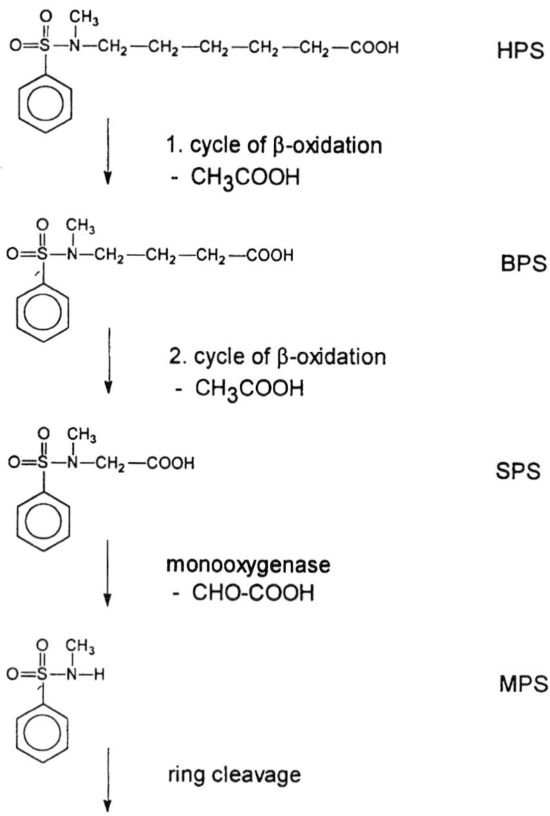

Fig. 9 Microbial degradation of 6-[methyl(phenylsulfonyl)amino]hexanoic (HPS) to sarkosind-*N*-(phenylsulfonyl)(*n*-methyl-*N*-phenylsulfonyl glycine) (SPS) via β-oxidation and further metabolism to *N*-methyl-*N*-phenysulfonyl (MPS) [16]

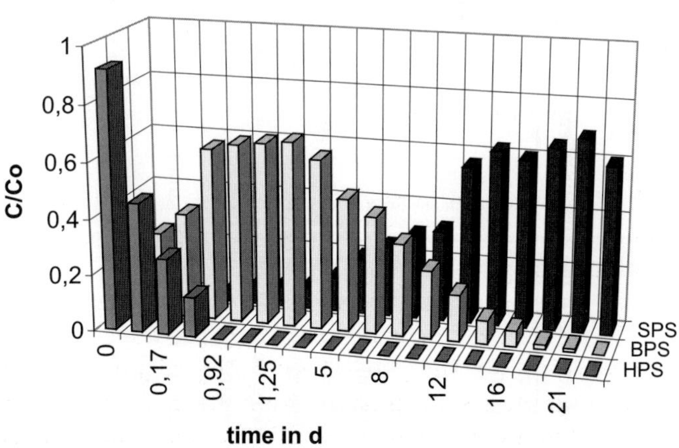

Fig. 10 Degradation studies performed by spiking the fixed bed bioreactor run with Rhine water with HPS

Table 5 Monitoring of the various phenylsulfonamides during the different steps of drinking water production out of surface water [16]

Sampling point	HPS		BPS		SPS		MPS
	Min–max	Mean	Min–max	Mean	Min–max	Mean	Mean
River Rhine	n.d.	n.d.	0.04–0.33	0.12	0.18–0.87	0.47	0.06
After sedimentation/aeration	n.d.	n.d.	n.d.–0.26	0.10	0.23–0.92	0.55	n.m.
After flocculation/rapid sand filtration	n.d.	n.d.	n.d.–0.19	< 0.03	0.04–0.99	0.43	n.m.
After activated carbon filtration	n.d.	n.d.	n.d.	n.d.	0.03–0.65	0.20	0.03
After underground passage	n.d.	n.d.	n.d.	n.d.	0.08–0.19	0.13	0.05
After slow sand filtration/disinfection	n.d.	n.d.	n.d.	n.d.	n.d.–0.07	< 0.03	n.d.

Concentrations in $\mu g\,L^{-1}$. All values except MPS are based on 24 samples taken randomly every 14 days in 1997, MPS is the mean value of $n = 3$, LOD = 0.03 $\mu g\,L^{-1}$,
n.d. not detected, *n.m.* not measured in this period

possible mother compounds for SPS other than HPS proved to be negative due to a different metabolism pattern and/or to a lesser production rate. HPS as source for SPS was confirmed through the detection of the other theoretically postulated intermediates, which were detected in almost all of the investigated sewage plant effluents and surface waters. HPS as the only source of SPS was furthermore confirmed by new measurements of SPS in the River Main in Germany. Accordingly, a significant reduction in SPS levels was monitored after closure of a HPS-producing factory that was discharging its sewage (after treatment) into the River Main [19]. SPS has been found to be poorly biodegradable and drinking water relevant and thus is recommended to be included in the group of environmentally relevant compounds.

4 Conclusion

Even in the industrial wastewater of chemical industries with well-documented production lines, less than 10% of the emitted DOC is normally known, and as such there is still a need for analytical methodology that permits identification of new metabolites and byproducts of chemical synthesis. In the case of the evaluation of complex structures of xenobiotics in difficult matrices, parallel application of a range of mass spectrometric methods are required in order to achieve complete structural characterization. With the combined structural information gained from GC-MS-spectra, following derivatization to the methylester, and LC-ESI-MS and ESI-TOF-MS-spectra it was possible to elucidate the structures of several other previously unknown polar pollutants.

The enormous investigative efforts of the isolation and characterization of byproducts of chemical synthesis is more than worthwhile, because only with available reference compounds will it be possible to study their impact on the aquatic environment, as well as their fate. Especially in the River Rhine, where almost all major European chemical industries emit their WW, the impact of these almost unknown chemicals is quite high.

Besides analytical approaches, the studies with the BIOL-isomers show that additional different toxicological tests still have to be performed since almost nothing is known about their risk to the population and environment.

Acknowledgements The experimental help of Ocke Rörden and Jutta Müller as well as the proof reading of Heike Weil is deeply acknowledged. We also thank the ARW and the VCI for financing most of the research.

References

1. Jahresberichte der Samenwerkende Rijn en Maaswaterleidingbedrijven (1995) RIWA, Niederlande, Amsterdam
2. Jahresberichte der Arbeitsgemeinschaft Rhein-Wasserwerke e.V. (1996) ARW, Köln, Germany
3. Jahresberichte der Arbeitsgemeinschaft Wasserwerke Bodensee-Rhein AWBR (1996) Schweiz, St. Gallen
4. Sontheimer H, Volker E (1987) Veröffentlichungen des Bereichs und des Lehrstuhls für Wasserchemie und der DVGW-Forschungsstelle am Engler-Bunte Institut der Universität Karlsruhe, 31, Karlsruhe
5. Linder K, Knepper, TP, Müller J, Karrenbrock F, Rörden O, Brauch HJ, Sacher F (2000) Entwicklung von Verfahren zur Bestimmung und Beurteilung der Trinkwassergängigkeit von organischen Einzelstoffen. Abschlußbericht des ARW/VCI Forschungsvorhabens, IAWR Rheinthemen 3, ISBN 90-70671-26-3
6. Weissermel K, Arpe HJ (1988) Industrielle Organische Chemie. Wiley, Weinheim p 284
7. Linder K, Knepper TP, Karrenbrock F, Rörden O, Brauch HJ, Lange FT, Sacher F (1996) Erfassung und Identifizierung von trinkwassergängigen Einzelsubstanzen in Abwässern und im Rhein. Abschlußbericht des ARW/VCI Forschungsvorhabens, IAWR Rheinthemen 1, ISBN 90-6683-080-8
8. Nielsen ST (1957) J Chem Soc 79:2518
9. Häusermann M (1951) Helv Chim Acta 34:1482
10. Knepper TP, Müller J, Wulff T, Maes A (2000) J Chromatogr A 889:245
11. Linder K, Knepper, TP, Müller J, Karrenbrock F, Rörden O, Juchem H, Brauch HJ, Sacher F (2003) Bestimmung und Beurteilung der mikrobiellen Abbaubarkeit von organischen Einzelstoffen bei umweltrelevanten konzentrationen in Gewässern. Abschlußbericht des ARW/VCI Forschungsvorhabens, IAWR Rheinthemen 5, ISBN 90-70671-32-8
12. Karrenbrock F, Knepper T, Sacher F, Lindner K (1999) Vom Wasser 92:361
13. Bücker T (2005) Untersuchungen zum mikrobiellen Abbau von Triphenylphosphinoxid und seinen Metaboliten. Diplomarbeit FH Aachen
14. Haberer K, Knepper TP (1993) GWF Wasser Abwasser 134:526
15. Knepper TP, Sacher F, Lange FT, Brauch H-J, Karrenbrock F, Roerden O, Lindner K (1999) Waste Management 19:77
16. Knepper TP, Kirschhoefer F, Lichter I, Maes A, Wilken R-D (1999) Environ Sci Technol 33:945
17. Knepper TP (2002) J Chromatogr A 974:111
18. Hogenboom AC, Niessen UA, Brinkmann T (2000) Mass Spectrom 14:1914
19. Knepper TP (2000) Environ Sci Technol 34:919

Hdb Env Chem Vol. 5, Part L (2006): 255–306
DOI 10.1007/698_5_020

Published online: 18 November 2005

Natural and Artificial Radioactivity in the Rhine and Its Tributaries

Sascha Wisser[1] (✉) · Jens Hartkopf[2]

[1]Rhine-Main Water Research Ltd., Kurfürstenstr. 6, 65203 Wiesbaden, Germany
sascha.wisser@hessenwasser.de

[2]Landesamt für Umwelt, Wasserwirtschaft & Gewerbeaufsicht, Amtsgerichtsplatz 1, 55276 Oppenheim, Germany
jens.hartkopf@luwg.rlp.de

Abstract Presently, natural and artificial radioactivity in the Rhine River does not present a radiological hazard to the population. On the contrary, the levels of radionuclides in water and sediments from the Rhine River are comparably low and can only be detected with sophisticated analytical procedures. The main reason for these comparably low levels of natural radioactivity is the dilution of concentration by the river water. Furthermore, the discharge from nuclear facilities and nuclear medicine does not lead to significant levels of artificial radioactivity in the river water. However, nuclear weapons tests and the accident at the Chernobyl nuclear power plant in 1986 lead to a temporary increase of the radioactivity levels present in the Rhine River. The key isotope ^{137}Cs from the Chernobyl event is still detectable in river sediments. This applies as well for ^{90}Sr, which mainly stems from nuclear weapons tests and to a smaller extent from the Chernobyl accident.

This section deals with the main properties of the most important natural and artificial radioisotopes and provides information about their occurrence in the Rhine River. Since natural radioisotopes are not regularly monitored, the emphasis has been set on artificial radionuclides occurring in river water and sediments.

Keywords Chernobyl · Monitoring · Radioactivity · River Rhine · Water

Abbreviations

A	Activity
Bq	Becquerel
Gy	Gray
λ	Decay constant
N	Number of atoms present
Sv	Sievert
Q	Quality factor
BfG	The Federal Institute of Hydrology, Germany
DGSNR	Directorate General for Nuclear Safety and Radiation Protection, France
EAWAG	Swiss Federal Institute for Environmental Science and Technology
EC	European Commission
ECURIE	European Community Urgent Radiological Information Exchange
EU	European Union
Euratom	European Atomic Energy Community
EURDEP	European Radiological Data Exchange Platform
HPGe	High-purity germanium
IAEA	International Atomic Energy Agency
ICP-MS	Inductively coupled plasma mass spectrometry
ICRP	International Commission on Radiological Protection
IMIS	Integrated Measuring and Information System, Germany
INES	International Nuclear Event Scale
IRSN	Institute for Radioprotection and Nuclear Safety, France
JRC/IES	Joint Research Center, Environmental Institute in Ispra, Italy
KFÜ	Nuclear Reactor Remote Monitoring System, Germany
KTA	Safety Standards of the Nuclear Standards Committee, Germany
LLD	Lowest limit of detection
MCA	Multi-channel analyzers
NAZ	Swiss National Emergency Operations Center
NMR	National Radioactivity Monitoring Network, Netherlands
PMT	Photo multiplier tube
REI	Guidelines of Emissions and Immissions Monitoring for Nuclear Facilities, Germany
REM	Radioactivity Environmental Monitoring, European Commission
ROI	Region of interest
SSK	German Radiation Protection Commission
UNSCEAR	United Nations Scientific Committee on the Effects of Atomic Radiation
WHO	World Health Organisation
Federal Law Gazette	BGBl, Bundesgesetzblatt, Germany
Federal Bulletin	Bundesanzeiger, Germany
Joint Ministerial Bulletin	Gemeinsames Ministerialblatt, Germany

1 Introduction

Natural radioactivity has existed for billions of years and has probably been a major contributor to the development of complex organic life on planet Earth. Radioactivity occurs in all kinds of natural waters, such as ground-, mineral- and surface waters. Since surface waters are an important source for human water supply and irrigation, the knowledge of radioactivity in those waters is very important. Today, the levels of natural radioactivity in surface waters are comparably low and the input of artificial radionuclides is mostly dependent on the distribution of nuclear facilities. However, the consequences of nuclear accidents, such as Chernobyl in 1986, and nuclear weapons tests are still detectable in most European rivers. Only a few investigations of the levels of natural radioactivity in surface waters have been performed, whereas a multitude of determinations of artificial radionuclides have been carried out.

The main topic to be introduced deals with the main properties of the most important natural and artificial radioisotopes and provides information about their occurrence in the Rhine River. In addition, the monitoring of radioactivity in the Rhine is introduced and the health effects of radionuclides present in the river will be discussed. Furthermore, the detection of radioisotopes by different means will also be described. Finally, an overview of the activity concentrations of the most important artificial radionuclides present in the Rhine will be discussed with respect to their contribution to the radiation exposure of humans.

2 General Aspects of Radioactivity

Radioactivity is a natural process, defined as the spontaneous transformation of an atomic nucleus into a different nuclide by the emission of particles or photons. The decay of a radioactive nuclide is a random process and is subject to the laws of statistics. This process is referred to as a transformation, a decay or a disintegrations of an atom. A common feature in all modes of decay is that ionizing radiation is released, i.e., the energy of the released particles or photons is sufficiently high enough to remove electrons from atoms and molecules, respectively.

Several modes of decay have been investigated since the discovery of radioactivity. The following section deals exclusively with the types of radioactive transitions, which occur within the three natural decay chains. Other natural modes of decay, such as spontaneous fission, have negligible importance with respect to radioactivity present in river waters and are not of concern here.

There exist three main types of ionizing radiation, defined after the first three letters of the Greek alphabet, as introduced in the following section.

2.1 Alpha Radiation

Alpha decay is usually restricted to elements with mass numbers higher than 170 and atomic numbers greater than 70. These heavy isotopes have excess protons and decay by the emission of α-radiation [1]. The α-particle is identical to a ^{4}He nucleus consisting of two neutrons and two protons. Alpha decay reduces the number of nucleons by four mass units and decreases the atomic number by two units. For example, alpha decay of ^{238}U produces ^{234}Th as the daughter isotope:

$$^{238}_{92}\mathrm{U} \rightarrow {}^{234}_{90}\mathrm{Th} + {}^{4}_{2}\mathrm{He} + 4.198\ \mathrm{MeV} \quad (1)$$

2.2 Beta Radiation

Electron emission (β^-) occurs when the ratio of neutrons to protons in the nucleus is too high. This process decreases the number of neutrons by one and increases the number of protons by one. Since the number of protons in the nucleus determines the element, electron emission leads to the formation of a different element. For instance, ^{40}K decays to ^{40}Ca by the emission of an electron:

$$^{40}_{19}\mathrm{K} \rightarrow {}^{40}_{20}\mathrm{Ca} + {}^{0}_{-1}\mathrm{e} \quad (2)$$

Another mode of beta decay is electron capture. The nucleus captures an electron from an inner orbital of the electron cloud and increases the ratio of neutrons to protons by this process. ^{40}K is not only a characteristic electron-emitting radionuclide; it may also decay to ^{40}Ar by electron capture under the release of a neutrino (ν) and the characteristic gamma-radiation of the daughter nuclide (h):

$$^{40}_{19}\mathrm{K} + {}^{0}_{-1}\mathrm{e} \rightarrow {}^{40}_{18}\mathrm{Ar} + h\nu \quad (3)$$

A third form of beta decay is the so-called positron emission (β^+). This process does not occur in nature and is consequently of no interest for natural waters, such as river waters. However, positron emission might be of importance for surface waters if artificial radioisotopes are released from nuclear power plants or other nuclear facilities.

2.3 Gamma Radiation

The daughter nuclides produced by α-decay and especially by β-decay are often obtained in an excited energy state. The energy associated with this excited state is released when the nucleus emits a photon of high energy.

Most frequently, the γ-radiation is emitted within 10^{-16} to 10^{-13} seconds after a particle emission [1]. In some cases, gamma decay is delayed, and a meta-stable radionuclide is formed. For example, the meta-stable ^{241m}Pa nuclide has a half-life of only 1.17 min. Since electromagnetic radiation does not carry any charge or mass, the product of γ-ray emission by ^{241m}Pa is the isotope ^{241}Pa.

$$^{241m}_{91}\mathrm{Pa} \rightarrow {}^{241}_{91}\mathrm{Pa} + \gamma \tag{4}$$

2.4 Radiation-Related Terms

This section will introduce the main terms and units encountered in the field of radioactivity and radiometry.

2.4.1 Decay Law

The decay rate or activity of a radionuclide source is described by a linear differential equation, the so-called decay law, where the number of decaying atoms (N) per unit time is proportional to the number of atoms that persisted decay at the moment t, assuming λ is the decay constant.

$$-\,\mathrm{d}N/\mathrm{d}t = \lambda N \tag{5}$$

Integration of Eq. 5 provides the following equation for the decay law [1]:

$$N(t) = N_0 \mathrm{e}^{-\lambda t} \tag{6}$$

This exponential expression may be useful for correcting the radioactive decay if short-lived radionuclides are investigated. The decay constant (λ) is dependent on the half-life ($T_{1/2}$) of the respective radionuclide and can be calculated from the following equation:

$$\lambda = \frac{\ln 2}{T_{1/2}} \tag{7}$$

Statistically, λ refers to the probability of the number of nuclei that will undergo radioactive transformation in time t and is characteristic for each radioactive isotope.

2.4.2 Half-Life

The half-life ($T_{1/2}$) is the length of time it takes for exactly one-half of the radioactive nuclei present in a radioactive sample to decay. Consequently, after one half-life 50% of the initially present atoms will remain. Substituting $T_{1/2}$

into the radioactive decay law and setting the result equal to $N_0/2$ will establish the relationship between the half-life and the decay constant (λ):

$$T_{1/2} = ln2/\lambda \qquad (8)$$

The half-lives of natural and artificial radioisotopes vary from some microseconds to billions of years.

2.4.3 Activity

The standard SI-unit of activity is the Becquerel (Bq), defined as one disintegration per second.

$$A = \lambda N \qquad (9)$$

Radioactivity is often expressed in larger units like: thousands (kBq), millions (MBq) or even billions (GBq) of a becquerel.

2.4.4 Absorbed Dose

The Gray (Gy) is a unit used to measure a quantity called absorbed dose. This relates to the amount of energy actually absorbed in some material, and is used for any type of radiation and any material. One gray is equal to one joule of energy deposited in 1 kg of a material. The unit gray can be used for any type of radiation, but it does not describe the biological effects of the different radiations. Absorbed dose is often expressed in terms of hundredths of a gray, or centi-grays.

2.4.5 Equivalent Dose

The Sievert (Sv) is a unit used to derive a quantity called equivalent dose. This relates the absorbed dose in human tissue to the effective biological damage of the adsorbed radiation. Not all radiation has the same biological effect, even for the same amount of absorbed dose. Equivalent dose is often expressed in terms of micro-sievert (μSv) or mili-sievert (mSv). The adsorbed (Gy) dose is multiplied by a quality factor (Q) to determine equivalent dose (Sv), that is dependent on the type of incident radiation.

2.5 Health Effects of Ionizing Radiation

Exposure to ionizing radiation represents a potential hazard to humans and radioactivity is known to trigger deterministic and stochastic effects. Deterministic effects are those of which generally a threshold level of dose exists [2]. Above this threshold, the severity of the effect is dependent on the radiation dose. The level of the threshold dose is characteristic of the particular health effect but also depends, to a limited extent, on the exposed individual. For instance, deterministic effects include erythrodermia and acute radiation syndrome. On the contrary, stochastic effects take the form of radiation-induced health effects, the probability of occurrence is greater for a higher radiation dose and the severity is independent of dose. Stochastic effects may be somatic effects or genetic effects, and generally occur without a threshold level of dose. For example, cancer and hereditary effects belong to the stochastic effects of ionizing radiation [2].

3 Natural Radionuclides in the Environment

Radionuclides occur naturally in our environment and have existed since 4.5 billion years ago when Planet Earth was formed. In nature, over 80 radionuclides can be found in detectable concentrations [3] and they can generally be grouped into three categories based on how they were formed:

Primordial – creation during nucleosynthesis, before the Earth was formed
Cosmogenic – formed by cosmic ray interactions in the upper atmosphere
Anthropogenic – artificial radionuclides produced through human processes

The primordial radionuclides which exist today are the oldest with half-lives that are at least comparable to the age of the Earth. Secondary radionuclides are derived from radioactive decay of the primordial isotopes. While cosmogenic radioisotopes are continuously produced in the Earth's atmosphere due to interactions of stable isotopes and cosmic rays.

Presently there exist three naturally occurring decay series beginning with the long-lived ^{238}U, ^{235}U, and ^{232}Th which were already present when the Earth was formed (Table 1). The most significant decay series with regard to human radiation exposure is the ^{238}U series, which contains 8 alpha and 6 beta decays to form stable ^{206}Pb.

The natural decay series of ^{232}Th, ^{238}U, and ^{235}U consists of 45 radioisotopes. The occurrence of natural radioisotopes in nature depends on several environmental processes. In waters, the existence of radioisotopes depends on the occurrence of rocks from which they stem and the processes, which concentrate the radionuclides. Hence, it is essential to have sufficient knowledge on the occurrence of source-rock materials containing elevated levels of radionuclides to understand the physical and geochemical processes that concentrate radionuclides. The following chapters are presented for a better understanding of the geochemical properties of the most important natural radionuclides.

Table 1 Properties of the three natural-decay series of the actinides

Name of series	Initial radionuclide	Half-life (years)	Stable end-member
Uranium	^{238}U	4.470×10^{9}	^{206}Pb
Actinium	^{235}U	7.038×10^{8}	^{207}Pb
Thorium	^{232}Th	1.405×10^{10}	^{208}Pb

3.1 Geochemistry of Uranium and Thorium

Uranium and thorium belong to actinides (Z 89–102). The actinides are the heaviest naturally existing elements in the cosmosphere. All actinides occur in association with varying amounts of their radioactive decay products that are isotopes of radium, radon, polonium and lead. The differences in the chemical and physical properties of an actinide and its daughters may lead to fractionation of the mentioned radionuclides. As a consequence, disequilibria between these radioisotopes are often observed in the natural environment. Uranium-series disequilibria are frequently used in earth sciences for different applications, for example in geochronology, hydrology and planetology.

The actinides and radium are electropositive elements and have a tendency to form strong ion bonds. On the contrary, radon is a noble gas and is inert under environmental conditions. Polonium, bismuth and lead are amphoteric and may form ionic bonds with some covalent character in the environment. The actinides and radium are lithophile elements and stable isotopes of bismuth and lead have chalcophile properties [4].

Uranium ($Z = 92$) is the heaviest naturally occurring element and exists as a silvery-white, ductile and slightly paramagnetic metal, which is pyrophoric when finely divided. It is slightly softer than steel and easily oxidizes and becomes coated with layer of oxide [5]. The Earth's continental crust contains uranium in an average concentration of about 2.7 ppm [6]. Uranium occurs naturally in the +2, +3, +4, +5, or +6 valence state, but most commonly found in the hexavalent form. In nature, hexavalent uranium is dominantly associated with oxygen as the uranyl ion, ${UO_2}^{2+}$. Naturally occurring uranium (U_{nat}) is a mixture of three isotopes as provided in Table 2.

There are six naturally occurring Thorium ($Z = 90$) isotopes: ^{232}Th, the parent isotope of a natural decay chain, and the intermediates ^{234}Th, ^{231}Th, ^{230}Th, ^{228}Th, and ^{227}Th. When in pure form, thorium is a silvery-white metal that is air-stable and retains its luster for several months. When contaminated with oxide, thorium slowly tarnishes in air, becoming gray and finally black [8]. Thorium oxide has a melting point of 3300 °C, the highest of all known oxides [9]. The average thorium content of the Earth's continental crust is about 8.5 ppm [6].

Table 2 Natural uranium isotopes and their properties [7]

	Abundance (%)	$T_{1/2}$	α-decay energy, MeV (probability)	Specific activity (Bq/g)
^{238}U	99.2745	4.470×10^9 a	4.198 (79.0%) 4.151 (20.9%)	12 455
^{235}U	0.7200	7.038×10^8 a	4.3987 (55%) 4.3661 (17%) 4.2147 (5.7%) 4.5964 (5.0%)	80 011
^{234}U	0.0055	2.455×10^5 a	4.7746 (71.38%) 4.7224 (28.42%)	2.31×10^8

3.2
Relevant Natural Isotopes in Surface Waters

Under consideration of the main properties of the most important radionuclides from the natural decay series, it becomes obvious that only a few radioisotopes are of any relevance in surface waters. These radionuclides have to possess a sufficient half-life and they must be transportable within surface waters. Since the members of the ^{238}U series are quite common in natural waters, radioisotopes of this decay chain are of major concern in water supply if they become enriched during water abstraction. In addition to that, a few radionuclides from the other natural decay series are of importance as well (refer to Fig. 1).

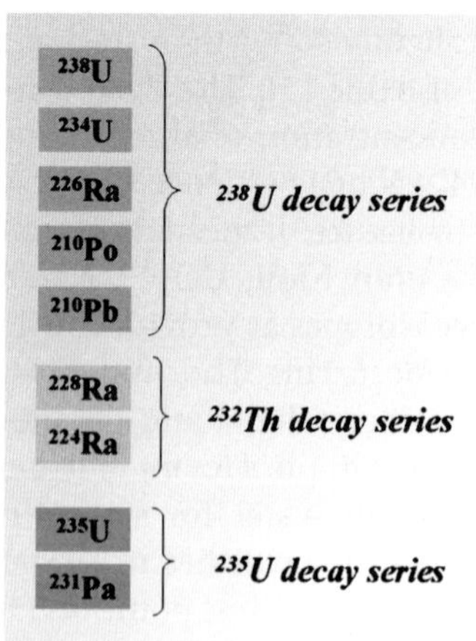

Fig. 1 Significant natural radionuclides in surface waters

Only these nine natural radionuclides and ^{40}K may occur in river waters in elevated concentrations. However, these radionuclides are part of a small fraction of the approx. 80 natural radionuclides that are present in the environment in measurable concentrations. Nonetheless, the occurrence of these most-widespread radioisotopes in potable waters may lead to serious health impacts to the population. It has to be acknowledged that the Rhine is the main drinking water source in Europe for about 20 million residents.

In summary, only some isotopes of uranium and radium are of any relevance in river waters. Additionally, long-lived radon decay products (^{210}Pb and ^{210}Po), ^{40}K and ^{231}Pa may be enriched in river waters as well.

4
Sources of Artificial Radioactivity in the Rhine River

4.1
Nuclear Power Plants and Other Nuclear Facilities With Direct Influence to the Radioactivity in the Rhine River

The Rhine River flows through the following countries: Switzerland, Liechtenstein, Austria, France, Germany, and the Netherlands. The complete distance from its source until its end in the North Sea is 1320 km. Its catchment area includes 185 000 km^2, where almost 50 million people are presently living [10].

Austria and Liechtenstein do not currently operate any nuclear power plants. In Switzerland, four nuclear power plants are in operation, where the one in Leibstadt is directly located at the Rhine River and the three others in Mühleberg, Gösgen and Beznau which all are situated at the Aare River that mounts into the Rhine. The French nuclear power plant Fessenheim is located at the Rhine River and Cattenom is situated at the River Moselle, which also attributes to the Rhine. In Germany, there are seven nuclear power plants of interest: three located directly at the Rhine River (Philippsburg, Biblis, Mülheim-Kärlich), two at the Neckar River (Neckarwestheim, Obrigheim) and two at the Main River (Grafenrheinfeld, Kahl). The reactor in Kahl has been shut down and the reactor in Mülheim-Kärlich is not in operation and is in the process of being decommissioned. Other than the nuclear power plants mentioned, there exists more nuclear reactors and nuclear facilities in Germany, in which radioactive material is handled. In the Netherlands, the nuclear power plant Dodewaard (which has been shut down) is located directly at the Rhine River. Figure 2 shows the course of the Rhine and the location of the mentioned nuclear power plants [11–14].

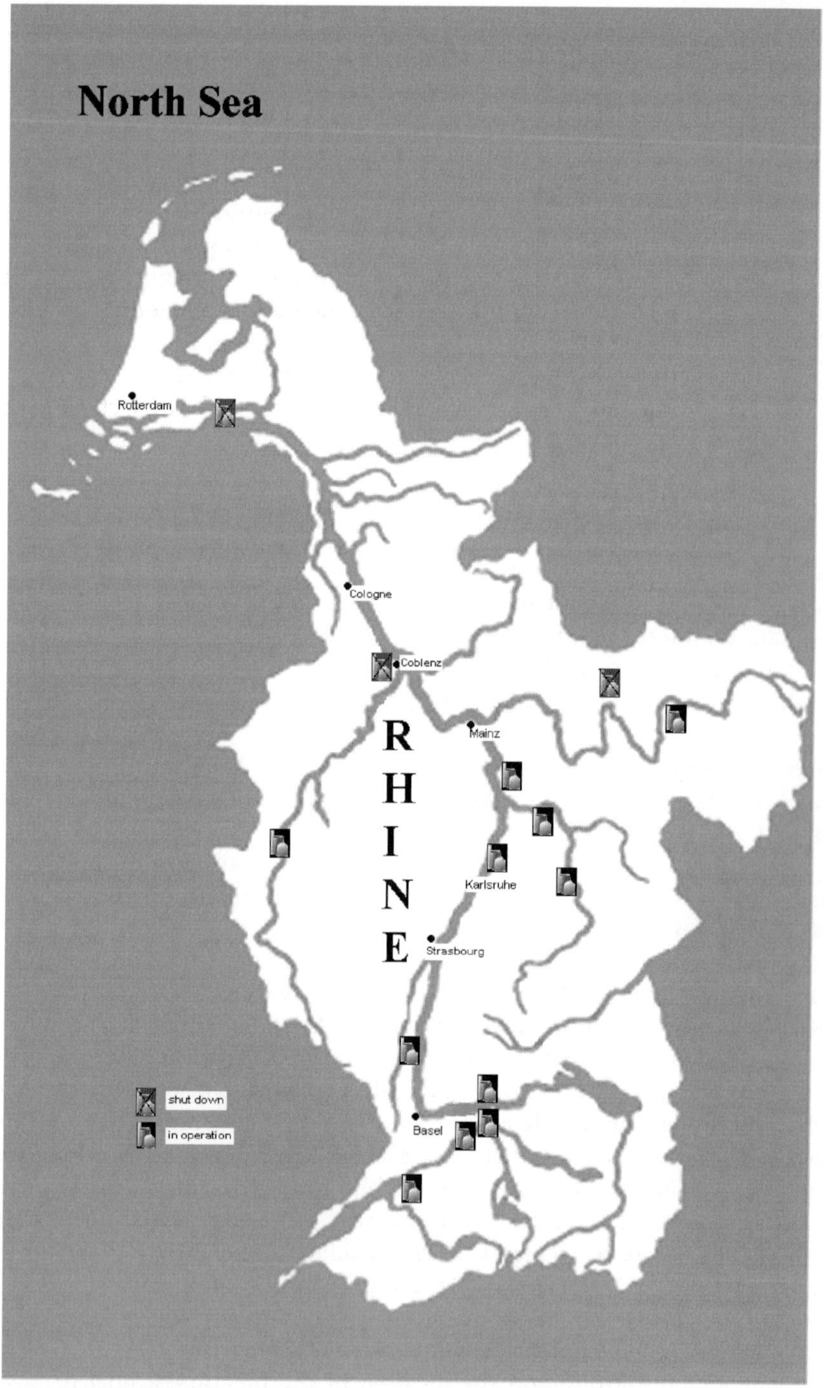

Fig. 2 Location of the nuclear power plants along the Rhine River

4.2 Sources of Radiation in Nuclear Reactors

The radiation emitted from a nuclear reactor has different sources:

- The radioactivity of uranium itself (is of minor importance)
- The fission of the nucleus of ^{235}U by neutrons resulting in: e.g., ^{131}I (refer to Sect. 5.5) along with the emission of neutrons and gamma radiation
- The mutation of ^{238}U by scavenging of a neutron and decay into ^{239}Pu. This creates a chain fission reaction with the release of energy and is the source of the fission products
- The radioactivity of the fission products, e.g., ^{89}Sr and ^{90}Sr (refer to Sect. 5.4), ^{103}Ru and ^{106}Ru, ^{124}Sb, and ^{125}Sb, ^{137}Cs (refer to Sect. 5.6). The radiation effects of isotopes of ruthenium and antimony towards humans are only of minor importance
- Isotopes of neptunium, americium and curium which are comparably radiotoxic alpha emitters
- The radioactivity of activation-products. By absorption of a neutron, inactive nuclides can mutate into radionuclides: e.g., ^{51}Cr, ^{54}Mn, ^{55}Fe and ^{59}Fe, ^{58}Co and ^{60}Co (refer to Sect. 5.3), ^{63}Ni, ^{65}Zn. These particular radionuclides can reach the environment (especially the surface water) via the sewage of nuclear power plants [15, 16].

The most important radionuclides in effluents of nuclear facilities are: Tritium, ^{58}Co, ^{60}Co, ^{131}I, ^{134}Cs, and ^{137}Cs, whereas only tritium can be measured in water samples with acceptable effort [17].

4.3 Nuclear Weapons Tests

The most catastrophic nuclear event in human history was the release of atomic bombs on Hiroshima and Nagasaki in 1945. Since that year and mainly between 1954 and 1962, numerous nuclear weapons tests have been carried out in different countries. Until 2002, 541 above-ground and 1867 underground bombs were detonated. There are three mechanisms that control the deposition of radionuclides onto the ground: The "local fallout" contains about 50% of the total activity and quickly contaminates an area of about 100 km around the site of the explosion. Smaller parts can reach the upper troposphere and can be transported in large distances along the latitude and will be deposited as "intermediate fallout" after some weeks. However, finest parts of dust stay in the stratosphere over years and will be transported as "global fallout" to the surface all over the earth [11, 15, 16].

The most important radioactive substances in a fallout resulting from atmospheric nuclear weapons tests are:

- fission products (^{89}Sr and ^{90}Sr, ^{95}Zr, ^{95}Nb, ^{103}Ru and ^{106}Ru, ^{129}I and ^{131}I, ^{137}Cs, ^{140}Ba),
- not used fission material (^{235}U, ^{239}Pu),
- induced construction material (^{54}Mn, ^{58}Co and ^{60}Co),
- activated soil material (^{54}Mn, ^{55}Fe and ^{59}Fe, ^{65}Zn),
- radiocarbon (^{14}C),
- and tritium [15, 16, 18].

The peak of nuclear testing has passed for more than 40 years. Most of these radionuclides with short half-lifes have already decayed and are without any impact on the environment in Central Europe. Only traces of long-lived radionuclides such as ^{90}Sr, ^{129}I and ^{137}Cs besides tritium can still be determined in environmental samples and identified as products from nuclear weapons tests. The amount of the activity of radioactive cesium released from the weapons tests corresponds closely to the total emission of ^{134}Cs and ^{137}Cs from the Chernobyl accident [15, 16, 18].

4.4
The Accident at the Chernobyl Nuclear Power Reactor

The most momentous accident during the peaceful use of atomic energy happened on the 26th of April 1986 at the Chernobyl nuclear power reactor which was located in the Ukraine (former USSR). This accident was caused by an incorrectly planned experiment and which was carried out with an inherent unsafe reactor. Due to this, the reactor block four exploded and large quantities of graphite were burning inside for about 10 days. The roof of the reactor building was completely destroyed [19–22].

The consequence of this accident was that a huge part of the reactor interior was released to the environment—in the amount of approximately 1.9×10^{18} Bq, which mainly consisted of volatile fission and activation products [15, 19, 23]. The debris was mainly deposited in a 30-km zone around the reactor. Due to explosion and burning of the building and blessed with the weather situation, the radioactive substances also reached the upper layers of the atmosphere and were transported far away. A large-scale radioactive contamination was found in Scandinavia, and also in Eastern and Central Europe [16, 20–22]. After this event, there was even a significant increase of radioactivity observed in the air, rain water and deposition of radionuclides in Japan, Canada and the United States. Only the southern hemisphere remained free of pollution. At the beginning of May 1986, the air upon Central Europe had been loaded with variable amounts of radioactivity from southeast to northwest. This was evaluated especially from the determination of ^{131}I and ^{137}Cs as key isotopes [15, 22].

The details about the composition of the substances in the radioactive air masses vary in literature. The dominating isotopes were: the short-lived noble gas ^{133}Xe (half-life 5.2 days) with high activity and the long-lived noble gas ^{85}Kr (half-life 10.7 years) with low activity. However, the contamination of lakes and rivers from these noble gases was negligible. Except for the very short-lived radionuclides, the most important isotopes of the radioactive air masses with respect to their relative proportion were ^{131}I, ^{132}Te, ^{132}I, ^{137}Cs and ^{134}Cs, ^{103}Ru and ^{106}Ru, and ^{140}Ba. Fortunately, the extremely radiotoxic nuclide ^{239}Pu (less than one millionth part of ^{137}Cs) was only liberated in small amounts. The release and influence of ^{89}Sr and ^{90}Sr was lower compared to ^{137}Cs. In the 2 months after the Chernobyl accident, ^{131}I had the greatest relevance of all radioactive substances which were discharged. However, since the half-life of ^{131}I is relatively short (8 days), the long-time effects of this isotope were only of minor importance. Therefore ^{137}Cs and ^{134}Cs, which were released from the Chernobyl reactor with an initial ratio of 2 to 1, received the final attention. However, ^{137}Cs became the long-lived key isotope of the Chernobyl accident because of its half-life of 30 years. For a few years after the accident, ^{134}Cs was detectable in environmental samples. However, it possesses a half-life of only 2 years, and thus the effect of this isotope is negligible today [15, 19–25].

The contamination in soil and surface water was mostly influenced by the weather, to be exact whether it was rainy or draught season during the event. The intake by precipitation (washout) predominated compared to dry deposition (fallout). As a characteristic in consequence of the different washouts, precipitations (with yields) strongly varied from place to place and resulted in short-distance variations of the surface contamination [15, 19–22]. Nevertheless, during that event, the level of contamination and the resulting dose values were well below the limits for the implementation of protective measures of emergency management in case of nuclear catastrophes [20].

4.5 Other Accidents in Nuclear Power Plants

Influential nuclear accidents in other power plants also took place in the Windscale reactor, Great Britain, in 1957, in the Three Mile Island reactor in Harrisburg, United States, in 1982, and in the nuclear reprocess facility Tokaimura, Japan, in 1999. These accidents were more isolated and of minor consequences compared with the Chernobyl accident.

4.6 Other Contributors to the Artificial Radioactivity in the Rhine River

There are more sources that can influence radioactivity in surface water. They can be medical, industrial, or originating from other civil facilities in which radioactive material is handled.

In hospitals and other medical institutions, several radioactive substances are used as tracers for diagnostic and radiotherapy. Radioactive waste or excreted radioactive substances that have been administered to patients can reach the rivers directly or indirectly as effluents from the sewage water treatment plants. The main component of these wastes is ^{131}I, which is often used for medical diagnostic and therapy [16]. A statement of the committee "Radiation Protection in Medicine" of the German Radiation Protection Commission (SSK) declared that the use of ^{131}I in diagnostic is no longer justified. The short-lived nuclides ^{99m}Tc, ^{18}F or ^{123}I have therefore been mostly recommended to replace ^{131}I [26].

In addition to this, effluents from industry and scientific institutions in which radioactive material is handled make a significant contribution to the radioactive contamination of surface water. Tritium, for example, is an important substance in the emissions from industrial effluence as a result of the clock and watch manufacture [12, 16].

Finally, it has to be mentioned that radioactive emission of a coal-fired power station is slightly higher than that of a trouble-free operating nuclear power plant due to the emission of the isotopes resulting from the decay of uranium, which is present in the fired coal [21].

5 Details on the Most Important Artificial Radionuclides

In this section, facts and values are provided for artificial radionuclides which are present in the environment and in the waste water released from nuclear power plants. This may be helpful for understanding and performing estimations in the field of environmental radioactivity. Table 3 gives an overview of the properties of some important artificial radionuclides.

5.1 Tritium

Tritium (^{3}H), the radioactive isotope of hydrogen, is produced by the interaction of cosmic radiation with compounds of the air in the upper atmosphere. Only traces of natural tritium can be measured in water: 1 tritium-atom occurs in about 10^{17} hydrogen-atoms [28]. Tritium originated as well from the above-ground nuclear weapon tests and is widely dispersed. Today, the content of tritium in surface water mostly stems from effluents of nuclear reactors, research reactors and reprocessing plants for nuclear fuel [15, 18]. Tritium is also used as a component in thermonuclear (fusion) weapons. It is also an important tool in life science research, in drug metabolism studies and in dating of ground water. In industry, tritium is used as fluorescent color for example in luminous signs, gauges indicator and for luminous dial,

Table 3 Properties of some artificial radionuclides [18, 27]

Name	Atomic Number	Chemical Symbol	Radiation Type Alpha	Beta	Gamma	Half-Life	Stable Daughter
Tritium (Hydrogen 3)	1	^{3}H		X		12.3 a	Helium 3
Carbon 14	6	^{14}C		X		5736 a	Nitrogen 14
Cobalt 60	27	^{60}Co		X	X	5.3 a	Nickel 60
Strontium 89	38	^{89}Sr		X		50.5 d	Yttrium 89
Strontium 90	38	^{90}Sr		X		28.5 a	Zirconium 90
Iodine 131	53	^{131}I		X	X	8.0 d	Xenon 131
Cesium 134	55	^{134}Cs		X	X	2.1 a	Barium 134
Cesium 137	55	^{137}Cs		X	X	30.1 a	Barium 137
Plutonium 239	94	^{239}Pu	X	X	X	24 130 a	various

face and hand of watches. The liquid effluents from this industry are another source of tritium in surface water [12, 23, 29]. Most of the tritium formed in the atmosphere combines with oxygen to form ^{3}HHO and can therefore rapidly take part in the global water cycle. In contrast to most other radionuclides, tritium is not strongly bound in soil and can therefore enter the groundwater easily [15, 30].

Tritium can also enter the human body rapidly as HTO from food and drinking water into the stomach and bowels or from water steam in the air into the lungs and the skin. Tritium is equally distributed in the human body water and possesses a biological half-life of approx. 10 to 12 days [21, 28]. However, tritium is a weak beta emitting isotope ($E_{max} = 18$ keV) which has a short effective half-life, thus the radiological toxicity of tritium is considered to be comparably low [16].

Tritium can be analytically determined by liquid scintillation spectroscopy [31].

5.2 Radiocarbon

Just like tritium, ^{14}C is produced naturally due to the interactions of cosmic radiation and nitrogen in the air through neutron scavenging and artificially from nuclear power plants and reprocessing plants for nuclear fuel. ^{14}C is a constituent of carbon dioxide in the air and of hydrogen carbonate in water [15, 28]. ^{14}C is used in natural sciences for the radiocarbon dating method of organic matter [32].

In the year 2000, the European Community estimated that only 5% of the whole activity of ^{14}C in the air originated from nuclear power plants. The radiological toxicity of radiocarbon (dioxide) is very low, since it is a week beta emitter (E_{max} = 158 keV). Nevertheless, exposure to ^{14}C results in an increasing risk for having the occurrence of a cancerous disease. It can be considered that the presence of ^{14}C will spread nearly homogenous all over the living organism resulting in a homogeneous radiation exposure to the body [15, 16, 18].

Similar to tritium, ^{14}C can also be analytically determined by liquid scintillation spectroscopy [31].

5.3 Cobalt 60

As an activation product, ^{60}Co is produced in nuclear reactor operations, resulting from the exposure of structural materials to neutron radiation. Since ^{60}Co emits a strong gamma radiation, it is one of the most common radionuclides used in industrial applications. It is also used in medical radiotherapy and for sterilization of medical equipment and, in some countries, for the sterilization of special foodstuffs, for example spices. Since ^{60}Co is exclusively an artificial radionuclide, it can only reach into the environment from nuclear power plants and other facilities, which use this isotope [18].

The radiological toxicity of ^{60}Co is high—being exposed to its strong gamma radiation increases the risk of developing a cancer. After the ingestion of contaminated food and drinking water or from inhalation, some ^{60}Co accumulates in the blood, liver and kidney. However, it is also quickly excreted in the feces [18].

Since ^{60}Co is a gamma emitter, it can be easily determined with a modern high-resolution gamma spectrometer [31].

5.4 Strontium 90 and Strontium 89

In the case of nuclear fission performed in nuclear power plants and weapons tests, the short-lived ^{89}Sr and the long-lived ^{90}Sr including the short-lived daughter nuclide ^{90}Y (half-life 2.67 days) will be produced. Both strontium isotope and the daughter are pure beta emitters and decay into their stable products without emitting gamma-rays [15, 28]. ^{90}Sr is used as radioactive tracer in medical and agriculture studies [18].

Among the monitored radionuclides, ^{90}Sr is of great importance because it has the tendency to be adsorbed to the bones and teeth. It also has both a long physical and biological half-life. The latter is estimated to be 50 years in humans. Strontium will be incorporated into the bones especially among children because it is chemically similar to calcium. This results in a long-lasting radiation exposure of the bone marrow which may cause leukaemia. Due to this reason, the radiotoxicity of strontium is considered to be high [18, 28].

Since the extent of released activity of ^{90}Sr from nuclear bomb test series was high, it was one of the most important radionuclides in the past. Nevertheless, strontium has been very well determined and a multitude of data has been collected for its contribution and behaviors in water and on the exposure paths to humans [15, 28].

Since both strontium isotopes do not emit gamma radiation, the analytical determination of ^{89}Sr and ^{90}Sr by proportional counters is only possible after an extensive radiochemical analysis and a long separation course [31]. The resonance ionization mass spectrometry (RIMS) is another method used to determine ^{89}Sr and ^{90}Sr in environmental samples [33].

5.5 Iodine 131

In the case of nuclear fission, several iodine isotopes with very different half-lives will be produced. Radioactive iodine will also be produced during nuclear weapons tests. Despite its relatively short physical half-life (of about 8 days) the most important isotope from this group is ^{131}I. From a damaged fuel rod of a nuclear reactor, this nuclide can reach the primary cooling cycle of nuclear power plants and may subsequently released to the air. It can rapidly disperse in air and water. ^{131}I is also often used in medicine, especially in radioiodine therapy and formerly for thyroid gland diagnostic [15, 28].

Iodine or iodide can be rapidly and completely incorporated into the bowels after inhalation or ingestion from foodstuff and drinking water. Subsequently, ^{131}I will be transported to the thyroid gland, accumulated therein and exposes this organ with ionizing radiation. The remaining fraction will be excreted as urine [15, 16, 18].

Due to the extent of released activity of ^{131}I from the Chernobyl accident and its high volatility, in the first weeks after this event, ^{131}I was the main contamination compound in the environment and in foodstuffs, especially in milk [28]. However, its long-time effect is only of minor importance and provides no hazard to ground water, because ^{131}I possesses a relatively short half-life [15].

^{131}I emits gamma rays and can easily be determined with modern high resolution gamma spectrometry [31].

5.6 Cesium 137 and Cesium 134

^{137}Cs and ^{134}Cs are artificial radionuclides that are also created by nuclear fission. While ^{137}Cs is a fission product which is frequently formed, ^{134}Cs is produced in the reactor by means of activation with neutrons. The activity of ^{134}Cs increases in correspondence to the running time of the reactor and will have about half of the activity of ^{137}Cs after a period of 1 year. The ratio of $^{137}Cs/^{134}Cs$ is useful to estimate the running time of a reactor and also to interpret the origin of a radioactive contamination measured for example in environmental samples. ^{137}Cs is used in numerous industrial processes, for example in moisture density gauges, levelling and thickness gauges or devices to measure and control the flow of liquids [18].

Cesium incorporated into the body from foodstuff and will almost be completely ingested by the gastro-intestinal system. For its chemical affinity to potassium, it will then spread into the muscle tissue, resulting in an all over body exposure. The biological half-life of ^{137}Cs is in average 110 days. The radiotoxicity of ^{137}Cs is high, and this isotope is therefore one of the most important artificial radionuclides [16].

As mentioned previously, ^{137}Cs was mainly produced during the atmospheric nuclear weapons tests and dispersed worldwide and was the long-life key isotope of the Chernobyl accident [15, 18, 28]. Today, ^{137}Cs is effectively the only radionuclide derived from the Chernobyl accident which can still be determined in some foodstuffs such as the meat of game or mushrooms and in some media in the environment such as soil and sediments. Since cesium has a very slow migration velocity in soil, a contamination of groundwater resulting from radioactive fallout will practically not occur [15, 28].

^{137}Cs and ^{134}Cs emits gamma radiation and can be easily determined with a modern high-resolution gamma spectrometer [31].

5.7 Plutonium

Plutonium is a heavy metal that belongs to the group of the actinides. It is the heaviest element existing naturally, but only in traces, produced under highly unusual geological circumstances [18]. Otherwise, it is mostly produced in a nuclear reactor, by neutron capture of ^{238}U which decays into ^{239}Pu. During the above-ground nuclear weapons tests at the end of the 1950s and at the beginning of the 1960s, three to four tons of plutonium had been spread all over the world which resulted in a deposition with an activity of approximately 10^{16} Bq. All plutonium isotopes are radioactive and emit alpha radiation and posses relatively long half-lives [18, 28].

Due to their alpha radiation, plutonium-isotopes are very radiotoxic, other than being chemically toxic. Alpha emitters are extremely dangerous when incorporated into the body from breathing, foodstuffs or from open wounds. What will happen is that the organs will be directly exposed to radiation. Plutonium remains in the lungs for different periods of time depending on the solubility of the compound. The period of stay in the lungs can last up to 4500 days. Other tissues which might accumulate plutonium are bone cells and the liver, in which the mean period of stay for plutonium is estimated to be 50 years or longer. There exists, however, no threshold dose of the ineffectiveness of incorporated plutonium. Intake of plutonium by inhalation enhances the risk of cancer to the lungs, bones, liver and leukaemia [18, 34].

Plutonium isotopes will form the insoluble tetravalent oxide under natural conditions that is complex-bound in the soil. The migration velocity of plutonium compounds is very low, thus representing a low danger of mobilization from the subsurface layers into drinking water and into the food chain [15, 28].

The analytical determination of the plutonium isotopes with alpha spectrometry is only possible after a radiochemical separation course [31]. Another analytical method used to determine plutonium is the application of a resonance laser ionization mass spectroscopy which is very effective and sensitive but also very costly [33].

6 Monitoring of Environmental Radioactivity

6.1 International Networks of Monitoring and Information Systems

In 1957, the "Treaty for the Founding of the European Atomic Energy Community, (Euratom Treaty)" was laid down to state the requirement for an establishment of facilities in each member state, the continuously monitor environmental radioactivity in the air, water and soil (Article 35) and for the regular report of the monitoring results from the member states to the Commission so that the public is kept informed about the level of radioactivity to which it is exposed (Article 36) [35].

Monitoring systems for environmental radioactivity were set up in countries such as Austria, France and Germany based on the individual radiation protection philosophies and technical resources. In the process of this, international cooperation was practically nonexistent [36]. The Chernobyl accident in 1986, however, has shown that coordination of the measuring and information systems at international level is an urgent necessity, in order to exchange of at least the most important data, in case of such an event. This knowledge led to the establishment of information systems by the EC (European Commission) and the IAEA (International Atomic Energy Agency) located in Vienna who, as a legal requirement, are to be kept informed by the member states and countries who have signed the treaty [36, 37]. This was based on agreements on early notification and speeding up information exchange in the case of a radiological emergency situation [38, 39].

Required techniques for monitoring environmental radioactivity are:

- Measuring techniques (and equipment) for the determination of environmental radioactivity,

- Simulation techniques for prognoses of transport processes, and for the determination of the exposures resulting from the measured radioactivity levels,
- Data processing (supporting) techniques for processing and transferring the measuring and calculation results [36].

The accreditation of the institutions of the Environmental Control Authority is based on the Good Laboratory Practice and the EN ISO/IEC 17025. This European Standard specifies the general criteria for the technical competence of contributing laboratories. The laboratory shall have—besides management and organization—sufficient personnel, having the necessary education, training, technical knowledge and experience. The laboratory shall be furnished with all equipment required for the correct performance of the tests and measurements. The laboratory shall have adequate documented manuals of all relevant equipments and for handling samples [40]. The analytical operation procedures of the nuclear analytical techniques in Germany are based on special measuring instructions, which are working procedures for the measurement of different radionuclides in e.g., water, air, soil and foodstuffs [31].

The EU member states provide the results of their monitoring systems to the European Union, or to be exact to the Joint Research Centre, Environmental Institute in Ispra, Italy (JRC/IES), where all results are collected in the database of "Radioactivity Environmental Monitoring" (REM), which is placed at the disposal of institutions of radiation protection and the general public. The REM project supports the European Commission in its responsibility to inform the European Parliament and the EU member states on the radioactive pollution in the environment [41, 42].

On the legal basis of the mentioned council decision, closely related projects were developed from the JRC-IES/REM with the scope of making early and reliable radiological information available to the EU member states in case of major nuclear accidents: ECURIE (European Community Urgent Radiological Information Exchange) is responsible for informing the competent authorities of the member states on the actual status of an emergency and its consequences. EURDEP (European Radiological Data Exchange Platform) has the aim to make radiological monitoring data available from most European countries on the Internet in real-time [37, 41, 43].

There exists an intensive international cooperation which deals with concerns related to nuclear safety and radiation protection by running discussions, exchange and comparison of the empiric data from observations and experiments. This also includes driving the state-of-art of science and technology forward. Among other tasks, the IAEA also performs many other scopes in relation to radiation protection [44].

The EC recommendation on the 8th of June 2000 on the application of Article 36 of the Euratom Treaty is with regards to the monitoring of radioactivity level in the environment, of which the Member States are obliged to perform under the terms of Article 35 in the Euratom Treaty. The information based on the results has the purpose of assessing radiation exposure to the population as a whole. The EC hereby recommends which of the requirements should be met by the Member States in order to discharge their obligations [45].

In most European states, the radiation protection limitation values and standards for ionizing radiation are laid out according to international regulations. Recommendations for radiation protection are developed by the ICRP (International Commission on Radiological Protection) based on exposures and risk assessments reports by for example UNSCEAR (United Nations Scientific Committee on the Effects of Atomic Radiation). These recommendations are taken up by the EU or by the IAEA where essential parts of them are implemented into directives or recommendations. As an immediate consequence, the EU directives regarding radiation protection have to be implemented directly into national laws, ordinances and other regulations which have to be obeyed [44].

In contrast to the radiation protection, the EC has currently neither authority in the field of nuclear safety nor direct influence on how the individual states should handle the supervision of their respective nuclear facilities. However, the EC intends to receive further-reaching power in this area [44].

6.2 Monitoring in the Surrounding of Nuclear Facilities in Germany

Due to the Euratom Treaty, the EU member states had to take over responsibility for the monitoring of radioactivity in air, water and soil and for a regular reporting of the results. The focus of monitoring shifted from the environment in general to the monitoring of the surroundings of civil nuclear facilities and the monitoring of the emissions from these facilities when nuclear energy began to be more often used in the end of the 1960s and the beginning of the 1970s [36, 46].

In Germany for example, this monitoring task is based on the legal standards of the "Atomic Energy Act" [47] and the "Radiation Protection Ordinance" [48] and is exactly described in the "Guidelines of Emissions and Immissions Monitoring for Nuclear Facilities" (REI) [49]. The task of monitoring is in accordance with the REI and is related to the respective facility. Continuous monitoring of emissions and immissions is performed in the surrounding of nuclear facilities, where, in the case of an event, contamination levels are to be obtained. On the basis of the measured data—especially the emissions data—radiation exposure of man is evaluated by means of model calculations [36, 46]. Measures that have to be taken in the case of a nuclear accident are also described in the REI [49]. The "Outline Recommendations for Emergency Management in the Surroundings of Nuclear Power Plants" recruit the measuring teams and dictate the necessary measurements and their implementation [36].

Of course, this monitoring has to be adjusted to a new situation in the state of a nuclear facility. For example the nuclear power plant in Mülheim-Kärlich is in the process of being shut down and all fuel rods were transported away in 2002. Therefore monitoring of fissions products such as ^{90}Sr is negligible and the extent of the supervision of the plant could be reduced [50].

A self-surveillance program carried out by the operator of the facility performs the emissions monitoring program, applying in the process the "Safety Standards of the Nuclear Standards Committee" (KTA). In addition to the surroundings monitoring, from the end of the 1970s onward, a "Nuclear Reactor Remote Monitoring System" (KFÜ) was set up by the supervisory authorities of the Federal States. The aim of this system is to continuously evaluate the state of operational security of the nuclear power plant through the compilation of data consisting of safety-relevant operational parameters, emissions data, meteorological parameters and immissions data. When abnormal operating conditions occur, the system operates independently, and indicates this action by means of alarm signals [36, 51].

6.3 General Environment Monitoring in Germany

In 1986, the Chernobyl accident resulted in a lack of accurate and immediately available data of the radioactive immissions. An lack of communication between the environmental monitoring facilities involved in many countries was observed, which led to a confusion in the public. These defaults provided the impulse to make a large-scale monitoring of environmental radioactivity more effective [15, 36].

In Germany, on the basis of the "Precautionary Radiation Protection Act" [52] passed in 1986, a comprehensive nationwide program of the monitoring of environmental radioactivity in accordance with the "Integrated Measuring and Information System" (IMIS) was initiated in 1988. Two guidelines were carried out with measuring programmes for routine and intensive operation. The routine program runs continuously and the intensive program takes over in the case of an event [53, 54]. IMIS is operated under the responsibility of the Federal Ministry for Environmental, Nature Protection and Nuclear Safety. The participants of this system are mostly installations and organizations of the Federation and the Federal States, which were set up following the requirements of the Euratom Treaty, which are to monitor the environmental radioactivity in air, radioactive fallout (precipitations), soil, surface water, ground and drinking water, sewage and waste water, agriculture products (plants) and foodstuffs [36, 46]. To be prepared for an emergency situation, two publications are of relevance: "The Overview about measures to reduction of radiation exposure (Measures Catalogue)" [55] and the "Guide for the radiation protection expert of the emergency staff" [56]. The basis of a part of the revision of this guide is a newly developed French–German model of atmospheric dispersion for emergences situation [57].

The IMIS system, which was put into operation in 1993, permits the online notification of the radiation protection authorities and a rapid overview on the current contamination situation in the entire area of the Federal Republic of Germany representatively. The data that have been processed are transferred to the superior Federal and Federal State authorities and, in case of a nuclear incident or accident, accompanied by recommendations [36, 46]. The aim of monitoring environmental radioactivity is to obtain extensive information on contamination and radiation levels in all relevant areas of the environment, and to ultimately determine the levels of radiation exposure to man. For this radiological protection, limit values are set to avoid detectable damage. The most important goal of the limit values is to keep radiation levels as low as possible [36].

The supervision of the Rhine River and other federal waterways is assigned to the Federal Institute of Hydrology (BfG) situated in Koblenz and the monitoring of drinking water is delegated to the Federal Office for Radiation Protection. The BfG operates a radioactivity monitoring and warning network that has the purpose of monitoring radioactivity in the federal waterways. This network has been incorporated as an early warning system into the IMIS. In the measuring stations, the total gamma and the total beta activity of the water are continuously and simultaneously measured with NaI(Tl) scintillation probes and proportional counters. These methods give an overview of the radioactive contamination, but they do not provide information regarding the detailed composition of the radionuclides. Therefore, the kind and the extent of the isotopes has to be determined by specific nuclide analysis. Furthermore, the equipment to determine the gamma activity is projected to be installed twice and to be upgraded to a multiple channel measurement. Due to its trouble susceptibility, the total-beta measuring will not be continued. The counting rates and information on the operation status are stored intermediately by a local computer and after transfer, validation and converting of the counting rates to activity concentrations will be performed by a central computer in a database [23, 58, 59]. In a nuclear emergency situation, the supervision of radioactive contamination in inshore waters becomes an essential task. This task consists of sample collection and the measurement of radioactivity level in water, suspended matter and sediment as well as the calculation for the forecasts on the behavior of radionuclides along the river course [23, 60].

6.4 Monitoring of Environmental Radioactivity in Austria

Similar to Germany and other EU member states, a large scale monitoring program of radioactivity is operated since 1957 in Austria. The legal basis for this is the Radiation Protection Act, which was passed in 1969 [61]. This routine program consists of the world wide first Radiation Early Warning System with on-line operation, real-time data collection and data communication to the authorities. It contains 336 gamma dose rate monitors and numerous nuclide specific air contamination monitors along with measuring radioactive contamination of air, precipitation, surface water, emissions from sewage works, soil, ground water and foods [30, 62]. Since 2003, the Early Radiation Warning System is assigned to the Austrian Environmental Federal Office [63].

The most important aims of the monitoring program are:

- The continuous measurement of radioactive contamination and the resulting radiation exposure of the population,
- The early recognition of increasing radioactivity in case of nuclear accidents,
- The complete overview and the supply of counter-measures in such a case [30].

As already mentioned earlier, no nuclear power plants exist in Austria, and since the few nuclear research installations have no direct influence to the Rhine, their surveillance and inspection shall not be discussed in this context.

Among other locations, at the Rhine River at Bangs and at the Lake Constance at Bregenz random sampling of surface water are periodic collected every month. The subsequent analysis is done by the Austrian Agency for Health and Nutrition Safety, that is the analysis of the gamma emitting isotopes and in most samples as well as the determination of tritium and the gross alpha and gross beta activity [30].

In case of a radiological event of major importance, the State Crisis Management coordinates the precautionary measures. A team of experts in a coordination committee gives the government advice, guarantees the exchange of information, coordinates the actual precautionary measures and ensures an unified action of all areas of administration. In analogy to the International Nuclear Event Scale (INES scale), eight warning levels were fixed in order to realize a classification of the actual danger and the necessary measures, which are carried out on the base of the "Frame recommendations of the establishing and implementation of measures for the protection of the population against ionizing radiation in case of large-scale radioactive contamination" [64]. It is projected that the Radiation Early Warning System will be a part of a comprehensive decision support system for nuclear emergencies [62].

6.5
Monitoring of Environmental Radioactivity in Switzerland

The regular surveillance of environmental radioactivity in Switzerland started in 1956. In the beginning, the tasks were restricted to the supervision of the fallout originated from the above-ground nuclear weapons tests. Later, the Swiss monitoring program shifted more and more to the monitoring of the nuclear facilities, of industrial, medical and other civil facilities in which radioactive material is handled. The legal basis for this monitoring is the Swiss "Radiation Protection Ordinance" [65]. The measurement program is coordinated by the Swiss Federal Office of Public Health, who is also responsible for collecting and evaluating the data and for the annual publication of the results. Twenty different institutions are involved in the general monitoring in Switzerland. For example, the supervision of the aquatic system (waters, fishes, sediments) is carried out by the Swiss Federal Institute for Environmental Science and Technology (EAWAG), the Institute Forel and the Laboratory Spiez. The supervision of nuclear facilities is performed by the Main Department of Nuclear Facilities Safety and the Environmental Radioactivity Section at the Swiss Federal Office of Public Health [12, 66]. At three locations (Hagneck, Klingnau, Pratteln) at the Rhine and the Aare Rivers, aggregate monthly water samples as well as sediment samples are collected [12].

Similar to other participating countries of the radioactivity monitoring, Switzerland has three goals:

- The pursuit of the development of a large-scale natural and artificial radioactivity in order to realize changes, accumulations and trends just in time,
- The monitoring of nuclear facilities and other institutions which deals with radioactive material, in order to inspect if the limitation values are kept, for example the exemption limits for water laid down in the Radiation Protection Ordinance,
- The determination of the radiation doses of the population [12, 66].

In the case of increasing radioactivity, where necessary steps for the protection of the population must be taken, the Swiss National Emergency Operations Center (NAZ) in Zurich has to be put into operation. The NAZ, which is the first element of the Swiss Emergency Organisation Radioactivity, own the logistic equipments and the personnel which are necessary in such events. After a nuclear accident, the NAZ will use the results from dispersion models to assess the situation in Switzerland [12, 67].

6.6 Monitoring of Environmental Radioactivity in France

The French nuclear safety and radiation protection organization is based on the principle of responsibility, which lies within those who perform hazardous activities (for example operator of a basic nuclear installation), supervised by a public authority. In 2002, an extensive institutional reorganization affecting nuclear safety and radiation protection was carried out in various decrees. One aim of this reformation was the separation of control authority and technical assessment. The Nuclear Safety Authority is the regulatory organization of monitoring nuclear safety in the basic nuclear installations. This regulatory organization consists of the Directorate General for Nuclear Safety and Radiation Protection (DGSNR) and decentralized departments. Beside others, two important tasks of the DGSNR are: to organize nationwide radiological monitoring of the environment and to organize public and media information in terms of nuclear safety and radiation protection. Thus, the DGSNR is responsible for the supervision of the liquid and gaseous effluents from nuclear installations. The Environment Code states the right be informed with regards to the environment, including radioactive substances [14].

The Nuclear Safety Authority obtains a consultative and technical support by the Institute for Radioprotection and Nuclear Safety (IRSN). The IRSN has the responsibility to assess the risks due to ionizing radiation. One of the main task of the IRSN is the protection of human health and the environment caused by radioactive pollution. This will be done by a nationwide permanent supervision of environmental radioactivity. In order to carry out these monitoring tasks, three measurement networks were established: Fully automated alert stations, observatories and 300 sampling stations, located generally near nuclear sites. Every year, more than 36 000 samples are taken from air, water, soil and foodstuffs. The laboratory analyses focus on the different alpha, beta and gamma emitters [68]. The Regional and Department Health and Social Action Directorates take part in radiological monitoring of drinking water [14].

The organizational provisions of the authorities and the operators are responsible for taking precautionary steps towards a nuclear emergency situation. This situation is defined in directives and emergency plans distinguished by decisions, action and coordination in national and international level. The Nuclear Safety and Radiation Protection Departments have to monitor the site and support the department Prefect to protect the population. The Prefect, who acts within the framework of an off-site emergency plan, is also supported by the Regional and Department Health and Social Action Directorates. The INES severity scale was adopted as well in France to inform the public and the media about the safety significance of nuclear incidents and accidents [14]. In a radiological emergency, the IRSN is able to rapidly provide information about the accidental situation to nuclear safety authorities [68]. The French-German model of atmospheric dispersion for emergences situation was implanted into the manuals and programmes used by the technical crisis center operated by IRSN [57]. An appropriate organization is available for emergency situations, which includes an on-call team, a technical crisis center and a mobile unit that is responsible for local measurements. Similar to other countries, crisis exercises are regularly carried out in particular as training for emergency teams and to practise and ensure the cooperation between the various institutions. Special computer tools are able to assess the possible development of an accident and the possible consequences on human and environment [68].

6.7 Monitoring of Environmental Radioactivity in the Netherlands

In the Netherlands, the National Radioactivity Monitoring Network (NMR) is a monitoring system consisting of a network of 163 ambient dose equivalent rate monitors and 14 aerosol monitors nationwide. The NMR is a facility under the authority of the Ministries of Internal Affairs and Environmental Affairs. The National Institute of Public Health and the Environment has to coordinate the exploitation, management and maintenance of this network. The measured data of this system are directly transferred to the national system in The Hague [69, 70].

The data on surface water from the main inland waters are obtained and determined by the laboratory of the Institute for Inland Water Management and Waste Water Treatment. One sampling point of the monitoring program for the determination of radioactive nuclides in surface water is the Rhine River at Lobith, which is located at the Dutch/German border. The determined radionuclides in these samples are tritium in surface water and ^{137}Cs in suspended solids [69].

Measuring equipments for radiation protection has been installed in the nuclear facilities. The Nuclear Research and consultancy Group has developed a software for the connection of different equipments to one centralized warning system, which enhances the effectiveness of the alarm system significantly. The NMR provide actual dates continuously and set off alarms, which were transferred to regional working stations at fire brigade departments and the national accident management organization. With this, quick decisions and security measures can be taken as soon as possible, whenever necessary [69, 70].

7 Detection of Radionuclides in Natural Waters

Measurement systems for radioactive isotopes are primarily based on the interaction between ionizing radiation and matter. Presently, three categories of radiation detection systems are mainly used for radionuclide determination:

- **Semiconductor detectors** (α-/β- and β-spectrometry)
- **Scintillation counters** (α- and β-measurements)
- **Proportional counters** (α- and β-measurements)

Counting techniques are still highly applicable for the determination of radioisotopes with half-lives of less than a few thousand years. However, inductively coupled plasma mass spectrometry is the current state-of-art instrumentation for measuring multiple elements at trace level. Unfortunately, this outstanding method is presently only applicable for stable isotopes or long-lived primordial isotopes, such as ^{232}Th and ^{238}U.

The following sections provide details of measurement techniques for the determination of radionuclides and present a brief summary of the advantages and disadvantages of the respective counting method.

7.1 Semiconductor Detectors

The use of solid detection medium is of great advantage for the detection of high-energy electrons, heavy particles and γ-rays. In the first place, detector dimension can be kept smaller due to the high density of semiconductors. In addition to that, a superior energy resolution can be achieved by using a solid semiconductor detector. There are different semiconductor materials available today, such as silicon and germanium. The fundamental information carriers are electron-hole pairs created in the crystal lattice by primary or secondary charged particles along their path to the detector [71]. The movement of electron-hole pairs through the crystal generates the basic electrical signal from the detector.

Since the intensity of the electric pulse is proportional to the energy of the penetrating particle, semiconductor detectors contribute full energy information. Multi-channel analyzers (MCA) provide the energy resolution and the primary analog signal is transformed into digital units by use of an analog-digital-converter.

Compared with other counting techniques, semiconductor detectors possess advantages, such as excellent energy resolution, good stability, excellent timing characteristics and simplicity of operation [71].

7.1.1 Alpha-Spectrometry

Alpha-spectrometry is an extremely useful and sensitive method for the detection of α-emitting radionuclides in a variety of materials. One of the main reasons for the sensibility of this method is the very low background radiation. High-purity silica surface barrier detectors have become the detectors of choice for the measurement of α-particles. The relatively small size of a silicon detector might be a limitation, especially when a large surface area is required. Modern α-spectrometers are equipped with vacuum chambers to achieve a high-energy resolution of 20–30 keV. The obtained α-spectrum can easily be analyses by using simple region of interest (ROI) settings. In addition, the energy resolution is generally sufficient to separate the energy peaks of the most important natural and artificial radionuclides present in waters. Unfortunately, extended measurement times (2–3 days) are necessary to achieve low detection limits in the range of several mBq/l.

7.1.2 Gamma-Spectrometry

High-purity germanium (HPGe) detectors are preferably used for complex γ-ray spectrometry. This is due to the lower melting point of germanium compared with silicon, which makes the exclusion of impurities in the refining process much easier [71]. Additionally, the extremely high density of germanium (5.33 g/cm^3) leads to a higher adsorption probability for γ-rays [71]. Another advantage of gamma-spectrometry is the relatively low self-absorption of γ-rays within the sample, which results in a simple sample preparation. However, the counting efficiency of HPGe detectors is comparably low and the naturally occurring γ-background radiation might be problematic when trying to achieve low detection limits.

These three effects of radiation interactions with matter are fundamental of γ-spectrometry:

- **Photo effect** (detector-atom absorbs the energy of the incident photon entirely)
- **Compton effect** (incident photons are scattered by electrons of the detector)
- **Pair production** (positron-electron pair is created by the incident photon)

The operation of germanium detectors at room temperature conditions is impossible. This is because of the thermally induced leakage current that would occur at room temperature. Hence, germanium detectors must be cooled with liquid nitrogen to reduce the leakage current to the point that the associated noise does not spoil their excellent energy resolution [71]. Usu-

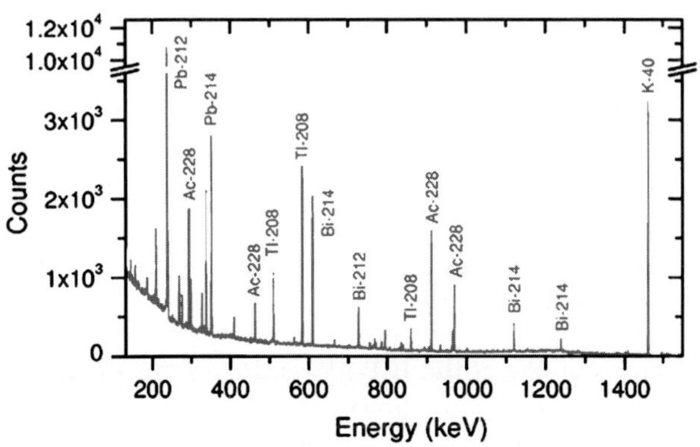

Fig. 3 Gamma-spectrum of a sample containing natural radionuclides

ally, the temperature is reduced to –196 °C by using an insulated vessel in which liquid nitrogen is kept in thermal connection with the detector. Figure 3 provides an example for the excellent energy resolution of an HPGe detector.

7.2
Liquid Scintillation Counting

The liquid scintillation method is one of the oldest and most useful method for detection and spectroscopy of ionizing radiations [71]. It is mainly utilized for the detection of low-energetic β-emitting radioisotopes, such as ^{3}H and ^{14}C, but α-emitting radioisotopes can also be detected with high efficiencies. The approach involves dissolving the sample directly into the liquid scintillator. Problems relating to sample self-adsorption or beta-backscattering from the detector are completely avoided [71]. The scintillation technique is based on the phenomenon of fluorescence and detects photons produced by the scintillator material. These light flashes are caused by exited states of the scintillator molecules after the absorption of energy from radioactive decay. The photons are converted in electrical pulses by a photo multiplier tube (PMT) and can be analyzed by electronic equipment. The liquid scintillation cocktail is composed of a primary and a secondary scintillator—that are able to transform the decay energy to light flashes—and an organic solvent (e.g., toluene) as carrier substance. Thus, the scintillator molecules can be regarded as the actual radiation detector in a liquid scintillation counter.

Modern liquid scintillation counters are able to distinguish between the different types of radiation and even between the different energies of the same type of radiation. The discrimination between α- and β-radiation in water samples is applicable for determining gross-alpha and gross-beta activity concentrations. One major disadvantage of liquid scintillation counters is their relatively poor energy resolution. Furthermore, quench effects may reduce the counting efficiency and background radiation might negatively affect the measurement. On the other hand, the counting efficiency for α- and β-particles can potentially be close to 100% in unquenched samples [71].

7.3
Inductively Coupled Plasma Mass Spectrometry

Inductively coupled plasma mass spectrometry (ICP-MS) provides a rapid and sensitive technique for determining long-lived radionuclides and stable isotopes. The principal use of the ICP-MS is that of a mass spectrometer—to detect the mass of elements according to their mass to charge ratio (m/z).

As the analytes enter the plasma, they are stripped off to their ionic form. This is only possible due to the high temperature of the plasma, which goes up to approximately 6000–10 000 K [72]. There are different types of mass separation devices, which can be used in a mass spectrometer. The most common ICP mass spectrometers are quadrupole-based instruments (ICP-QMS), which allow isotope ratio measurements with a precision for short-term isotope ration measurements from 0.1% to 0.5% relative standard deviation [73]. The quadrupole consists of four parallel conducting rods of which controlled potentials are applied to. This enables only ions of specific masses/charges to go through. The positively charged ions will be attracted to the negatively charged dynode of the detector and the number of ions is displayed as count per second (cps).

The achievable lowest limit of detection (LLD) of an ICP-MS for the determination of uranium isotopes is about 0.1 ng/l – without further sample preparation and sample concentration, respectively.

Major advantages of ICP-MS are the small sample sizes, high sample throughputs, and short measurement times with only few sample preparation steps [74]. Not only that, ICP-MS offers efficient and sensitive multi-elemental analysis in trace or ultra-trace levels [73]. However, the instrument itself is rather costly and it requires high maintenance. High sample matrix concentration may interfere with the determination of low levels of long-lived radioisotopes, but solid-phase extraction may overcome these problems [75].

8
Extent of Radioactivity Levels in the Rhine River

This chapter deals with a presentation of the contents of natural and artificial radionuclides in the Rhine and its main tributaries, respectively. There are different sources and sinks for the concentrations of natural and artificial radionuclides in the River Rhine, as introduced in Fig. 4.

Significant contaminations of the environment from artificial radionuclides were influenced by two historical events: The nuclear weapons tests and the Chernobyl accident. This was reflected in two different periods, where the concentration of ^{137}Cs evidently increased, especially in sediment samples, first in 1955 and second later in 1986 [25]. These periods are illustrated in Fig. 5, which shows the depth profile of the ^{137}Cs values in a sediment nucleus from the Lake Constance. The share of ^{137}Cs originated from the nuclear bomb tests is detectable in a depth of about 12 cm from the surface of the lake, whereas the intake from the Chernobyl accident is clearly separated from this in a depth of about 2 to 3 cm [29].

Today, the artificial radioactivity in European rivers is still affected from historical events besides the influence of nuclear power plants and other facilities in which radioactivity is handled. Other common artificial radionuclides are tritium, ^{60}Co, ^{90}Sr, and ^{131}I [13].

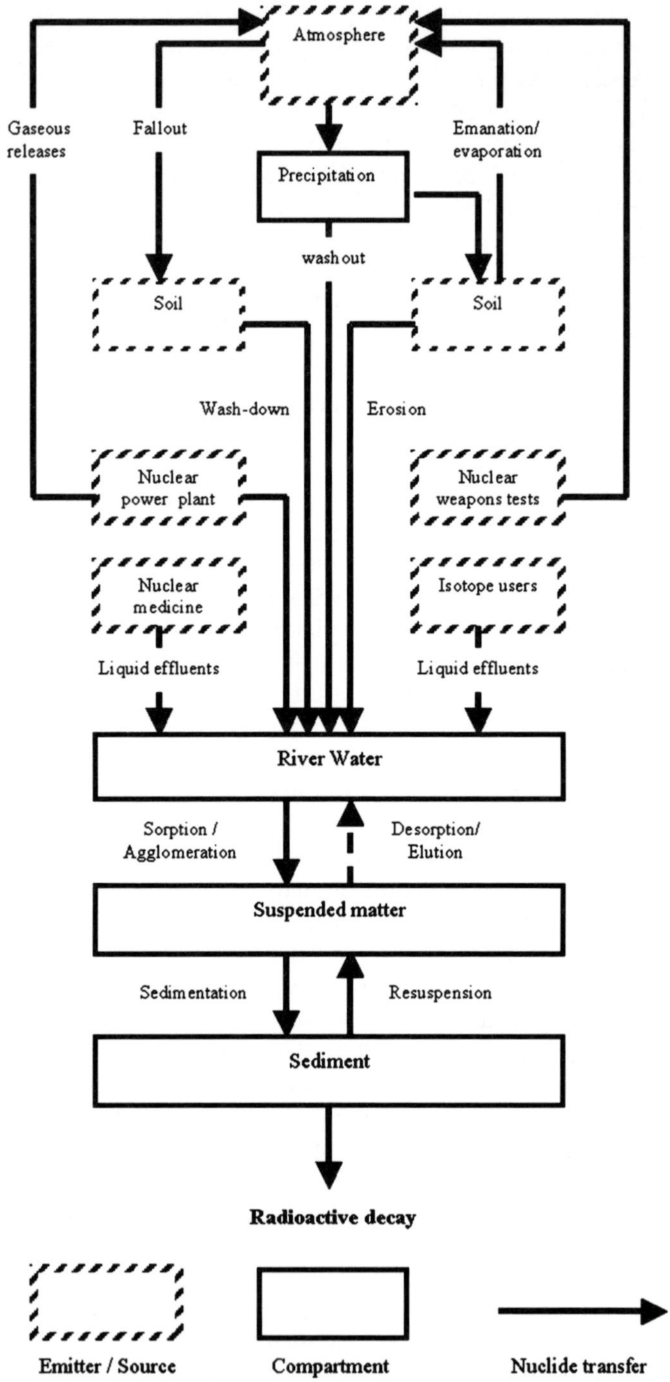

Fig. 4 Main pathways of radioactive substance to the Rhine River (modified after [76])

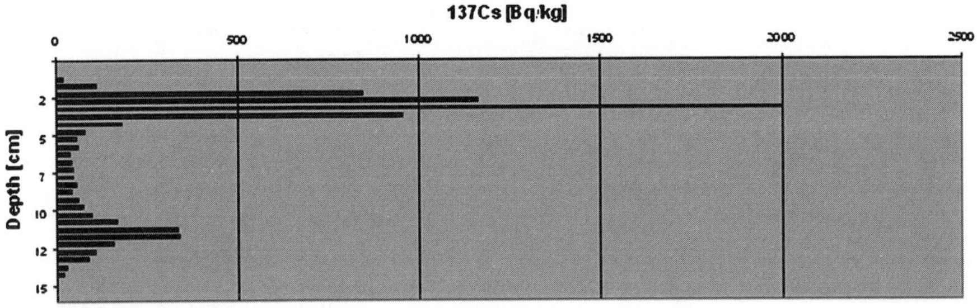

Fig. 5 Depth profile of the ^{137}Cs values [Bq/kg dry matter] in a sediment core from the Lake Constance (data from State Institute for Environmental Protection Baden-Württemberg [29])

8.1 Levels of Natural Radioactivity

The concentrations of natural radionuclides in the Rhine are usually very low compared to the levels of natural radioisotopes in ground- and mineral waters. This is mainly due to the dilution of concentrations by the river water and tributaries. Natural radionuclides in the river water stem from precipitation (rainwater) and the discharge of groundwater into the river under effluent conditions. Furthermore, the weathering of radionuclide-bearing rocks is another important source for the input of natural radioisotopes.

In general, the isotopes of the natural decay series are not highly enriched in water and sediments of the Rhine. Therefore, these isotopes are not regularly monitored by the competent authorities. The only isotope which is usually covered by the official monitoring of water and sediment is ^{40}K. Is has to be mentioned that ^{40}K mainly stems from the weathering of potassium-bearing rocks and minerals and from the usage of potassium-containing fertilizer in agriculture.

Natural radionuclides contribute to the gross-alpha- and gross-beta activity which is regularly monitored by official authorities. It can be assumed that ^{40}K is a major contributor to the gross-beta activity in the Rhine River, mostly in the range of about 0.1–1.0 Bq/l. However, the levels of natural uranium and radium isotopes in the river water are usually only in the range of a few mBq/l. This applies for radon as well, since radon degasses very quickly to the atmosphere.

Natural radionuclides could be determined in sediments from the River Rhine, whereas the concentrations are comparably low as well. It has to be kept in mind that the concentrations of natural radionuclides in a common soil are usually higher than in the river sediments. Figure 6 provides data concerning natural radioisotopes in sediments from the Rhine River as published by the North Rhine-Westphalia Environment Agency, Germany [77]. It has to be acknowledged that the following data is representing the activity concentration with respect to the dry-matter of the sediments.

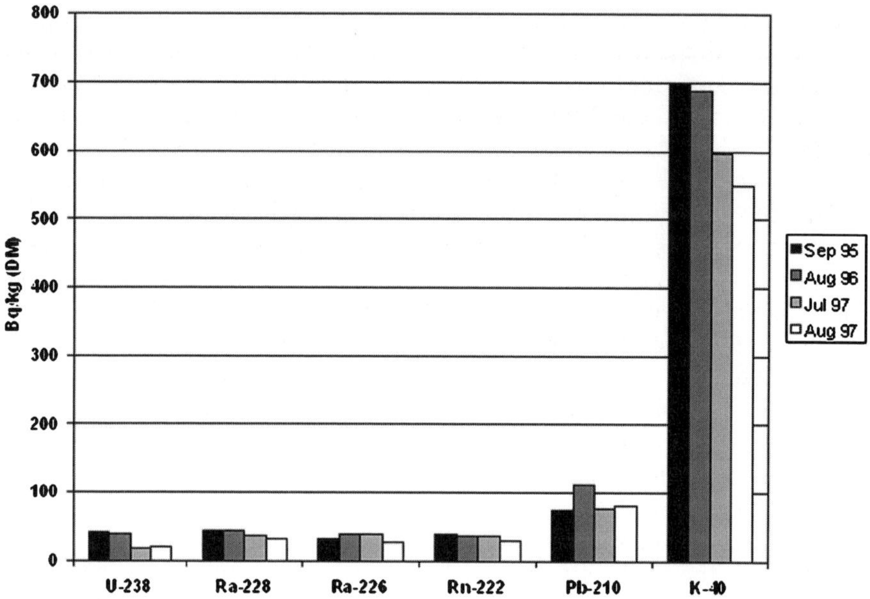

Fig. 6 Natural radionuclides in sediment from Cologne harbor (dry-matter; raw data taken from [77])

It is obvious, that ^{40}K is the most important natural radionuclide in the measured samples. The other natural radioisotopes are approx. 10 times less enriched than ^{40}K. There is a tendency that the activity concentrations decrease from September of 1995 to August of 1997. Since the levels of natural radionuclides in the Rhine are not regularly monitored and their levels are usually very low, the main part of this section deals with the contribution of artificial radioisotopes to the radioactivity of the river.

8.2 Radioactive Contamination Originating From Nuclear Weapons Tests

Following the atmospheric nuclear weapons tests at the end of the 1950s and early 1960s, radioactive fallout was appreciably enhanced all over the world, which lasted for years. Due to the long duration of this fallout, the contamination of the environment was nearly evenly spread. The long-lived radionuclides such as ^{90}Sr and ^{137}Cs were especially dominating, whereas the short-lived isotopes such as ^{131}I were rarely detectable due to the long period of stay of the radioactive pollution in the stratosphere and troposphere [15].

After the above-ground bomb test series, radioactive pollution in the environment differed greatly in terms of the duration of the fallout, the kind of radionuclides, and the extent of the contamination, whereas the Chernobyl accident was a singular event. The deposition of the radiotoxic nuclides ^{90}Sr and plutonium originated from the fallout of the bomb test series were much higher compared with the reactor accident. The whole activity of ^{90}Sr released from nuclear tests amounts to about 6×10^{17} Bq even until today. In Central Europe, on the other hand, the deposition of radioactive cesium after the Chernobyl accident has been in the same order but with greater variations [15].

Until the beginning of the nuclear weapons tests period, the gross beta activity level in European rivers was below 0.4 Bq/l but had increased significantly from 1957 to the middle of 1959 and again from the end of 1961 to 1964. In the Rhine River, the maximum activity concentration was over 1.0 Bq/l. Nevertheless this value is still clearly less than the level in other smaller rivers. The measured activity showed appreciable short-term variations caused by the influence of radioactive contaminated precipitations. Single specific radionuclides were rarely determined, since the analytical equipment was limited at that time [15].

During the bomb test series, large amounts of tritium were released into the atmosphere, resulting in an increase of the tritium activity concentration in the Moselle River, which is the largest tributary of the Rhine, to values of about 100 Bq/l [78, 79]. The mean activity level in terms of tritium measured in the Rhine River were about 20–25 Bq/l in the 1970s and about 10 Bq/l at the beginning of the 1980s followed by a further decrease in the years afterwards. This fact confirmed the assumption that this measured tritium activity concentration was chiefly caused by the bomb tests. Other than the removal of bomb tritium to the oceans, the radioactive decay also reduced the input of tritium [15, 77, 78]. Between 1970 and 1990, tritium concentrations in the Rhine have decreased exponentially, with rate constants up to 0.15 per annum [78].

Today, only very small amounts of ^{90}Sr, ^{137}Cs and tritium originating from these atmospheric nuclear weapon tests are still detectable in the environment, for example in soil and sediments of inshore waters [13].

8.3
Radioactive Contamination Originating From the Accident at the Chernobyl Nuclear Power Reactor

After the Chernobyl accident two periods in terms of radioactive effects were established: The initial period, directly after the event, in which short-time effects occurred where ^{131}I was particularly the most relevant radionuclide at the time. The secondary period is the long period contamination of the sediment which resulted in having ^{137}Cs as the indicator nuclide [19, 22, 24].

Due to the various extent of radioactive contamination, a differentiation of surface water, suspended matter and sediment will be described in the following paragraph. The differences between the determined composition of the radionuclides in these three media may be effected by: The extreme different half-lives of the nuclides, the fractionation during transfer and the distinction of the volatility and solubility of the nuclides [15, 19].

Surface water: The airborne activity caused by the Chernobyl nuclear accident on 26 April 1986 resulted in a pollution of surface water that even reached Central Europe. Most of this contamination originated from wet precipitation ("washout") and only a small part from dry deposition ("fallout") [19, 20]. In samples of rainwater, collected on 3 May 1986 in Koblenz, activity levels up to 3500 Bq/l in terms of ^{131}I and 320 Bq/l in terms of ^{137}Cs were detected. However, the contamination of the rainwater decreased rapidly in the days after [19]. Since the reactor accident, numerous radioactivity measurements on surface water samples were carried out. In Central Europe, the highest contents of artificial radionuclides were observed in the Danube River. In its upper layer, contents of ^{131}I reached up to 351 Bq/l and in terms of ^{137}Cs 25 Bq/l, which were determined at the beginning of May 1986 [15]. After this maximum of the radioactive pollution of the flowing water, a slow decrease followed. Nevertheless the period of this decrease was much faster than the radioactive decay. Therefore, the contamination of a river is dependent primarily on the extent of the fundamental radioactive pollution and the precipitation of radioactive contaminated clouds onto its catchment area, and secondarily on the ratio of precipitation to the draining away and the processes of dispersion and sedimentation [15, 19].

In contrast to the conditions of the Danube River, the contamination of the Rhine was clearly lower. As the first signal of a radioactive pollution, the gross gamma activity measured at the upper Rhine was significantly enhanced on 30 April 1986, because some of the most important radionuclides of the Chernobyl accident, e.g., ^{131}I, ^{137}Cs and ^{103}Ru, were gamma emitters. Since these nuclides emit beta particles as well, the determined gross beta activity in the river also significantly increased, but both gamma and beta activity lingered only for a short time. Since these are only monitoring methods, the kind and the extent of the radionuclides had to be determined by specific nuclide analysis. Since the dilution in the Rhine was higher and the ratio of precipitation to stream flow was favorable, the activity concentration of ^{131}I in surface water of the central part of this river was mostly below 10 Bq/l. The contents of ^{137}Cs, which were amounted onto the river were still minor [15, 19, 23]. Five years after the accident the activity of ^{137}Cs in the river water was mostly less than 0.005 Bq/l [25]. The concentrations of ^{89}Sr and ^{90}Sr were only insignificantly enhanced. In terms of alpha emitters and tritium, no influence to the surface water was detected [15, 19, 23].

No artificial radioactivity was detected in drinking water in most cases, and only very low concentrations of artificial radionuclides were found in some ground waters. An exception was water from cisterns, from reservoirs used for the production of drinking water and from bank filtrate of surface waters, which slightly exceeded the detection limits but only for a short time [15, 80]. Most of the fission nuclides—with the exception of long-lived radionuclides—have decayed completely before reaching the ground water. The function of the suspended matter as a filter and the good retention capacity of the subsoil, that is able to hold back radionuclides and other pollutants, protected the ground water against the enrichment of radioactivity. Nevertheless, the drinking water supply in Central Europe is well protected from radioactive pollution [15].

Suspended matter: The enrichment of radionuclides, which are included in inshore waters, mainly occur in the fraction of the fine grains of suspended matters, that means $\leq 20\,\mu m$ grain size diameter [19]. This contaminated matter can be rapidly transported far away from the area where the radioactive fallout has taken place and then reach the sedimentation in still water areas, for example harbors, branches, and barrages. Due to its high mobility and good affinity to radionuclides, the suspended matter can be called a "short-time monitor" and can express the actual contamination level of inshore waters [13, 19, 23].

After the Chernobyl accident, the enhancement of the radioactive contamination of the suspended matter is comparable to that of the surface water in the Rhine River, but nevertheless with an obviously higher extent. First, the content of ^{137}Cs increased up to two orders of magnitude higher than the former background level and afterwards decreased continuously. In May 1986, the maximum activity level in terms of ^{137}Cs measured was 5000 Bq/kg in dry matter, and in July 1986, an activity value of 600 Bq/kg was determined in dry matter [15, 19]. It was also remarkable that the contamination of the suspended matter was decreasing along the river course from Weil at the Swiss-German border to Bockum, north of Cologne. This reflects the general gradient of the radioactive pollution from southeast to northwest in Central Europe after the Chernobyl accident, which was confirmed by the determination of a significantly higher content of cesium in the Danube River [19].

Five years after the accident, the content of ^{137}Cs in the Rhine was below 100 Bq/kg. The suspended matter phase seems to be the most suitable medium to assess the discharges of radionuclides from various sources into the river [25].

Sediment: In contrast to soil pollution, the content of radionuclides in flowing surface water is permanently variable due to the processes of transport and dispersion in the three media: surface water, suspended matter and sediment. The contamination of suspended matter by radionuclides is a much faster process than the exchange between the suspended matter and the sediment. The rapid transport of the river water causes a pollution of sediments in areas originally not exposed to the radioactive washout or fallout. Therefore after the Chernobyl accident, no good correlation between contamination in suspended matter and sediment collected at the same place was observed [19, 23]. The sediment can be called a "long-time monitor", because it reflects the radioactive pollution of rivers and lakes in the long term [13, 19, 23].

The highest activity level of ^{137}Cs was also detected in the River Danube, which was measured to be up to 7000 Bq/kg in dry matter. High accumulations up to 6000 Bq/kg in dry matter also occurred in the Moselle River.

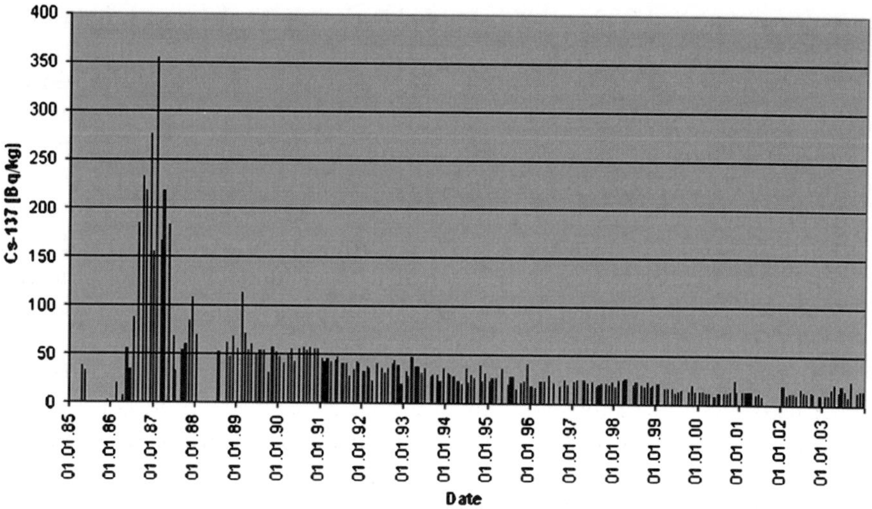

Fig. 7 Content of ^{137}Cs [Bq/kg dry matter] in sediment samples originated from Mainz, (496 km from the Rhine River) in 1985 until 2003 (data from Landesamt für Wasserwirtschaft Rheinland-Pfalz, Germany [81, 82])

In contrast to these values the contamination in the Rhine amounted significantly lower [19, 80]. These values are remarkable because of the fact that the maximum activity level was determined some months after the Chernobyl accident [19].

As mentioned earlier, ^{137}Cs is the key radionuclide that represents the long-term effects of the Chernobyl accident. A typical diagram demonstrating the development of the radioactivity is illustrated in Fig. 7, showing the decrease of the specific content of cesium 137 in the years of 1985 until 2003 measured in sediment samples from Mainz (496 km from the Rhine River). After the Chernobyl accident, the content of ^{137}Cs increased slowly until a maximum value of 355 Bq/kg in dry matter was measured in February 1987, 10 months after the event. This value then decreased extremely in the beginning and much slower later. During the latest years, a value less than 20 Bq/kg in dry matter was determined. This figure obviously demonstrates the influence of the Chernobyl accident to the content of ^{137}Cs in the Rhine [81, 82]. It is also noteworthy the fact that the activity of ^{137}Cs in sediment samples is often fluctuating considering the heterogeneous dispersion of the sediment phase [25].

8.4 Radioactive Contamination Originating From Nuclear Power Plants, Nuclear Facilities and Other Sources

In Austria, Switzerland, France, Germany and in the Netherlands, the contamination of the aquatic systems with artificial radionuclides and the radiation doses resulted from anthropogenic sources always are clearly below the permitted limits [12, 13, 30, 68, 69].

Relatively high levels of artificial radionuclides are emitted by liquid effluents of various sources. Table 4 provides a compilation of the annual discharges of fission and activation products and tritium by nuclear power plants into the Rhine in the years of 2000 and 2001 [12, 13, 29].

No artificial radionuclides have been detected in the Rhine River, upstream of nuclear facilities from the confluence of Aare River in Switzerland and Austria [12, 30]. The exception is ^{137}Cs, which has been remobilized from the Chernobyl accident and can occasionally be determined in very low concentrations [12].

In the last years, the mean activity level in terms of ^{137}Cs and ^{90}Sr has been below 0.01 Bq/L for Swiss and German surface waters in general [12, 13, 29, 77, 81, 83, 84]. The most important artificial radionuclide in suspended matter and sediment of inshore waters has been ^{137}Cs, which activity level has been determined up to 100 Bq/kg in dry matter [13]. In the Rhine River, the values of ^{137}Cs has the amount of mainly below 25 Bq/kg in dry matter [13, 29, 69, 81]. Low activity level of the ubiquitous ^{137}Cs have also been determined in effluents from garbage dumps and incinerating plants. The

Table 4 Compilation of the annual discharges of fission and activation products and tritium by nuclear power plants into the Rhine in the years of 2000 and 2001 [12, 13, 29]

Nuclear Power Plant	Year	Fission and Activation Products [Bq]	Tritium [Bq]
Leibstadt	2000	$< 2.0 \times 10^8$	$\sim 2.0 \times 10^{12}$
	2001	$< 2.0 \times 10^8$	1.1×10^{12}
	2002	2.0×10^8	1.6×10^{12}
Fessenheim	2000	$\sim 4.0 \times 10^9$	$\sim 4.0 \times 10^{13}$
	2001	$\sim 2.0 \times 10^9$	$\sim 3.0 \times 10^{13}$
Philippsburg 1	2000	1.6×10^8	4.8×10^{11}
	2001	1.3×10^8	6.5×10^{11}
Philippsburg 2	2000	3.5×10^8	1.8×10^{13}
	2001	4.9×10^8	1.3×10^{13}
Biblis block A	2000	1.6×10^8	1.6×10^{13}
	2001	9.6×10^7	7.7×10^{12}
Biblis block B	2000	2.0×10^8	1.5×10^{13}
	2001	2.6×10^8	1.1×10^{13}
Mülheim-Kärlich*	2000	8.1×10^6	1.1×10^{11}
(* not in operation)	2001	1.1×10^7	5.3×10^9

^{137}Cs activity in effluents from sewage works has mostly been below the detection limit, which is claimed to be 0.1 Bq/l [13, 29].

The content of ^{137}Cs in Lake Laach was quite remarkable, where approx 0.04 Bq/l in surface water and approx 100 Bq/kg in dry matter in sediment were measured in the past years. This lake is situated northwest of Koblenz and less than 10 km away from the Rhine, but without a water connection to the river. The activity of ^{137}Cs is probably the result of the Chernobyl accident and the fact that there is only a small water exchange between other water bodies [81, 83].

Isolated traces of ^{131}I that resulted from medical applications in sludge and effluents from sewage works were determined [12]. In Germany, the general median values in terms of ^{131}I amounted to approx 0.2 Bq/l in waste water from sewage works and approx 40 Bq/kg in dry sludge in the last years. In the rivers, a level is resulted up to 0.1 Bq/l in surface water and 50 Bq/kg in suspended matter [13].

In the Swiss part of the Rhine, downstream of the mouth of the Aare River, traces of ^{58}Co and ^{60}Co were detected in surface water. These isotopes obviously originated from Swiss nuclear power plants. Levels of ^{137}Cs (up to 22 Bq/kg), ^{60}Co (up to 56 Bq/kg), ^{58}Co (up to 4 Bq/kg) and ^{54}Mn (up to 6 Bq/kg) were detected in sediment samples [12]. The nuclides ^{58}Co and ^{60}Co originating from the discharges of nuclear power facilities are also detectable in the German and French part of the Rhine but only in traces [13, 29, 85]. These discharges of ^{58}Co and ^{60}Co can be used to register the particulate nuclide transport in flowing waters and to describe their propagation and sedimentation behaviors, whereas by means of tritium, the propagation behavior within the surface water can be investigated [17].

During the last years, the mean annual content of tritium in the Rhine River has been below 5 Bq/l [12, 13, 30, 69]. A tritium level of 8 Bq/l has been determined in the Rhine in April 2002, downstream of Basel caused by the nuclear power plant in Gösgen. However, this level was still within its permitted limits, which amounts to 12 000 Bq/kg [12]. In 2002, the maximum tritium level in the liquid effluents from the Philippsburg nuclear power plant amounted to 130 Bq/l from block II and 40 Bq/l from block I [85].

An exception constituted the tritium levels in samples from individual water supplies near by an old branch of the Rhine in the vicinity of the Research Centre Karlsruhe, which were 230 Bq/l (1998), 140 Bq/l (1999), 74 Bq/l (2000), and 78 Bq/l (2001) [13].

Local emitters in the industrialized regions, especially the clock and watch manufacturers, and the incineration of waste containing tritium has caused a detectable value of tritium in some rivers. However, the contamination with tritium has been decreased in the last years. The reasons behind is that on the one hand there is a declined use of tritium as fluorescent colors and on the other hand waste containing tritium has been collected [12, 29].

Since tritium in form of ^{3}HHO behaves like water, it is an optimal key substance to investigate the dispersion of solved pollutants in running water. Investigations at the Rhine showed that the tritium levels decreased in an exponential function in the course of the river. After a flowing distance of 600 km the tritium levels amounted to about 60% of the initial emission [86].

Figure 8 demonstrates the tritium level in the Rhine, to be exact from Koblenz in the Middle Rhine area and Bimmen at the German/Dutch border, from 1981 to 2001 [10].

As a comparison, Fig. 8 also shows the development of the tritium contents in water samples collected at two locations at the Moselle River that attributes to the Rhine [10, 81]. The background level resulted from the nuclear bomb tests and the cosmic tritium is also shown from 1981 until 1986. As the French nuclear power plant Cattenom went into operation between the end of 1986 to 1991, the content of tritium originating from the reactor waste had increased appreciably in the river water [23, 81, 87]. However, the radiological toxicity of tritium is relatively low (refer to Sect. 5.1). The tritium activity decreased

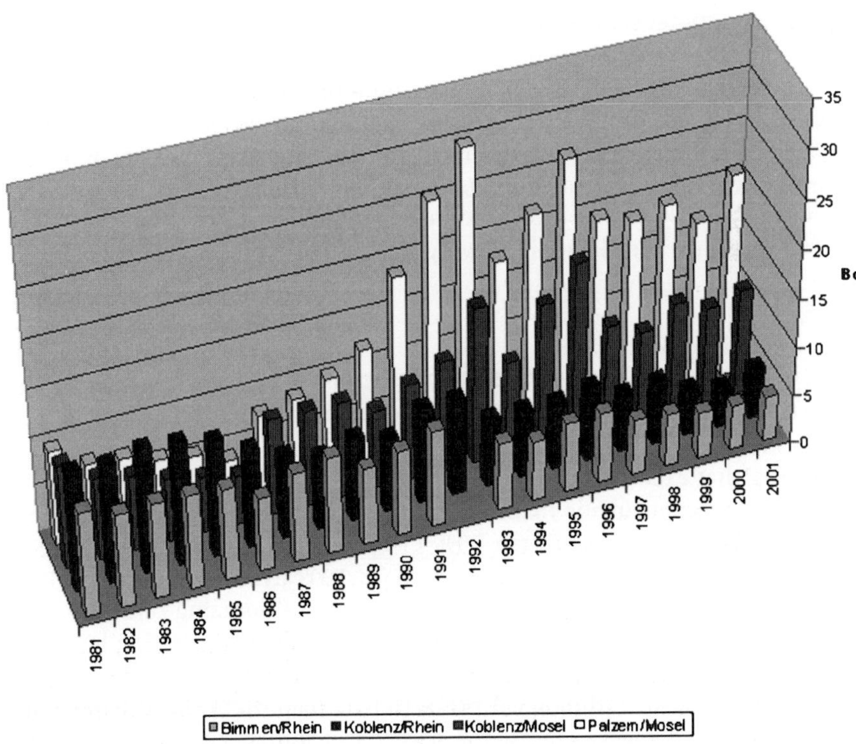

Fig. 8 Tritium levels in the Rhine and the Moselle Rivers from 1981 to 2001 (data from Federal Institute of Hydrology and Landesamt für Wasserwirtschaft Rheinland-Pfalz [10, 81])

in the course of the Moselle River, but is still detectable in Koblenz, where the Moselle River attributes to the Rhine [88]. This is obviously illustrated in Fig. 8 by comparison of both levels in Koblenz.

Nevertheless, the artificial emission of the nuclear power plant in Cattenom is surpassed by the input of natural radionuclides, as an investigation showed in suspended matter [79]. The discharges of the artificial ^{60}Co and ^{137}Cs by this nuclear facility amounted to 2.6 and 1.0 GBq/year in 1992, against the annual input of the natural ^{7}Be and ^{40}K were approximately 21 and 130 GBq/year. Nevertheless, the extent of the emission of ^{137}Cs by the nuclear power plant in Cattenom was clearly exceeded by the preload from the Chernobyl accident and in lower extent from the atmospheric weapons tests [79].

The aim of a long time study, which was carried out in the Middle Rhine area, was to investigate the distribution of the stable and radioactive isotopes of cobalt and cesium in the system surface water/suspended matter and surface water/sediment, respectively. In comparison to the artificial radionuclides ^{60}Co and ^{137}Cs, the naturally occurring stable isotopes ^{59}Co and ^{133}Cs are more firmly bound in the solid phase, whereas the distribution coefficients of the two systems are quite comparable. These distribution coefficients are important parameters for the estimation of the radiation exposure caused by the pathway "exposure on spoil fields", as discussed in the following chapter [89].

In the routine monitoring programs, fishes were also analyzed, which came from the rivers especially from the surrounding of nuclear facilities. Only traces of ^{137}Cs (meanly 0.2 Bq/kg) that originated from the Chernobyl accident and ^{90}Sr (below 0.05 Bq/kg) from the above-ground nuclear weapons tests were able to be determined [13, 29, 90]. The activity level of ^{137}Cs in 2001 was only about 2% of that measured in 1986 [77]. The activity concentration of ^{137}Cs was one range lower in flowing waters than in lakes [13].

9
Effect of Natural and Artificial Radioactivity in the Rhine River to the Population (Exposure to Man)

Everything that exists on this planet is exposed to a variety of radiation. Human radiation exposure is mainly dominated by natural radiation (Fig. 9). Radon contributes to more than 50% of the annual radiation exposure of humans. Other forms of radiations such as terrestrial radiation and cosmic radiation are other important sources of natural ionizing radiation. The global average of the effective dose attributed to natural radiation has been estimated to be about 2.4 mSv per year [2].

The main goal of radiological protection is to protect the general public and exposed persons against the hazards of ionizing radiation. Among other activities, the institutions, which have the task of performing radiation protection, provide information on the mechanisms of biological effects of radiation, behavior of radionuclides in environment and the improvement of the dosimetric and monitoring methods. One of the most important regulations of the radiation protection is the minimization rule. This means that everyone who has to handle radioactive material should try everything possible in order to minimize the radiation exposure of human and the environment [16, 21].

The radioactive contamination of inshore waters can result as a radiation exposure to man in different pathways. In terms of a river: the exposure on spoil fields that is originated from sediment, the drinking water pathway and the exposure, which results from ingestion of radioactive substances by food influenced from the river water via the following pathways: water – fish, cattle drinking trough – cattle – milk or meat, irrigation – plant (– cattle – milk or meat) [48]. An unusual contamination of the inshore waters with natural or artificial radionuclides has to be supervised and the concerned group of people has to be protected [15].

One of the most prominent aims of the Euratom Treaty is the radiation protection of the population in the European Union. In the recommendation of 96/29/EURATOM, the limits for effective dose are laid down to 1 mSv per annum for members of the public and 20 mSv for people exposed to radiation as a result of their occupation [91]. The EU member states had to transfer this recommendation into their own national radiation protection laws with limitation values, to be exact maximal effective or organic doses which should not be exceeded in a certain time period. The Swiss Federal Radiological Protection Ordinance lays down quite the same maximum figures as in this EU recommendation. It also establishes limit values for ambient concentrations in the atmosphere and water in places accessible to the public [12]. In the

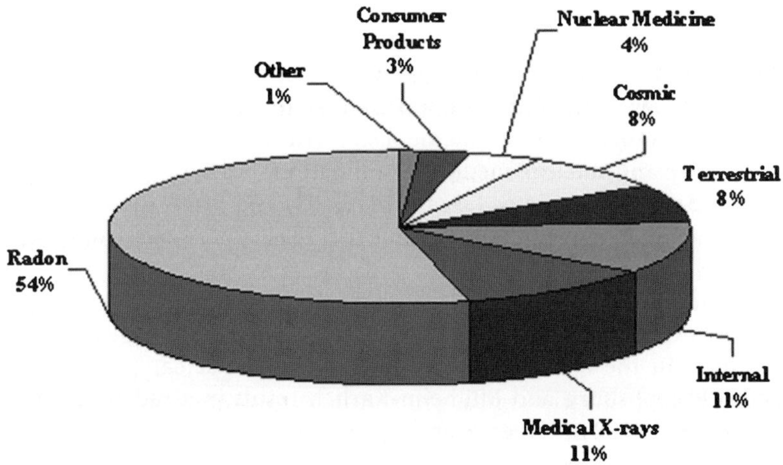

Fig. 9 Human radiation exposure (data taken from [2])

German Radiation Protection Ordinance, the maximum permitted effective dose level caused by draw out of radioactive substances in effluents or used air amounts to 0.3 mSv per year and person [48].

Except for natural radiation, inhabitants of Central Europe are exposed to an artificial radiation mainly caused due to the following applications described with the estimated annual doses, which include the radiation exposure of radionuclides deposited on earth, the inhalation and the ingestion of contaminated foodstuffs:

- The radiation exposure of the population in Central Europe caused by the fallout of the **above-ground nuclear weapons tests** resulted onto an effective doses of approx. 1.6 mSv in the period from 1960 until 1970 and approx. 0.4 mSv from 1970 until 2050. Therefore the actual dose amounts to about 0.005 mSv per year and person. That is only a small fraction of the legal allowed annual doses [16]. It has to noted that the ^{90}Sr values which originated mainly from the time of the nuclear weapons tests might be overestimated [15].
- In Central Europe, the effect of the **Chernobyl accident** actually cause an effective annual dose of approximately < 0.015 mSv/year, in comparison to 0.011 mSv in 1986 [46]. The component of the radiation exposure via rivers and lakes amounted to only 1 or 2 percent of the total doses resulting from the Chernobyl accident [22]. Directly after that event, the highest pollution had originated mainly from the ingestion of fresh vegetables exposed to the fallout, and also from milk and dairy products [28]. Even in 1991, the radiation exposure on the drinking water pathway resulted by the Chernobyl accident was lower than that of the natural ^{40}K and nearly negligible in comparison to the maximum permitted dose level. In contrast to this, a higher effect was determined from the so-called "exposure on spoil fields", which is the most sensitive exposure pathway of the aquatic system. On the base of the highest concentrations measured in samples of the River Danube, a theoretical radiation exposure of about 30% of the legal adopted value was estimated [25].
- The use of **radiation in medicine**, i.e., medical application for diagnostic or therapy and nuclear medical examinations, cause an effective annual dose of 2 mSv in Germany, 1.3 mSv in Austria and 1 mSv in Switzerland for every human in average. It is obvious that these doses vary from person to person [12, 13, 30, 46].
- In Germany, the regular operation of **nuclear power plants** results in a radiation exposure to an adult and an one-year-old child below 0.01 mSv per year. In the last years, the effluents of the nuclear power plants in Biblis, Philippsburg and Mülheim-Kärlich result in a radiation exposure of each ≤ 0.0002 mSv per year [13, 46]. In Switzerland, an annual dose resulted from effluents of nuclear power plants to < 0.001 mSv in the year 2002 [12]. Nevertheless, a coal-fired power station releases higher

levels of radioactive substances to the environmental than a trouble-free working nuclear power plant and cause an annual dose of approx. 0.002 mSv [16].

- The use of **ionizing rays in science, technique and household** cause an effective annual dose of approximately ≤ 0.01 mSv/year [16, 46].
- The **radiation exposure resulting from aircraft traffic**, which depends highly on the frequency and the kind of the flights, is estimated for a Central European to be in average 0.005 mSv per annum [16]. A flight from Frankfurt to New York would have an effective dose of 0.032 to 0.075 mSv dependent on the sun cycle, flight height and air-route [92].
- Due to its character, the **radiation exposure at work** is heterogeneous and cause a mean annual dose of approx. 0.03 mSv/year [21] to approx. 0.05 mSv/year [30].

Water–fish pathway: The dose, which resulted from the consumption of fish, is only a very small part of the whole radiation exposure [29, 85].

Drinking water pathway with the following assumption: Activity level of tritium 5 Bq/l, ^{90}Sr 0.005 Bq/l, ^{137}Cs 0.005 Bq/l in the Rhine and annual consumption of not purified drinking water 700 l would result in a theoretical annual dose of approx 0.21 μSv/year for an adult. In comparison to that, the annual dose for an adult caused by natural ^{40}K amounts to approx 1.7 μSv/year [13]. Thus, the exposition path of drinking water may nearly be neglected compared with other exposition routes [13, 93].

Pathway exposure on spoil fields with an analogous assumption: Activity level of ^{137}Cs 20 Bq/kg sediment of the Rhine River and annual length of stay 1000 hours would result in an annual dose of approx 2 μSv/year for an adult. Activity level of ^{60}Co 34.5 Bq/kg in direct effluents from a nuclear power plant, and an annual length of stay 1000 hours would cause an annual dose of approx 11 μSv/year for an adult. These values are still minor in comparison with the exposure caused by natural radioactivity [13].

Acknowledgements The authors would like to thank the colleagues who contributed to this article by permitting the use of their materials, such as results, graphs and figures. Furthermore, the tremendous support of Mrs. Marcella Liem during the conclusion of this paper is gratefully appreciated.

References

1. Lieser KH (2001) Nuclear and radiochemistry: fundamentals and applications, 2nd edn. Wiley, Berlin
2. UNSCEAR (2000) Sources and effects of ionizing radiation. United Nations Scientific Committee on the Effects of Atomic Radiation. United Nations Publication, New York
3. Siehl A (1996) Umweltradioaktivität. Verlag Ernst & Sohn, Berlin

4. Gascoyne M (1992) Geochemistry of the actinides and their daughters. In: Ivanovich M, Harmon S (eds) Uranium series disequlibrium. Oxford Science Publications, 2nd edn. p 34
5. Bleise A, Danesi PR, Burkart W (2003) Properties, use and health effects of depleted uranium (DU): a general overview. J Environ Radioact 64:93
6. Wedepohl KH (1995) The composition of the continental crust. Geochim et Cosmochim Acta 59:1217
7. Wahl W (2001) Radionuclide handbook for laboratory workers in spectrometry, radiation protection and medicine. Schliersee (Germany)
8. World Nuclear Association (2002) Internet publication. http://www.world-nuclear.org/info/info.htm
9. Webelements.com (2003) Thorium key information. Internet publication, http://www.webelements.com/webelements/elements/text/Th/key.html
10. International Commission for the Protection of the Rhine against Pollution (ICPR) (2004) Internet publication, http://www.iksr.org, data from http://had.bafg.de/iksr-zt
11. International Atomic Energy Agency (IAEA) (2004) Nuclear power plants information. Internet publication, http://www.iaea.org/programmes/a2/index.html
12. Swiss Federal Office of Public Health (ed) (2003) Environmental radioactivity and radiation exposure in Switzerland 2002. Fribourg
13. Trugenberger-Schnabel A, Peter J, Kanzliwius R, Bernhard C, Bergler I (ed) (2003) Environmental radioactivity in the Federal Republic of Germany 1998 until 2001 (in German). Federal Office of Radiation Protection (BfS), Salzgitter
14. Directorate General for Nuclear Safety and Radiation Protection (DGSNR) (ed) (2003) Nuclear Safety in France 2002, Paris
15. Haberer K (1989) Umweltradioaktivität und Trinkwasserversorgung, 1st edn. (in German). R. Oldenbourg Verlag, Munich, Vienna
16. Volkmer M (2003) Radioaktivität und Strahlenschutz, 2nd edn., (in German). Informationskreis KernEnergie, Berlin
17. Mundschenk H, Krause W (1993) Study of the long-range effects of radioactive effluents from nuclear power plants in the case of the River Rhine. Environmental Impact of Nuclear Installation, p 277
18. U.S. Environmental Protection Agency (EPA) (2004) Internet publication, http://www.epa.gov/radiation/radionuclides and http://www.epa.gov/radiation/understand
19. Mundschenk H, Wengler P (1987) Der Unfall im Kernkraftwerk Tschernobyl und seine Aus- und Nachwirkungen in deutschen Oberflächengewässern (in German). Federal Institute of Hydrology, Koblenz
20. Swiss Federal Office of Public Health (ed) (1996) Environmental radioactivity and radiation exposure in Switzerland 1995. Fribourg
21. Grupen C (2003) Grundkurs Strahlenschutz, 3rd edn, (in German). Springer, Berlin Heidelberg New York
22. Nuclear Energy Agency (2002) Chernobyl: assessment of radiological and health impact 2002 Update of Chernobyl: Ten years on. Internet publication, http://www.nea.fr/html/rp/chernobyl
23. Mundschenk H, Krause WJ, Dersch G, Wengler P (1994) Überwachung der Bundeswasserstraßen auf radioaktive Stoffe im Normal- und Ereignisfall – Konzept, Methoden und Ergebnisse (in German). Report BfG-0783, Federal Institute of Hydrology, Koblenz
24. Ministerium für Umwelt und Gesundheit Rheinland-Pfalz (ed) (1986) Informationen zu den Auswirkungen des sowjetischen Reaktorunfalls im Land Rheinland-Pfalz (in German). Mainz (additional 1st–6th supplement, 1986 until 1988)

25. Mundschenk H (1992) On long-term effects of the accident of the nuclear power plant in Chernobyl in the German Federal main waterways (in German). DGM 36:7
26. Radiation Protection Commission (SSK) (2002) Anwendung von ^{131}I in der Nuklearmedizin (in German). Internet publication, http://www.ssk.de
27. Erdtmann G, Soyka W (1979) The gamma rays of the radionuclides, 1st edn. Verlag Chemie, Weinheim, New York
28. Roempp Online. Internet publication, http://www.roempp.com
29. State Institute for Environmental Protection Baden-Württemberg (LfU) (ed) (2003) Radioaktivität in Baden-Württemberg, Jahresbericht 1998–2001 (in German). Karlsruhe
30. Österreichisches Bundesministerium für Land- und Forstwirtschaft, Umwelt und Wasserwirtschaft (Lebensministerium)/Österreichisches Bundesministerium für soziale Sicherheit und Generationen (ed) (2002) Radioaktivitätsmessungen in Österreich 2001, Daten und Bewertung (in German). Vienna, Austria
31. Federal Ministry for the Environment: Nature Conservation and Nuclear Safety (BMU) (ed) (2001) Messanleitungen für die Überwachung der Radioaktivität in der Umwelt und zur Erfassung radioaktiver Emissionen aus kerntechnischen Anlagen (Loose-leaf edition, in German). Urban & Fischer, Munich Jena
32. Faure K (1986) Principles of isotope geology, 2nd edn. Wiley, New York
33. Trautmann N, Passler G, Wendt KDA (2004) Ultratrace analysis and isotope ratio measurements of long-lived radioisotopes by resonance ionization mass spectrometry (RIMS). Anal Bioanal Chem 378:348
34. Federal Office for Radiation Protection (BfS) (2001) Wie gefährlich ist Plutonium? (in German). Internet publication, http://www.bfs.de
35. Euratom-Treaty – Treaty on the founding of the European Atomic Energy Community (Euratom) of 25 March 1957, Federal Law Gazette II 1957, pp 1014, 1679
36. Bayer A (ed) (1998) The monitoring of radioactive contamination and radiation exposure in the environment in Germany – tasks, techniques, realizations. Federal Office for Radiation Protection (BfS), ISH-IB-7-E, Neuherberg, Germany
37. De Cort M, de Vries G, Petit P (2001) Overview of EC projects related to European-wide exchange of radiological information during nuclear emergencies. In: Bayer A, Leonardi A (ed) Measurement and calculation in nuclear emergency management. Cologne, p 269
38. IAEA – Act on the agreement of 26 September 1986, for the early notification in the case of nuclear accidents and the provision of assistance in the case of these or radiological emergencies occurring (Act on the IAEA notification agreement and on the IAEO provision of assistance agreement) (in German). Federal Law Gazette Part II 1989, p 434
39. Decision of the Council of the European Communities of the 14 December, 1987, on the Council Agreement for the Accelerated Exchange of Information in the Case of a Radiological Emergency Situation (87/600 Euratom). Off J Eur Commun, 30.12.87, No L 371/76
40. Federal Ministry for the Environment: Nature Conservation and Nuclear Safety (BMU) (ed) (2002) Umweltpolitik: Umweltradioaktivität und Strahlenbelastung im Jahr 2001 (in German). Berlin, Germany, General requirements for the competence of testing and calibration laboratories (Trilingual version EN ISO/IEC 17025: 2000)
41. European Commission, Joint Research Centre (JRC) (2001) Radioactivity Environmental Monitoring (REM). Internet publication, http://java.ei.jrc.it
42. Bayer A (ed) (2001) The monitoring of radioactive contamination and radiation exposure in the environment in Germany – tasks, techniques, realizations (German Version). Federal Office for Radiation Protection (BfS), SH-IB-3, Neuherberg, Germany

43. De Cort M, de Vries G, Galmarini St (2003) European radiological emergency information exchange systems developed at EC-JRC/IES: current status and future developments. In: Federal Ministry for the Environment: Nature Conservation and Nuclear Safety (BMU) (ed) 12th Professional discussion for the monitoring of the environmental radioactivity. Bonn, p 617
44. Federal Office for Radiation Protection (BfS) (2003) International Co-operation and Committees. Internet publication, http://www.bfs.de/bfs/zusammenarbeit.html
45. 2000/473/Euratom: Commission Recommendation of 8 June 2000 on the application of article 36 of the Euratom Treaty concerning the monitoring of the levels of radioactivity in the environment for the purpose of assessing the exposure of the population as a whole (27/07/2000). Off J Eur Commun L 191, p 37
46. Federal Ministry for the Environment: Nature Conservation and Nuclear Safety (BMU) (ed) (2000) Die Leitstellen zur Überwachung der Umweltradioaktivität – Historie, Aufgaben und Perspektiven (in German). Berlin, Germany
47. German Atomic Energy Act (1985) (Act on the peaceful use of nuclear power and the protection against its dangers), revised version from 15/07/1985, Federal Law Gazette, Part I, p 1565
48. German Precautionary Radiation Protection Ordinance (Ordinance on the precautionary protection against damage caused by ionizing radiation) (2001) Federal Law Gazette, Part I, p 1714; (2002) p 1459
49. Guidelines of Emissions and Immissions Monitoring for Nuclear Facilities (REI) (1993) German Joint Ministerial Bull 29:502
50. Ministerium für Umwelt und Forsten Rheinland-Pfalz (2004) Technischer Umweltschutz – KKW Mülheim-Kärlich (in German). Internet publication, http://www.muf.rlp.de/index2.asp?bereich=106
51. Bayer A (ed) (1997) The Precautionary Radiation Protection Act and its regulations – Monitoring of environmental radioactivity and precautionary measures, Federal Office for Radiation Protection (BfS), ISH-IB-3-REV-3-E, Neuherberg, Germany
52. German Precautionary Radiation Protection Act (Act on the precautionary protection of the population against radiation exposure, StrVG). Federal Law Gazette, Part I, p 2610 (1986)
53. Guidelines for the monitoring of radioactivity in the environment, in accordance with the Precautionary Radiation Protection Act, Part I: measuring programme for routine operation (Routine measuring programme). Joint Ministerial Bull 45:929 (1994)
54. Guidelines for the monitoring of radioactivity in the environment, in accordance with the Precautionary Radiation Protection Act, Part II: measuring programme for intensive operation (Intensive measuring programme). Joint Ministerial Bull 49:261 (1995)
55. Federal Ministry for the Environment: Nature Conservation and Nuclear Safety (BMU) (ed) (1999) Übersicht über Maßnahmen zur Verringerung der Strahlenexposition nach Ereignissen mit nicht unerheblichen radiologischen Auswirkungen (Maßnahmenkatalog) (in German). Berlin, Germany
56. Radiation Protection Commission (SSK) (1995) Leitfaden für den Fachberater Strahlenschutz der Katastrophenschutzleitung bei kerntechnischen Unfällen, vol 13, 2nd edn. (in German)
57. Schnadt H (2003) Actual state of the French–German model and its usage in the in the revision of the guide for the radiation protection expert of the emergency staff. In: Federal Ministry for the Environment: Nature Conservation and Nuclear Safety (BMU) (ed) 12th Professional discussion for the monitoring of the environmental radioactivity. Bonn, p 351

58. Dersch G, Mundschenk H, Wengler P (1993) A stationary monitoring and warning network for the continuous surveillance of the radioactivity in the Federal waterways (in German). DGM 37:127
59. Dersch G, Krämer T, Krause WJ, Wengler P (2003) Concept for the development of the measurements techniques for the radioactivity monitoring network along the Federal waterways by the Federal Institute of Hydrology (BfG) (in German). In: Federal Ministry for the Environment: Nature Conservation and Nuclear Safety (BMU) (ed) 12th Professional discussion for the monitoring of the environmental radioactivity. Bonn, p 61
60. Dersch G, Krause WJ (2001) The surveillance of radioactive substances in inland waters in the case of an incident (in German). In: Bayer A, Leonardi A (ed) Measurement and calculation in nuclear emergency management. Cologne, p 93
61. Austrian Radiation Protection Act from 11/06/1969, Federal Law Gazette 227/1969, revised version from 20/08/2002, Federal Law Gazette 146/2002
62. Henrich E, Hohenberg JK, Haider W, Zimmel G (2001) The new Austrian radiation early warning system (in German). In: Bayer A, Leonardi A (ed) Measurement and calculation in nuclear emergency management. Cologne, p 204
63. Österreichisches Umweltbundesamt (2003) Entwicklung - Das Strahlenfrühwarnsystem im Wandel der Zeit (in German). Internet publication, http://www.umwelt bundesamt.at/umwelt/strahlenfruehwarnsys
64. Österreichisches Bundesministerium für Land- und Forstwirtschaft, Umwelt und Wasserwirtschaft (Lebensministerium) (2004) Atomenergie & Strahlenschutz - Rahmenempfehlungen. Internet publication, http://www.lebensministerium.at/umwelt
65. Swiss Federal Radiological Protection Ordinance (1994). Internet publication, http://www.admin.ch/ch/d/sr/c814_501.html
66. Völkle H (2002) Environmental radioactivity monitoring in Switzerland (in German). In: Michel R, Täschner M, Bayer A (ed) Praxis des Strahlenschutzes: Messen, Modellieren, Dokumentieren. TÜV-Verlag, Cologne, p 159
67. Blättler M (2001) Use of the different dispersion models in Switzerland: overview and principles (in German). In: Bayer A, Leonardi A (ed) Measurement and calculation in nuclear emergency management. Cologne, p 71
68. Institute for Radiation Protection and Nuclear Safety (IRSN) (2003) Annual Report 2002. Internet publication, http://www.irsn.fr
69. Knetsch GF (ed) (2002) Monitoring of radiation in the environment in the Netherlands - Results in 2001. National Institute for Public Health and the Environment, RIVM report 610056048/2002, Bilthoven
70. Nuclear Research and Consultant Group (ed) (2003) Annual report 2002. Internet publication, http://www.nrg-nl.com
71. Knoll GF (2000) Radiation detection and measurement, 3rd edn. Wiley, New York, p 802
72. Ponce de Léon CA, Montes-Bayón M, Caruso JA (2002) Elemental speciation by chromatographic separation with inductively coupled plasma mass spectrometry detection. J Chromatogr 974:1
73. Becker JS, Dietze HJ (2000) Precise and accurate isotope ratio measurements by ICP-MS. Fresenius J Anal Chem 368:23
74. Shen CC, Edwards RL, Cheng H, Dorale JA, Thomas RB, Moran SB, Weinstein SE, Edmonds HN (2002) Uranium and thorium isotopic concentration measurements by magnetic sector inductively coupled plasma mass spectrometry. Chem Geol 185:165
75. Unsworth ER, Cook JM, Hill SJ (2001) Determination of uranium and thorium in natural waters with a high matrix concentration using sold-phase extraction inductively coupled plasma mass spectrometry. Anal Chim Acta 442:141

76. Reprinted from Mundschenk H (1996) Occurrence and behaviour of radionuclides in the Moselle River – Part I: entry of natural and artificial radionuclides. J Environ Radioact 30:199, with permission of Elsevier
77. North Rhine-Westphalia State Environment Agency (LUA NRW) (ed) (2003) Gewässergütebericht 2001 (in German). Essen
78. Mundschenk H, Krause WJ (1991) Behaviour and radiological significance of tritium from nuclear power plants and other sources in the River Rhine Basin. J Environ Radioact 14:341
79. Mundschenk H (1996) Occurrence and behaviour of radionuclides in the Moselle River – Part I: entry of natural and artificial radionuclides. J Environ Radioact 30:199
80. Weller D (1987) Der Reaktorunfall von Tschernobyl in seiner Wirkung auf die rheinland-pfälzische Wasserwirtschaft (in German). Landesamt für Wasserwirtschaft Rheinland-Pfalz (report 203/87), Mainz, Germany
81. Hartkopf J (2003) Radioaktivität in rheinland-pfälzischen Gewässern in den Jahren 2001 und 2002 (in German). Landesamt für Wasserwirtschaft Rheinland-Pfalz (report 211/03), Mainz, Germany
82. Hartkopf J (2004) Landesamt für Wasserwirtschaft Rheinland-Pfalz (unpublished results), Mainz, Germany
83. Hartkopf J (2001) Radioaktivität in rheinland-pfälzischen Gewässern in den Jahren 1986 bis 2000 (in German). Landesamt für Wasserwirtschaft Rheinland-Pfalz (report 205/01), Mainz
84. Hessisches Landesamt für Umwelt und Geologie (ed) (2003) Kontamination der Umwelt mit künstlichen radioaktiven Stoffen (in German). Internet publication, http://atlas.umwelt.hessen.de/atlas/radioakt/kont_txt.htm
85. State Institute for Environmental Protection Baden-Württemberg (LfU) (ed) (2003) Überwachung der baden-württembergischen Umgebung kerntechnischer Anlagen auf Radioaktivität, Jahresbericht 2002 (in German). Karlsruhe
86. Krause WJ, Mundschenk H (1995) Zur Ermittlung von Fließzeiten und Kenndaten zur longitudinalen Dispersion mit Tritium als Tracer als Basis für Ausbreitungsprognosen am Beispiel des Rheins (in German). In: Federal Ministry for the Environment: Nature Conservation and Nuclear Safety (BMU) (ed) 9th professional discussion for the monitoring of the environmental radioactivity. Oberschleißheim, p 425
87. Landesamt für Umweltschutz Saarland/Landesamt für Umweltschutz und Gewerbeaufsicht Rheinland-Pfalz (ed) (2003) Umgebungsüberwachung des frz. KKW Cattenom – Jahresbericht 2002 (in German). Saarbrücken
88. Weller D (1996) Radionuklide in der Mosel (in German). Landesamt für Wasserwirtschaft Rheinland-Pfalz (report 215/96) Mainz, Germany
89. Mundschenk H (1985) On the exchange behaviour of radionuclides in contact with suspended matter/sediment of the River Rhine (in German). DGM 29:4
90. Hessisches Landesamt für Umwelt und Geologie/Landesamt für Umweltschutz und Gewerbeaufsicht Rheinland-Pfalz (ed) (2003) Jahresbericht 2002 der unabhängigen Messstellen zur Umgebungsüberwachung des Kernkraftwerks Biblis. Darmstadt
91. Council Directive 96/29/Euratom of 13 May 1996 laying down basic safety standards for the protection of the health of workers and the general public against the dangers arising from ionizing radiation. Off J Eur Commun L 159 (29/06/1996) p 1
92. Federal Office for Radiation Protection (BfS) (ed) (2003) Annual Report 2002 (in German). Langenhagen
93. Bünger T, Obrikat D, Rühle H, Viertel H (1999) Material volume 1996 for radioactivity in drinking water, ground water, sewage, sludge, remaining matter and waste (in German). Federal Office for Radiation Protection (BfS), BfS-ST-19/99, Berlin, Germany

Hdb Env Chem Vol. 5, Part L (2006): 307–368
DOI 10.1007/698_5_031

Published online: 6 October 2005

Assessment of Organic Compounds in the Rhine Estuary

R. W. P. M. Laane[1] · P. de Voogt[2] (✉) · M. H. Bik[3]

[1]National Institute for Coastal and Marine Management, P.O. Box 20907, 2500 EX The Hague, The Netherlands
r.w.p.m.laane@rikz.rws.minvenw.nl

[2]Earth Surface Processes and Materials-IBED, University of Amsterdam, Nieuwe Achtergracht 166, 1018 WV Amsterdam, The Netherlands
pdevoogt@science.uva.nl

[3]Royal Netherlands Institute for Sea Research/kNIOZ, P.O. Box 79, 1790 AB den Burg, Texel, The Netherlands
bik@nioz.nl

Abstract Over the past centuries the Rhine estuary has changed from a natural, hardly disturbed estuary into an anthropogenetically influenced area that extends into the North

Sea. The impact of the River Rhine on the adjacent coastal zone of the North Sea has thus increased dramatically, with relatively high loads of suspended matter and associated as well as dissolved compounds reaching as far as the Wadden Sea and further north. This chapter briefly discusses the historical changes in morphology and hydrology and reviews the concentrations of organic substances in the lower River Rhine and their possible effects on the estuarine and adjacent coastal zones. Gradually the loadings of these zones from riverine inputs have shifted from the more persistent chemicals such as HCHs and drins to PCBs and PAHs, and today to the more polar persistent compounds including surfactants and pharmaceuticals. Most of the concentrations of the persistent organic compounds have decreased, although a few, including the brominated flame retardants, show an increasing trend. Because of the large area involved and the technical problems of effect assessments in marine areas, a proper effect assessment can hardly be made for the area. Although effects related to specific exposure (e.g. organotins) have been observed occasionally, in general neither in vivo nor in vitro responses have been observed in the Dutch coastal zone of the North Sea, not even at dumping sites for harbour dredge. Regulatory bodies disagree remarkably about the compounds that should be prioritised, or the bioassay batteries that should be involved in evaluating actual exposure. Harmonisation of procedures is urgently needed in this respect.

Keywords Assessment · Dutch coastal zone · Organic compounds · Rhine estuary

Abbreviations

APEO	alkylphenol ethoxylate (surfactant)
BBP	butylbenzyl phthalate
BDE	brominated diphenyl ether
BpA	bisphenol-A
CALUX	chemical activated luciferase gene expression assay
COMMPS	combined monitoring based and modeling based priority setting
DBT	dibutyl tin
DDT	dichlorodiphenyl trichloroethane
DEP	diethyl phthalate
DEHP	diethylhexyl phthalate
DMP	dimethyl phthalate
DOP	di-*n*-octyl phthalate
DR-CALUX	dioxin responsive CALUX assay
d.w.	dry weight
E1	estrone
E2	17β-estradiol
E2-17α	17α-estradiol
EC50	effect concentration causing effect in 50% of test animals
EE2	17α-ethinylestradiol
EEQ	estrogen equivalent
ER-CALUX	estrogene responsive CALUX assay
GC	gas chromatography
GPC	gel permeation chromatography
HCH	hexachlorocyclohexane
K_{ow}	*n*-octanol/water partition coefficient
LC	liquid chromatography
MBT	monobutyl tin

MPRC	maximum permissible risk concentration
MS	mass spectrometry
NOEC	no (observed) effect concentration
NP	nonylphenol
NRC	negligible risk concentration
OSPAR	Oslo Paris Commission for the protection of the marine environment of the North-East Atlantic
PAH	polycyclic aromatic hydrocarbons
PCB	polychlorinated biphenyls
PBDE	polybrominated diphenylether
PFC	perfluorinated alkylated compounds
PFOA	perfluorinated octanoic acid
PFOS	perfluorinated octanesulfonic acid
RPS	relative penis index
SPE	solid phase extraction
TBT	tributyl tin
WFD	water framework directive

1 Introduction

Man has always preferably settled close to water. The reasons for this include the availability of drinking water, fishing, trade and dilution of waste. Dikes were built as a protection against natural forces, for example rising water. Conflicts of interest arose when the environment changed due to human activity. In general, these activities can be separated into three different types [1]. The first involve landscape interventions, such as the disappearance of various landscape types through urbanisation etc. The second type is contamination with natural and man-made synthetic (in)organic compounds, noise and energy. The last type of activity deals with extraction of materials including water, fish, sand and gravel.

The general principle behind changes caused by these activities is that the natural energy and matter fluxes through and in a system changed due to human activities. Earlier in history, with small settlements, changes in energy and matter fluxes due to human activity were relatively small. Also man tended to move from time to time to other places. As time progressed, the number of people living in one settlement increased and they stayed in one place for a longer period of time. However, the concurrent environmental impact was still restricted to discharges of easily degradable organic matter and the generation of bad smells. The energy used was based on man or horsepower only. Human and cattle manure was used to improve crop yields. In this way the concentration of natural compounds such as nutrients increased in surface waters around the settlements. The natural concentrations of metals in surface waters started to increase due to the mining of metal ore.

However, man's activities changed completely in the 20th century. European cities grew larger and the European chemical industry changed from natural products to the synthetic production of chemicals. The application of synthetic chemicals increased immensely and so did the matter flux to the aquatic systems. Environmental regulations became more and more necessary to separate conflicting uses of the same area or the same matter. To restrict rivers from their natural paths through the country and to use them for shipping, energy and materials were put into the water systems, for example through canalisation and the building of dikes. The developments described above all occurred within the Rhine catchment area. Presently, the Rhine catchment area is highly populated and industrialised and has changed tremendously due to human activities. These changes in the catchment area and river also have an impact on the adjacent coastal zone.

In this contribution the main focus is on the impact of chemical changes in the Rhine estuary due to human activities: more specifically the focus is on xenobiotic organic compounds. Results of different approaches to assess the changed chemical conditions in the Rhine estuary are described. First, the concentrations are compared with chemical targets. Second, results of bioassays are used to evaluate ecotoxicity, and lastly different negative effects on organisms observed in the field are discussed.

2 Historical Development

The River Rhine, with a lenght of 1025 km drains an area of about 220 000 km^2 covering Switzerland, France, Germany, and the Netherlands. Rainwater and melt water from Swiss glaciers constitute the sources of the river water. The discharges range between 620 $m^3 s^{-1}$ and more than 13 000 $m^3 s^{-1}$ with a long-term average around 2200 $m^3 s^{-1}$. The highest discharges are measured in February and March and the lowest in October and November [2, 3]. In the Netherlands, the River Meuse joins the Rhine water and together they pass through a common estuary, often referred to as the Delta, to discharge into the North Sea. The River Meuse drains an area of 33 000 km^2 and is a typical rain river [4]. Its length is 925 km and the discharges vary between 250–3000 $m^3 s^{-1}$ [4]. Hence, compared to the Rhine the Meuse is a moderate river [5].

The Rhine estuary is also affected by the River Scheldt in the southwest of the Netherlands. The River Scheldt drains an area of 10 505 km^2 and has a length of 280 km. The average water discharge is 103 $m^3 s^{-1}$ [6, 7].

2.1 Morphology

To protect himself from the risk of flooding and to use the river for economic purposes, during the last few centuries man has changed the natural path of the River Rhine through its catchment and the outflow path into the North Sea tremendously (Fig. 1). Presently, one can conclude that according to the classification of the new European Water Framework Directive (WFD), the River Rhine is definitely no longer a natural river system [8]. Dike building began in the Middle Ages and gradually forced the river within its present channels in the lower part of the Rhine. Later, canalisation and regulation of many of its parts fixed the river into a certain position. For instance, in the 19th century the River Rhine between Mannheim and Basel was shortened from 354 to 273 km. The positions of the Rivers Rhine and Meuse often changed in the 14th and 15th century because the dikes did not hold and the rising river and sea water flooded formerly enclosed land (Fig. 1b). Slowly the areas in the river mouth were reclaimed and became closely connected with the coastal dunes. In the 19th century an artificial channel was excavated to stop the river mouth becoming brackish. In this way Rotterdam became directly connected to the North Sea in 1868 via the New Waterway. After the disastrous flooding in 1953 in the southern part of the Netherlands the

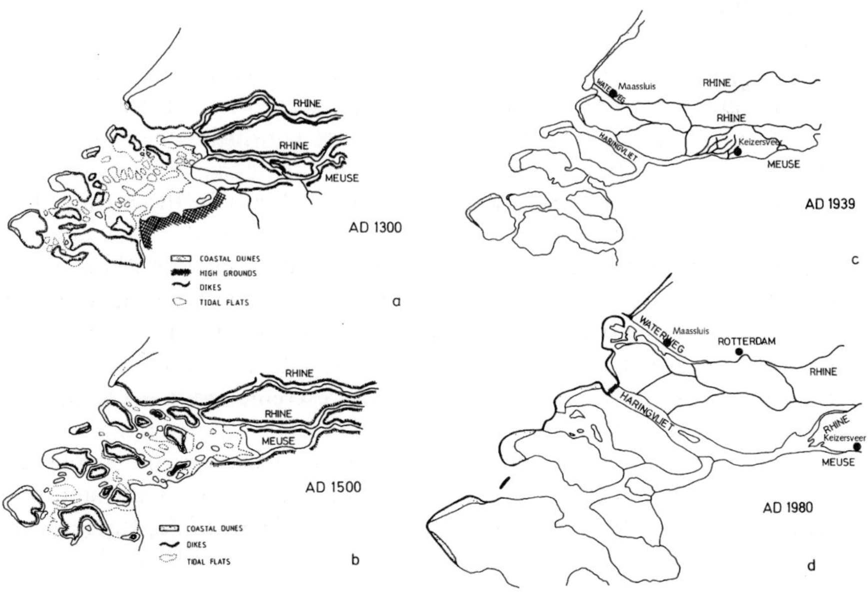

Fig. 1 Historical development of the mouth of the Rhine estuary in **a** ±1300, **b** ±1500, **c** 1939 and **d** 1980 (from Eisma et al. [14], with permission)

main branches of the Rivers Rhine and Meuse were closed by dams to protect against flooding from offshore storms. During the spring tide and with strong winds from the northwest seawater can pile up five meters above the normal level.

In 1971 the dam in the Haringvliet, formerly the main outflow of the Rhine, was closed with doors, which are only opened when the river discharge exceeds 1700 $m^3 s^{-1}$. To the south a dam closed the Grevelingen, and the dam in the Eastern Scheldt was semi closed to protect the brackish ecosystem. Presently, the only open outflow channel of the River Rhine to the sea is the New Waterway, which is continuously maintained at a certain depth by dredging to keep shipping lanes open to the harbours of Rotterdam.

To explain the transport and fate of compounds in the River Rhine it is necessary to have a general introduction to the behavior of the river water and the fluvial suspended matter in the Rhine estuary and Dutch coastal zone.

2.2 Salinity

Before the excavation of the New Waterway the Rhine water flowed through the Haringvliet and the other Delta branches to the North Sea. Although there are no salinity measurements from that time, it can be expected that the estuarine zone extended more inland than at present. Due to the dams in the Delta area and the New Waterway, today the estuary is pushed towards the Dutch coastal zone of the North Sea, giving rise to lower salinities along the coast. Indeed, the salinity distribution at the surface in the Dutch coastal zone is highly variable and complex and is mainly influenced by the river discharges and variations in wind field [9, 10]. It is also influenced by the general circulation pattern of the water in the Dutch coastal zone, which is anti clockwise, pushing the Rhine outflow northward along the Dutch coast. The maximum tidal currents along the Dutch coast are directed to the north and vary between 0.75 and 1.5 $m s^{-1}$ [11].

During calm weather conditions the less dense fresh water floats over the more saline seawater, and vertical stratification occurs over a distance of about 30 km from the mouth of the Waterway to the north [9]. This area can expand to 50 km offshore, to the south to the mouth of the Scheldt estuary, and nearly to Den Helder in the north under extreme weather conditions. The vertical density gradient in the Dutch coastal zone drives an onshore residual current near the bottom, giving rise to an estuarine-like circulation with an associated turbidity cloud close to the coast [10, 12]. Bennekom and Wetstijn [13] showed that under normal conditions, part of the Rhine water flows to the south along the coast and that at least 10% of the coastal water at the mouth of the Scheldt estuary originates from the Rhine.

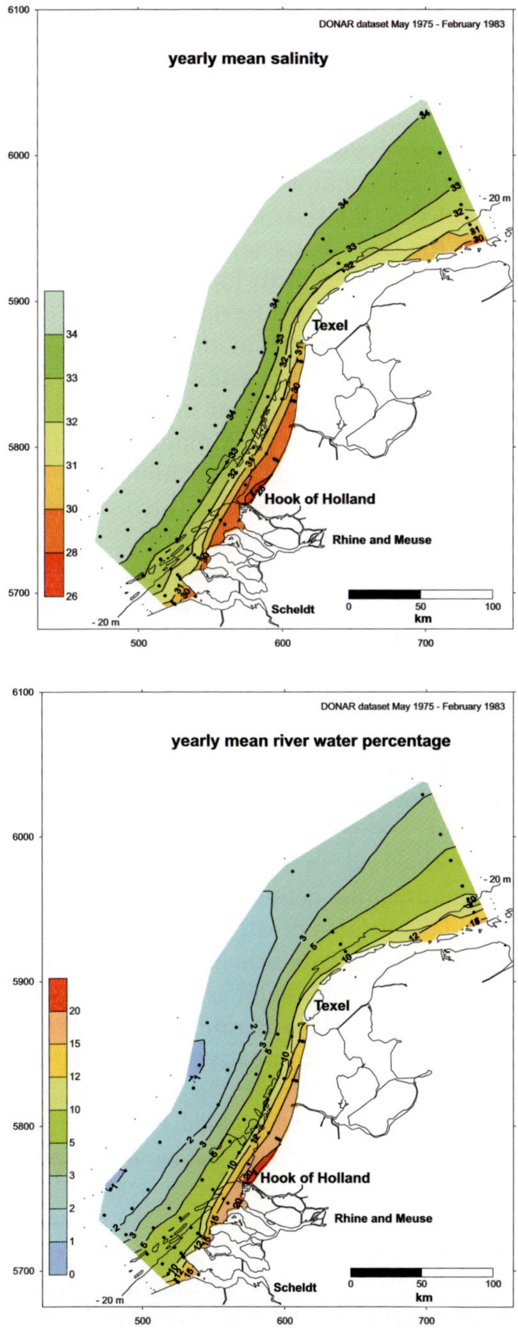

Fig. 2 Average salinity distribution at the surface in the Dutch coastal zone during summer and winter months in the period 1975 and 1983 (from Suijlen and Duin [182], with permission)

2.3 Suspended Matter

The suspended matter distribution in the Rhine estuary changed completely due to the morphological changes and the dams built in the Delta. One can assume that in earlier days due to relatively small water currents the fluvial suspended matter settled in the river itself and in the Delta, e.g. in the Haringvliet. Presently, as the Delta branches are closed the Rhine/Meuse water with its suspended matter flows directly and much faster to the Dutch coastal zone and is only partly trapped in the Haringvliet [14].

The fate of the suspended matter in the Rhine estuary is strongly influenced by estuarine circulation processes. According to Eisma et al. [14] the annual suspended matter supply from the Rivers Rhine and Meuse is 3.5–4.0 Mtonnes y^{-1}, of which about 1.5 Mtonnes y^{-1} is transported through the Rhine/Meuse estuary [15]. Model simulations suggest that 53% is deposited in the estuary and 47% reaches the Dutch coastal zone of the North Sea [16]. Large amounts of marine suspended matter are transported from the coastal zone of the North Sea into the Rhine/Meuse estuary: on average 3.94 Mtonnes y^{-1} [14, 16, 17]. On the basis of these results, the retention of fluvial suspended matter in the Rhine/Meuse estuary is estimated to be around 44–53%.

In general, the estuaries around the North Sea are importing estuaries for suspended matter: meaning that fluvial suspended matter is mainly trapped in the estuary and that sand and suspended matter are imported from the North Sea. The suspended matter concentration in the Dutch coastal zone (see Fig. 3) has various sources: the Channel and the Belgian coast to the south, and the River Rhine [15]. To the south the suspended matter originates from the Channel, the dredging activities at the harbour of Zeebrugge and associated dumping in the Belgian coastal zone [15]. The main part of the fluviatile suspended matter of the River Scheldt is retained in its estuary. Various suspended matter budgets of the Scheldt estuary have been made [18–22]. It has been estimated that industrial and domestic sources accounted for two thirds of suspended matter load to the River Scheldt in the 1970s [7]. Man-made loads have decreased considerably due to waste water treatment plants. Van Maldegen [23] constructed a suspended matter balance of the Scheldt estuary, with explicit distinction between fluvial and marine suspended matter: the transport of fluvial suspended matter is about 315 Ktonnes y^{-1}; on average 160 Ktonnes y^{-1} reaches the North Sea. The annual import of marine suspended matter is about 200 Ktonnes y^{-1} into the Scheldt estuary [24]. On the basis of these results, the estuarine retention of fluvial suspended matter is estimated to vary between 60–95% [25].

A variable amount of dredged material is removed annually from the harbours of Rotterdam; there are rather large differences between various years (Fig. 4). At the end of the sixties and in the beginning of the eighties the

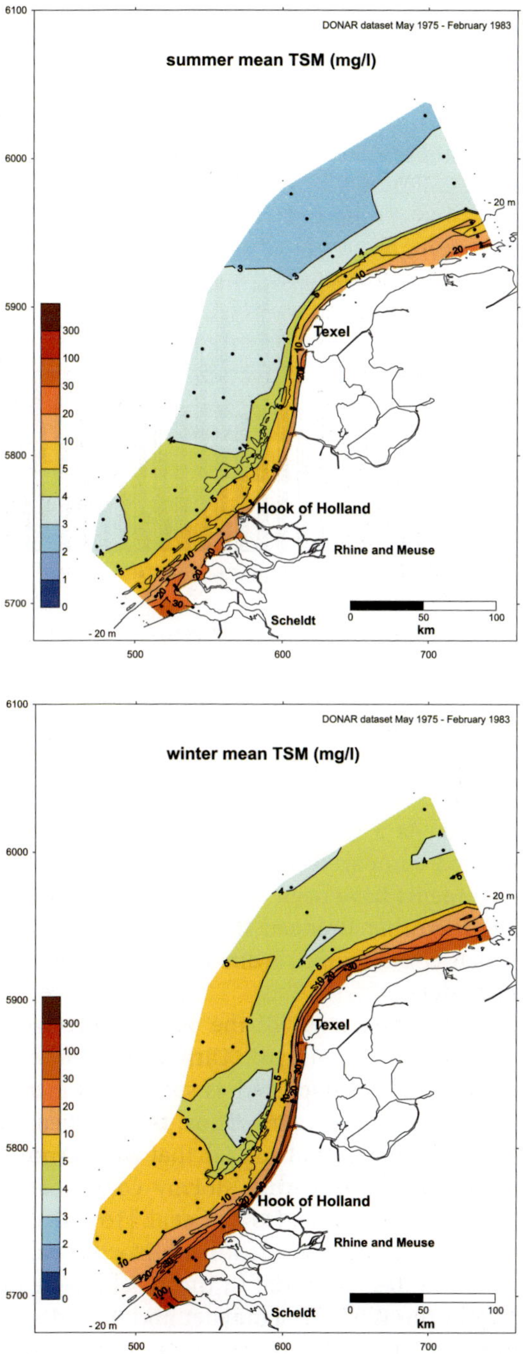

Fig. 3 General distribution of suspended matter in the Dutch coastal zone during summer and winter months in the period 1975 and 1983 (Suijlen and Duin [182], with permission)

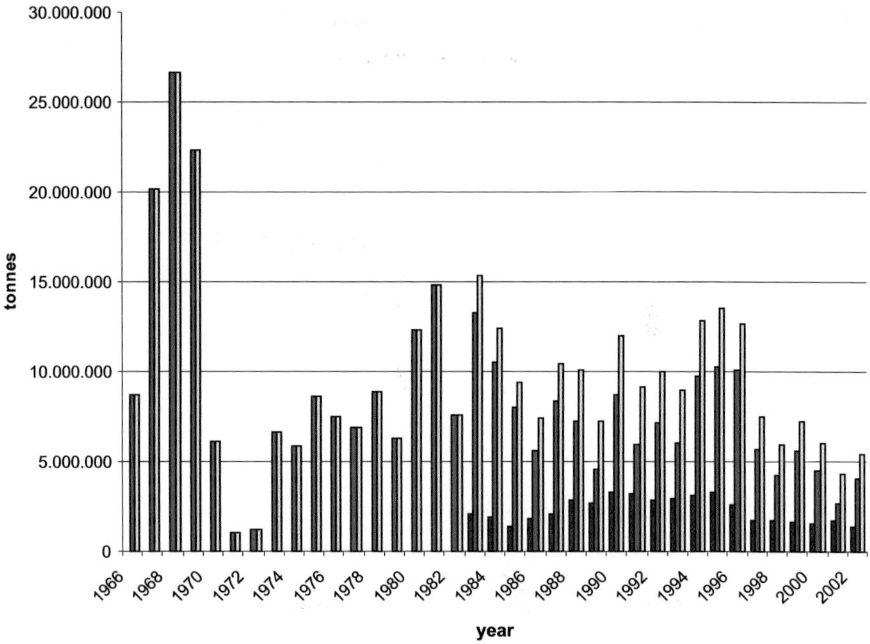

Fig. 4 Total amount (*light bars*) of dredged material (tonnes dry matter) from Rotterdam harbours and waterways during 1966–2002. The most polluted sediments, which are stored on land are indicated with *grey bars* (from Sandeh [26], with permission). The majority (*dark bars*) of the dredge spoil is deposited in the coastal zone

channel was deepened. As a result, during these years much more dredged material was produced compared to other periods. From 1983 on the most heavily polluted sediments have been stored on land in specially constructed basins. Up to 1996 the other dredged material was dumped at an offshore location in the Dutch coastal zone. From 1995 on a new location was used, situated further to the west.

It has been estimated that 40% of the suspended matter in the Dutch coastal zone is dredged material [16]. Ultimately, the major part of this dumped dredged material is transported as suspended matter by currents to the Wadden Sea and further north [26]. Thus, the Wadden Sea is a net deposition area of part of the suspended matter that is transported in the Dutch coastal zone [27, 28]. However, during stormy conditions the flux of suspended matter can be seawards [29]. De Kok et al. [16] calculated a net import of 4.3 Mtonnes y^{-1}: 1.73 Mtonnes y^{-1} from the North Sea, 1.89 Mtonnes y^{-1} from the dumping of dredged material in the coastal zone of the North Sea, 0.32 Mtonnes y^{-1} from the Rhine/Haringvliet and 0.41 Mtonnes y^{-1} from the IJssel Lake. From the suspended matter imported into the western Wadden Sea 80% is of marine and 20% of fluvial (mainly from the Rhine) origin. Molegraaf [30] derived a sedimentation rate of 3–3.5 Mtonnes y^{-1}. With an import of 4.35 Mtonnes y^{-1} and the fact that suspended matter from the IJssel Lake has a relatively high sedimentation rate because it is discharged in an area of low current velocities the overall retention of suspended matter in the Wadden Sea is 80–90%.

As part of the water and part of the suspended matter in the Wadden Sea are from the River Rhine, it is also likely that compounds present in the River Rhine enter the Wadden Sea. Trends in concentrations of (in)organic compounds in the Wadden Sea have been described by De Jong and coworkers [31] and Lozán and coworkers [32].

3 Human Use

The Dutch coastal zone is used in many ways: cities with associated industries, fishing, oil- and gas extraction, sand and gravel extraction, shipping, discharge of waste and contaminated dredged material, discharges of chemical compounds, and recreation [33].

Here the focus will be on the discharge, transport and fate of chemical compounds; more specifically on xenobiotic organic compounds. Contamination is defined as the increase of a concentration of a natural compound (such as metals and nutrients) above its natural background level, and above zero for man-made compounds [34]. Pollution is understood to mean a level of certain chemical compounds that adversely affects the natural environment. In other words: a too high concentration present in the wrong place and at the wrong time.

4 Problems

Problems induced by chemical compounds can be identified in different ways. Problems may arise when the concentrations of chemical compounds exceed a certain limit, e.g., for human consumption or exposure. Non-lethal and lethal effects on organisms in the field induced by chemical compounds can pose problems for the functioning of the ecosystem. Currently, in particular the non-lethal field effects are still difficult to detect in the (marine) area considered, because hardly any reliable detection method is available, which, in turn, is because only a few researchers have focussed on these problems up to now, and because the North Sea covers an area 16 times as large as the Netherlands. Moreover, it is difficult to distinguish qualitatively effects induced by chemical compounds from effects induced by other human uses, such as beam trawling [35], and effects induced by the natural variability of the physical, biological and chemical environment [36].

Another, easier way is to compare the concentrations in the field with a target value based on experimental toxicological knowledge. (Inter)national policymakers have agreed that measures must be taken when the concentration observed is higher than the target value. However, it must be recognised that the latter method gives an indication of the chance that a possible negative effect on 5% of the organisms present in the ecosystem will occur [37]. In other words, if the concentration observed is higher than the target value, this does not necessarily mean that real effects will occur in organisms in the field.

The impact of chemical compounds on various processes in organisms has been measured indirectly by transferring the organisms to the laboratory. Different biomarkers (e.g., cytochrome P450, metallothioneins [38]) and various bioassays have been developed and performed to obtain an indication of a possible effect in the field. Performing bioassay experiments directly in the field is still a difficult problem, because the surrounding environment cannot be kept constant as in the laboratory [39].

Another way to express a possible effect of chemical compounds is to determine their concentrations in the body and tissues of various organisms. Many measurements of this kind have been performed in the North Sea and Rhine estuary [40–47]. However, the presence of chemical compounds in biota does not directly mean that sub-lethal effects, such as retardation of growth and negative effects on reproduction, will occur [48, 49]. It does give a good representation of the bio-available fraction of the chemical compounds [50, 51].

Below we will focus on the field effects and bioassays, and the comparison of the concentrations observed in the field with a chemical target value.

4.1 Biological Effects in the Field

Some specific adverse effects of chemical compounds on organisms have occurred in the past 40 years in the Rhine estuary. For instance, numerous oil pollution incidents have led to dead seabirds in the Dutch coastal zone and on the beaches during the last few decades [52]. It has been reported that there is probably a decline in the number of dead oiled seabirds found on the Dutch beaches, which could be related to a decline in oil accidents [53].

In the 1960s the decline in some breeding populations and the large-scale bird mortality in the Dutch Wadden Sea could be related to the production and associated discharge of the pesticides telodrin and dieldrin in Rotterdam [54–56]. These pesticides were transported with the River Rhine water and suspended particulate matter to the western Wadden Sea [57–59]. The decrease in population was most severe for the sandwich tern (*Sterna sandvicensis*) and little tern (*Sterna albifrons*). Reorganisation and part closure of the factory in Rotterdam resulted in an increase in the population size of the different birds in the Wadden Sea.

The sharp decline of the seal (*Phoca vitulina*) population and the death of seals in the western Wadden Sea were correlated by Dreschler and coworkers [60, 61], and Reijnders [62–64] with too high PCB concentrations in seal tissue. Reijnders subsequently proved that PCB compounds affected the reproductive process in female seals [64]. The main source of PCBs in the western Wadden Sea is the Rhine estuary [65].

Kamp [66] was one of the first to describe an increase in vertebral abnormalities in herring (*Clupea harengus*) in areas in the North Sea where the chemical pollution was highest. In a review Möller [67] summarised the knowledge on observed diseases of dab in the North Sea and hypothesised that pollution could induce ulcers. Dethlefsen and coworkers [68–71] tried to link fish diseases to pollution in the North Sea. They found correlations between disease prevalence and high concentrations of metals and PCBs. They also suggested that low oxygen concentrations could lead to an increase in fish disease abundance. Van Banning [72] showed that the presence and prevalence of diseases vary strongly as a function of fish species, area and season. He found especially a relatively high prevalence of diseases around the Doggerbank area. However, it must be recognised that the interpretation of the results of these fish disease studies was rather difficult because at that time only few data on organic compounds were present. Consequently, the relation between pollution and fish diseases was difficult to identify.

Vethaak studied the internal and external lesions in flounder and dab in the Dutch coastal zone during 1983 and 1987 [73–76]. He found that the prevalence of diseases was also linked to the sex of the fish. But the most important results indicated that in the Dutch coastal zone more than 40% of dab and flounder aged 3 years or more were affected with liver nodules (> 2 mm). These lesions in flounder were correlated with the concentration of PAHs in surface sediments [77]. In large-scale mesocosm experiments with dredged material from the Rhine estuary, Vethaak [76] showed that contaminated sediments could induce liver tumours (neoplasia) in flounder. In time the occurrence of liver tumours in flounder along the Dutch coast decreased as did the PAH concentration in surface sediments (Fig. 5).

Embryonic development in fish can be used as another biological indicator of exposure to high concentrations of chemicals. High malformation rates in pelagic fish embryos were found in areas affected by rivers, including the near shore area of the Dutch coastal zone and the German Bight [70, 78]. For all species in all areas the same trend was found: intermediate incidences in 1984, maximum in 1987 and a significant decrease thereafter [71]. The same temporal trend was found for organochlorine compounds, such as DDT and HCHs, but also temperature appeared to be linked to malformation. The hypothesis was put forward that low temperatures may predispose developing fish embryos to the impact of organochlorines.

Parrett [79] concluded that there was no clear evidence that North Sea fish stocks were impacted by chemical contamination.

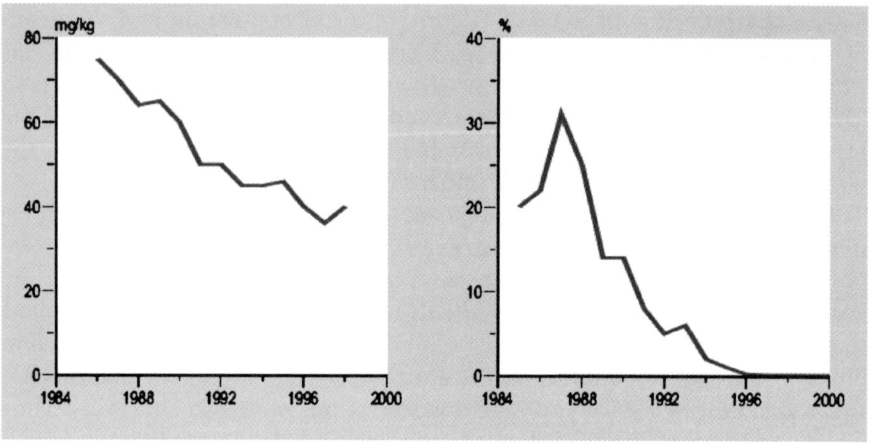

Fig. 5 Decrease in incidence of liver tumours in flounder (*right hand panel*, %) and decrease in the concentration of PAH in surface sediments (*left hand panel*, mg kg^{-1}) from the Dutch coastal zone (Courtesy A.D. Vethaak, RIKZ)

Imposex in marine snails has been found along the Dutch coast and in the open North Sea [80–82]. Imposex also occurred in all collected specimens of common whelk close to the mouth of the River Rhine [81]. It appeared that the female population of the dogwhelk (*Nucella lapillis*, L) and the common and red whelk (*Buccinum undatum* and *Neptunea antique*, respectively) obtained masculine properties. It is rather well established that these effects are induced by tributyltin compounds (TBT), which are classified as the most toxic chemical compounds in the marine environment [83]. These chemicals are applied to ships to prevent hulls from fouling.

The first studies in the Netherlands showed relatively high concentrations of TBT in the River Rhine, estuary and coastal zone and some harbours [84], as well as associated imposex for the local dogwhelk [85–87]. The induction of imposex in the dogwhelk in the Dutch coastal zone and the North Sea was established with transplanted dogwhelks in both a national [85] and an international study [88], and for the common whelk in laboratory experiments [89, 90]. Both studies showed that within a few weeks female dogwhelks had a RPI (relative penis index) of more than 50%. The juvenile stage of the common whelk was the most sensitive stadium for imposex induction. Although a recovery of the imposex has not yet been observed (C. Schipper, personal communication), during the last decade the concentrations of TBT have decreased significantly in the Dutch coastal zone.

Before an assessment is made of the possible impact of organic compounds in the Rhine estuary an overview will be presented of the occurrence of various chemical compounds in the Rhine estuary.

4.2 Organic Compounds

The historical development of the attention paid to compounds in the aquatic environment has been described by Laane et al. [91]. In the 17th and 18th century the well-known scientists Boyle and Lavoisier studied the content and composition of the different salts in seawater. In the beginning of the 20th century attention was focussed on chemical compounds that could be measured easily, e.g. oil and radioactive elements. Due to the application of the spectrophotometer it was possible to measure the concentration of nutrients at relatively low concentrations around 1930. After the publication of "Silent Spring" [92], attention became focussed on chlorinated pesticides. In 1966 Jensen, with the aid of mass spectrometry, was the first to identify the occurrence of the presently well-known group of PCB congeners in wildlife [93, 94]. There was a time lag of nearly 40 years between this finding and the start of the commercial production of PCBs [94–96].

The production of xenobiotic chemicals started around 1900. Since then the production has increased tremendously, both in numbers of different compounds and their amounts. Presently, it is estimated that around 150 000 different chemicals have been produced.

With the modern instrumental techniques and methods (e.g., GC-MS/MS and LC-MS/MS) it is possible at the moment to detect nearly all chemical compounds in the aquatic environment at relatively low concentrations ($< 10^{-9}$ g L^{-1}). Over time, the public and scientific attention for certain compounds or groups of compounds have shifted. Initially, the chlorinated pesticides such as lindane and DDT became the subject of scientific investigations, followed consecutively by the lipophilic PCBs, dioxins and furans, polybrominated flame retardants, and then gradually shifting to more water soluble compounds including surfactants and most recently the perfluorinated compounds [91, 97]. In the course of time, the attention became more and more focussed on persistent, bioaccumulating and toxic chemical compounds: the so called PBT compounds. Presently, about 150–200 compounds are measured regularly in the Rhine estuary by national monitoring programmes.

However, here it must be recognised that it is impossible to study for all these thousands of compounds the path, fate and impact in the aquatic environment. So, based on various criteria, selections and prioritisation mechanisms have been proposed by various (inter)national organisations. OSPAR [98] introduced the DYNAMEC (dynamic selection and prioritisation mechanism for hazardous compounds) methods. The EU [99] applied the COMMPS (combined monitoring-based and modelling-based priority setting) for the selection of chemical compounds, which need political attention in the Water Framework Directive [8]. The OSPAR and WFD priority lists do not contain the same compounds: there is a small overlap. Even the different national lists of priority compounds in Europe show little overlap [91, 100].

Below, we present examples from the screening of organic compounds and results of the first group of compounds that gained attention, viz. the isomers of HCH and some of the PAHs. In addition, the concentrations of newly emerging compounds including TBT, surfactants, azaarenes, phthalates, and perfluorinated compounds will be discussed.

4.2.1 Screening Organic Compounds

In 1997 a screening study was set up to identify so-called forgotten chemical compounds in the Dutch surface waters [101]. Forgotten chemical compounds were defined as potentially hazardous compounds which are not included in existing (inter)national priority lists. At different locations close to the mouth of the Rivers Meuse and Rhine and near discharge points of waste-water treatment plants water samples were taken, filtered, concentrated on a XAD column, extracted with dichloromethane (DCM) and analysed and identified with GC-FID and GC-MS [101, 102]. In this way, organic compounds were identified with a log K_{OW} value between 2.5 and 6 and a concentration above 0.01 $\mu g\,L^{-1}$. An example of a gas chromatogram is given in Fig. 6.

In total the presence of 151 organic compounds could be identified in this way [101]. The compounds with the highest concentrations and which were found nearly in all samples are presented in Fig. 7. In the Barreveld study [101], water samples from the River Rhine at the border with Germany (Lobith) were also analysed. It appeared that 83% of the 151 compounds were present both near Lobith and in the New Waterway at Maassluis. Forty-three compounds found to be present in Lobith were not observed in Maassluis samples, and 25 compounds found in Maassluis were not present in Lobith.

In 2001 water samples from four different locations in the Rivers Rhine and Meuse around Rotterdam were collected and analysed for forgotten compounds with GC-MS and LC-MS [103]. Compared to the study of Barreveld and coworkers [101], this study focussed more on the identification of different groups of compounds, such as surfactants, volatile compounds, synthetic musks, anilines, alkylphenols, PCBs, pesticides and brominated compounds. Various extraction techniques were applied to isolate specific groups of compounds, e.g. Solid Phase Extraction (SPE) for the surfactants nonylphenol ethoxylates, and acid/base with DCM extraction for different phenols (Fig. 8a,b). The concentrations of surfactants and some other chemical compounds were also determined in suspended particulate matter.

The most important dissolved compounds, found in more than 50% of the samples included: triphenylphosphate (softener), tetraglyme (tetraethyleneglycol) (solvent), dimethylether (solvent), benzenesulfonamide, *N*-ethyl-4-methyl (softener), phosphoric acid tributyl ester (softener), adipinic acid, mono-(2-ethylhexyl) ester (metabolite of softener), benzenemethanol and alpha-phenyl (intermediate of pharmaceutical and pesticide). In more than

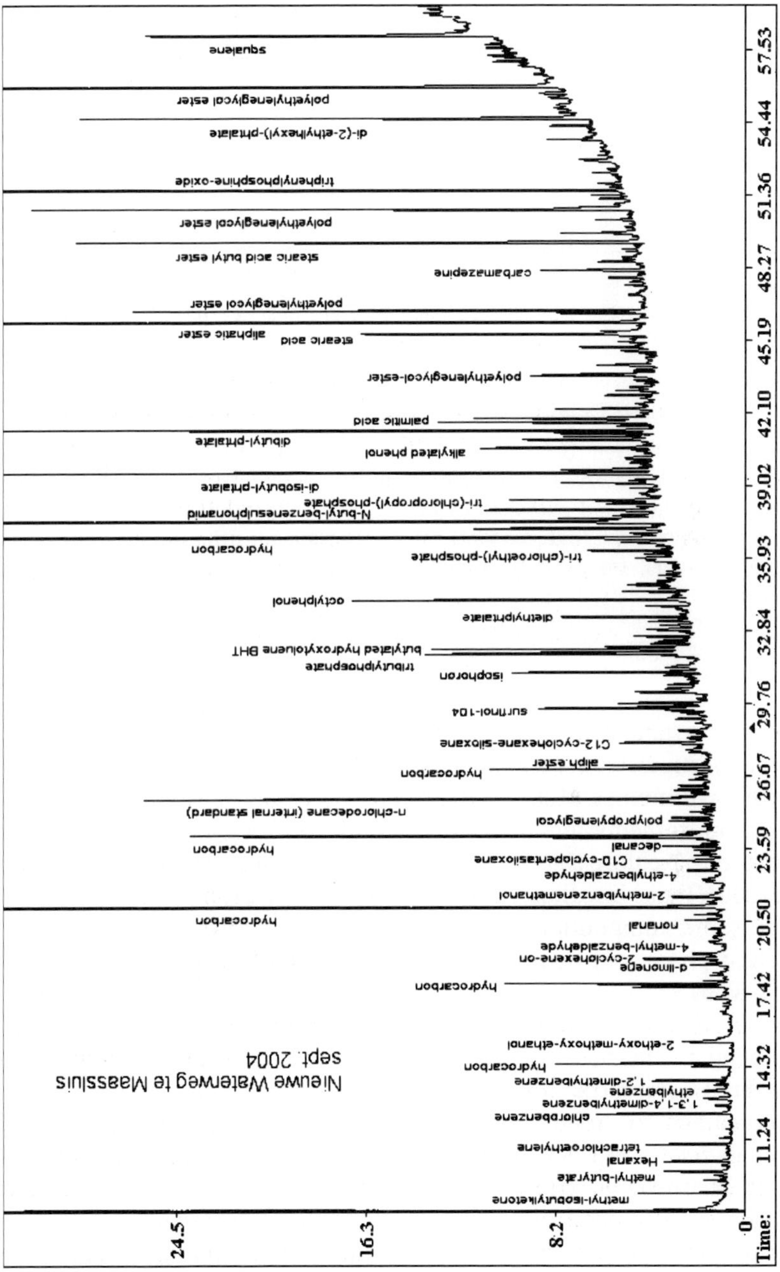

Fig. 6 Gas Chromatogram (Agilent-HP-MSD quadrupole mass spectrometer, DB-1 capillary column, 50 cm, linear temperature program 40–300 °C) of a water sample taken near Maassluis in the Rhine estuary in September 2004. Identification with NIST 2000 spectrum library in combination with AMDIS deconvolution program (Courtesy H. Barreveld, RIZA)

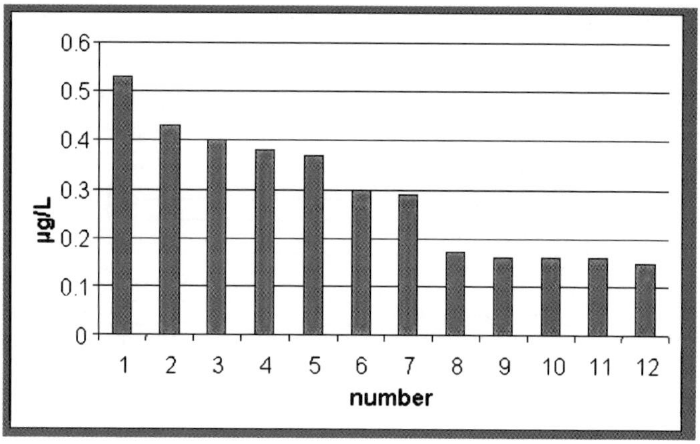

Fig. 7 Average dissolved concentrations (μg L^{-1}) of 12 organic compounds with the highest concentrations observed in 16 water samples in spring 1997 in the River Rhine near Maassluis. (1) diisobutyl-phthalate (softener); (2) 2-ethylhexanol (odour); (3) 2,6-di-*t*-butyl-4-hydroxytoluene (anti-oxidant); (4) cyclohexanone (nylon production); (5) TAED (washing powder); (6) 2-*t*-butyl-4-methoxyphenol (anti-oxidant); (7) dibutylphthalate (softener); (8) cyclohexanol (solvent); (9) tri-(2-chloro-isopropyl)-phosphate (flame retardant); (10) propyleneglycol methylether (solvent); (11) benzothiazole (odour); (12) butylstearate (odour) (taken from Barreveld et al. [101])

50% of the samples the following compounds were identified in the suspended particulate matter: 9*H*-xanthen-9-one, 2,7-dichloro-1-hydroxy-3,6-dimethoxy-8-methyl, 1,3,5,7-tetrachloronaphthalene, 7-chloro-10-hydroxy-3-(4-(trifluoromethyl)phenyl)-1-(aminoimino), 1,3,4,10-tetrahydro-9(2*H*)-acridinone, 7-chloro-3,4-dihydro-10-hydroxy-3-(4-(trifluoromethyl)phenyl)-1,9(2*H*,10*H*)-acridinedione, 1-(1*H*-benzimidazole-2-*yl*)hydrazone.

For some compounds there is an overlap between those found in the dissolved phase and those in the particulate phase (compare the numbers in Fig. 8a,b).

It can be concluded that none of the compounds identified in 2001 corresponded to the ones identified in the 1997 study. This is mainly due to procedural differences (see below).

Berbee and coworkers [102] identified forgotten compounds in the River Meuse in 2002 and 2003. In total 207 compounds could be identified. Most find their origin in consumer products that are used in housekeeping; e.g., odours from washing powder, anti-oxidants from food, and softeners from plastics. More than half of the number of compounds identified were lipophilic ($\log K_{ow} > 3$).

In the screening studies described above the aim is to isolate and identify as many compounds as possible. The isolation and extraction methods as well as the limits of detection of the analytical instrumentation are major determinants of which chemicals can be identified. The results described above

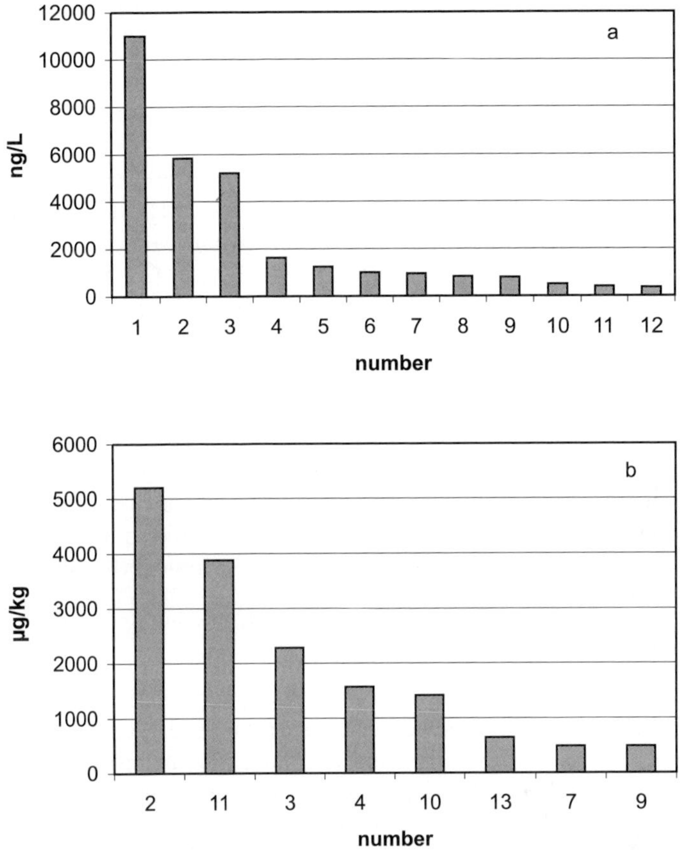

Fig. 8 Highest observed dissolved (**a**, $\mu g\,L^{-1}$) and particulate (**b**, $\mu g\,kg^{-1}$) concentrations of various compounds at different locations in the rivers Rhine and Meuse in 2001. (1) diisobutylphthalate (softener); (2) A_9PEO_{TOT} (surfactant); (3) A_9PEO_{3-16} (surfactant); (4) $A_9PE_{3-6}C$ (surfactant transformation product); (5) diisopropylether; (6) di-*n*-butylphthalate; (7) A_9PE_1C (surfactant); (8) methyl-*t*-butylether; (9) A_9PE_2C (surfactant transformation product); (10) A_9PEO_2 (surfactant); (11) nonylphenol; (12) acetone (solvent) 13) 4-chloro-3,5-xylenol. From Geerdink and Schrap [175]

are not consistent with each other because different extraction and isolation techniques were used.

In (inter)national monitoring programmes it is impossible to analyse all compounds present because this is too expensive. Moreover, as already said, not all compounds are toxic. Therefore, these monitoring programmes focus more on the persistent, bioaccumulating and toxic compounds. The selection criteria used and the starting list of compounds, of course, determine the ultimate selection of priority chemicals. It is remarkable that when three different international lists of priority compounds are compared, only six compounds appear on all three international priority lists [104].

4.2.2 Hexachlorocyclohexane

HCH is the chemical abbreviation for hexachlorocyclohexane. HCH has been applied world wide as an insecticide. During production different isomers are formed: α (60–70%), β (5–12%), γ (10–15%, lindane) and δ (6–10%). The dissolved concentrations of the different isomers of HCH have been measured at the river end of the Rhine and Meuse estuary since 1973 (Fig. 9). The mean dissolved concentration of HCH in the Meuse (Keizersveer) dropped from 18 ng L^{-1} in 1972 to below detection limit in 1977 and onwards [4]. However, there is an unexplained increase in the concentration during the last few years.

Values for the HCHs reported by Dijkzeul [2, 4] for Maassluis and Keizersveer are in close agreement with each other. Duinker and Hillebrand showed that during transport through the Rhine estuary the isomers of HCH behaved conservatively [105]. Ernst and coworkers [43] reported that HCHs were found everywhere in the North Sea and postulated that the earliest HCH data for the North Sea were from 1977. Concentrations of dissolved α- and γ-HCH in the Dutch coastal zone were around 1.5 and 2.5 ng L^{-1}, respectively, in 1979 [43, 106]. The earliest data of HCH isomers in the North Sea and Dutch coastal zone are from 1973 (Bik, unpublished). These results show that the surface concentrations of two HCH isomers, going from Rotterdam to Den Helder in the North, remained rather constant in 1973 (Fig. 10).

In 1974 a sharp gradient in dissolved surface concentrations of α-HCH and γ-HCH was observed when sampling across the North Sea in a westward direction going from the Dutch to the English coast (Fig. 11).

The yearly average dissolved concentrations of two isomers of HCH in the Dutch coastal zone between 1991 and 2002 are given in Fig. 12. In general, the concentration of the γ-isomer is higher than the α-isomer.

These concentrations in the Dutch coastal zone agree with the values reported in 1983 [107]. Ernst and coworkers [43] reported dissolved lindane concentrations in the seaward part of the Dutch coastal zone of 3 ng L^{-1} in 1986. The general distribution of dissolved α- and γ-HCH across the entire North Sea has been measured for the first time in 1986 [108]. The authors showed that the maximum value of the sum of the isomer concentrations of HCHs in the Dutch coastal zone ($>$ 5 ng L^{-1}), just north of the Wadden Sea ($>$ 6 ng L^{-1}), and in the German Bight ($>$ 7 ng L^{-1}) were higher than in the open North Sea (1.7–2.8 ng L^{-1} for γ-HCH and $<$ 1.0 ng L^{-1} for α-HCH). Lower concentrations were measured in the continental coastal zones of the North Sea in 1987: all values were below 5 ng L^{-1}.

Duinker and coworkers [3, 105] also analysed the concentrations of the different isomers of HCH in suspended matter and surface sediments in the Rhine (1974) and Meuse estuary (1976). Dissolved concentrations of α-HCH ranged between 3–4 and 2–9 ng L^{-1} and for γ-HCH between 6–8 and

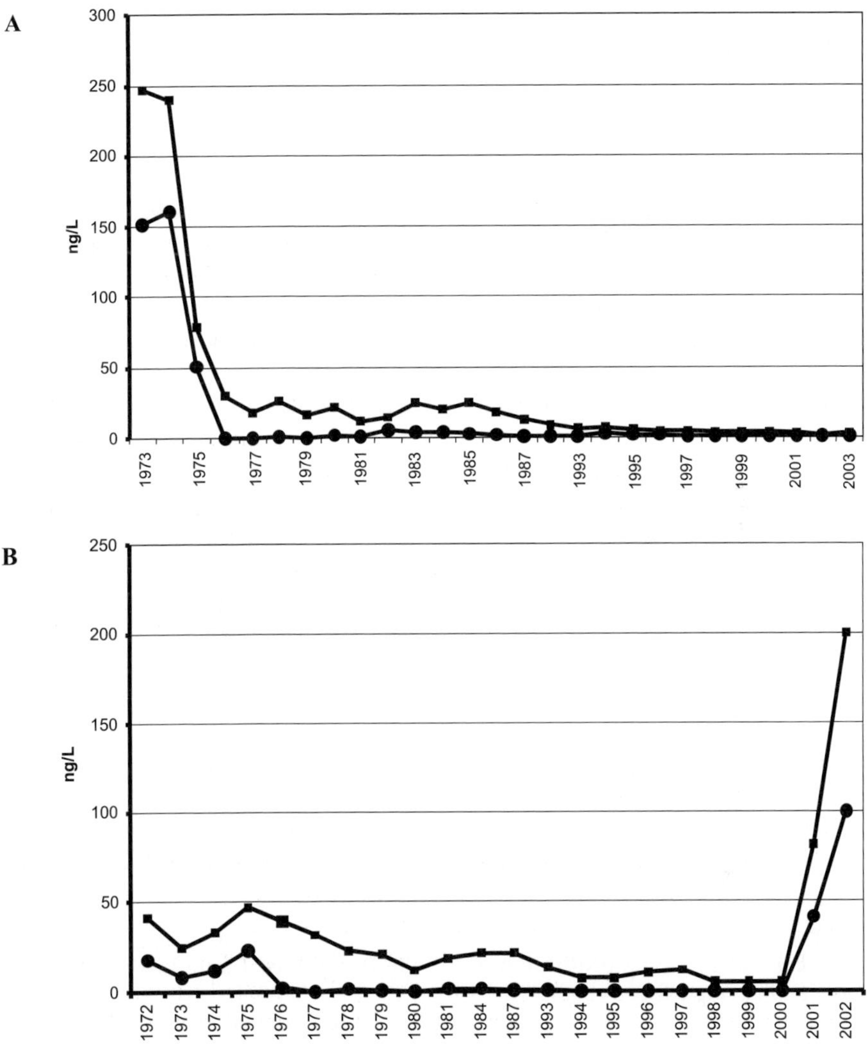

Fig. 9 Yearly averaged dissolved concentrations (in ng L^{-1}) of α-HCH (•) and sum of α and γ-HCH (■) in the River Rhine at Maassluis (**A**) and Keizersveer (**B**) from 1973–2003. Note shift in years at Keizersveer (data applied from Rijkswaterstaat database DONAR)

6–12 ng L^{-1} in the Rhine and Meuse estuary, respectively. Concentrations of α- and γ-HCH in suspended particulate matter were measured in the Meuse (13–50 and 75–100 ng g^{-1} respectively) and the Rhine estuary (40–60 and 145–170 ng g^{-1} respectively). In the sediments the concentrations of the isomers of HCH were below the limits of detection.

The concentration of the sum of the HCH isomers in sediments from the southern Bight of the North Sea varied between 5 and 1000 pg g^{-1} [109, 110].

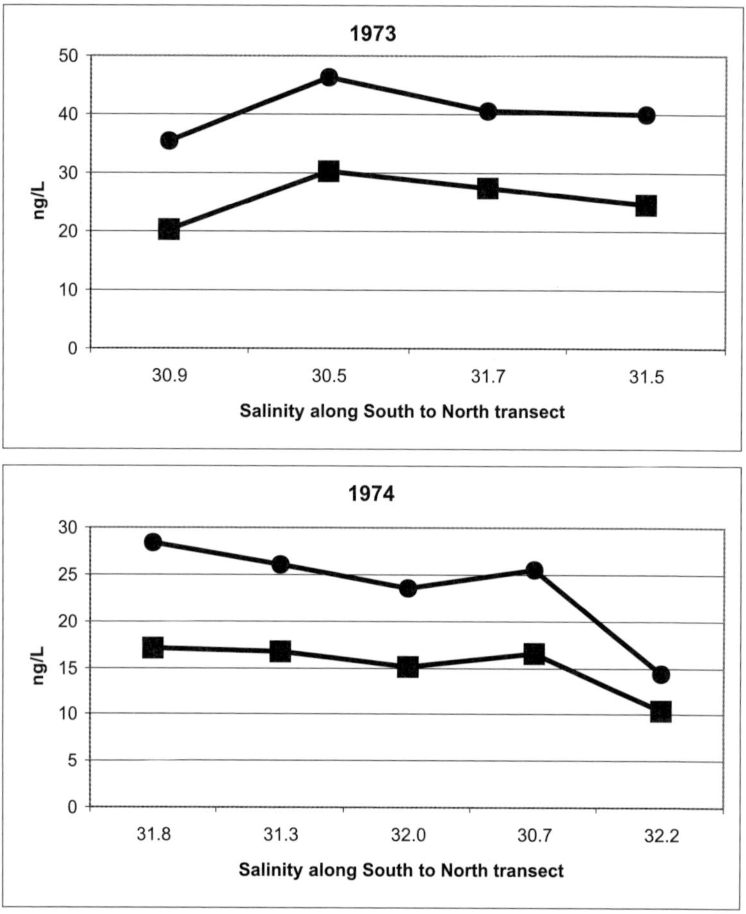

Fig. 10 Concentrations (in ng L^{-1}) of averaged dissolved α-HCH (•) and γ-HCH (■) versus salinity along a South-North transect in the Dutch coastal zone in 1973 and 1974. From M. Bik, unpublished results

It is remarkable that the concentration of lindane in surface sediments did not show the expected distinct distribution pattern in the North Sea in 1986, with relatively high concentrations along the coasts and lower in the open sea sediments [108]. Also, the amount of lindane in the hermit crab (*Pagurus ssp.*) in the North Sea during 1986 and 1987 did not show this pattern: in 1987 somewhat higher concentrations were found in the continental coastal zones [108]. It is suggested that this could be caused by the rather important atmospheric deposition of these compounds [111]. The author estimated the annual riverine input of the sum of the HCH isomers to be between 2–5 tonnes for the 1980s. He noticed also that the annual atmospheric deposition of the HCH

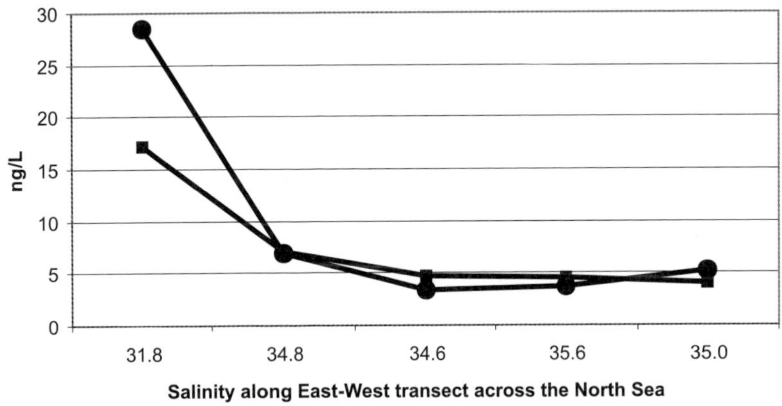

Fig. 11 Gradients in the averaged dissolved concentrations (in ng L^{-1}) of α-HCH (•) and γ-HCH (■) versus salinity along an East-West transect across the North Sea from the Dutch coast to the English coast in 1974. From M. Bik, unpublished results

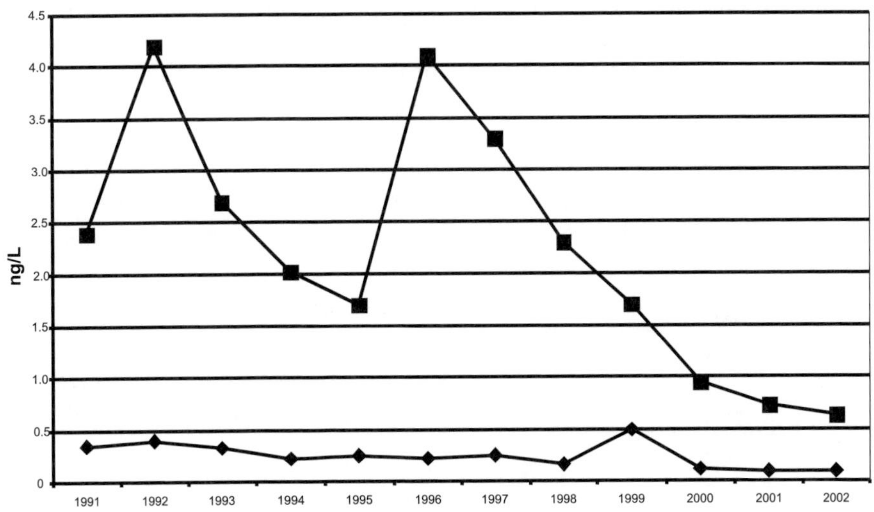

Fig. 12 Yearly averaged concentrations (in ng L^{-1}) of dissolved α-HCH (♦) and γ-HCH (■) in the Dutch coastal zone between 1991 and 2002

isomers is of the same order as the riverine input: 1.5 tonnes. However, he did not explain why the atmospheric deposition was so high in that particular year.

4.2.3
Polycyclic Aromatic Hydrocarbons

Polycyclic aromatic hydrocarbons (PAHs) is the classification applied to a very large group of chemical compounds that have in common a skeleton of at least two fused aromatic rings. There exist hundreds of PAH and PAH metabolites. Azaarenes are a special group in which an aromatic carbon of a PAH is substituted with a nitrogen atom. Most studies in the Dutch coastal zone have been focussed on the PAH concentrations because of the carcinogenic and mutagenic characteristics of some of these PAHs.

The spatial and temporal distributions of PAHs in the surface sediments of the Dutch coastal zone were studied by Laane and coworkers [112, 113]. The concentration of PAHs in the fraction smaller than 63 μm was measured in the surface sediments in the years 1986, 1991, 1996, 2000 and 2003. Whereas the concentrations of metals and PCBs decreased significantly in the Dutch coastal zone in that period, the concentration of PAHs remained nearly constant (Fig. 13a,b).

Sonneveldt and Laane [114] applied the sediment quality model, developed by Laane and coworkers [112], to describe and predict the concentration of PAHs in the surface sediments of the Dutch coastal zone. It appeared that 30–50% of the PAHs in the surface sediments originated from the River Rhine and from the dumping of dredged material in the coastal zone. Atmospheric deposition accounted for 40% of the PAHs in the surface sediments and the rest was transported to the area by the Channel water entering the North Sea in the south (see [115]). Here it must be recognised that only the

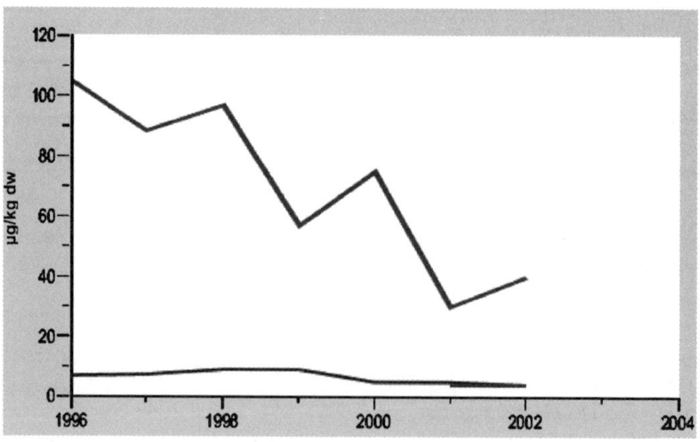

Fig. 13 Concentration of Tributyltin (TBT) in suspended particulate matter in the River Rhine, close to the Rhine estuary (Maassluis, upper graph) and at the German border (Lobith, lower graph (1996–2002)), and in the river Meuse at the border (Eijsden, lower graph (2001–2002)) (From RIKZ, with permission)

A)

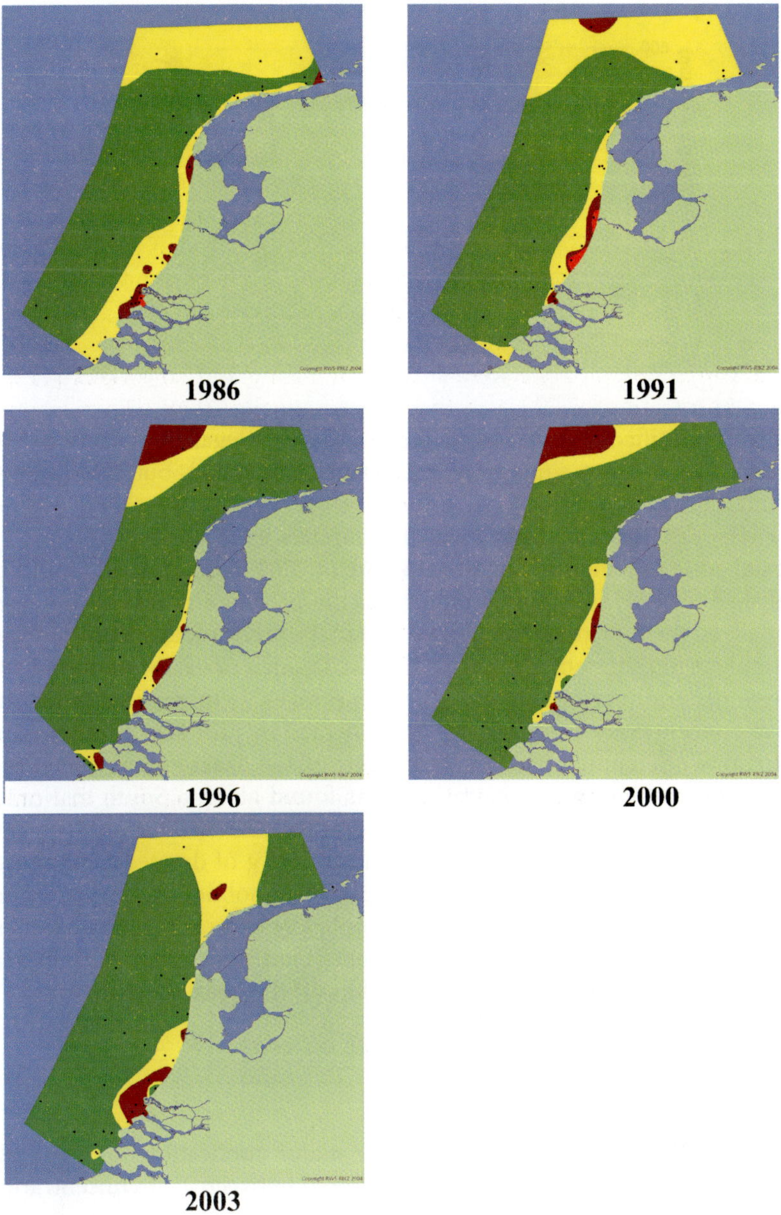

Fig. 14 Spatial distribution of PAH (sum of Borneff) in the fraction < 63 μm of surface sediments from the Dutch coastal zone in 1986, 1991, 1996, 2000 and 2003 (**A**), and the median of the sum of PAH concentration (six of Borneff) in surface sediments (< 63 μm) in three regions from the coastal zone (Southern coast: *squares*, Northern coast: *circles* (both extending to 20 km offshore), and the Central North Sea = 50 km offshore: *triangles*) versus the year of sampling (**B**). *Isolines* are drawn by hand, small dots = sampling locations (Courtesy: Hegeman and Laane [113])

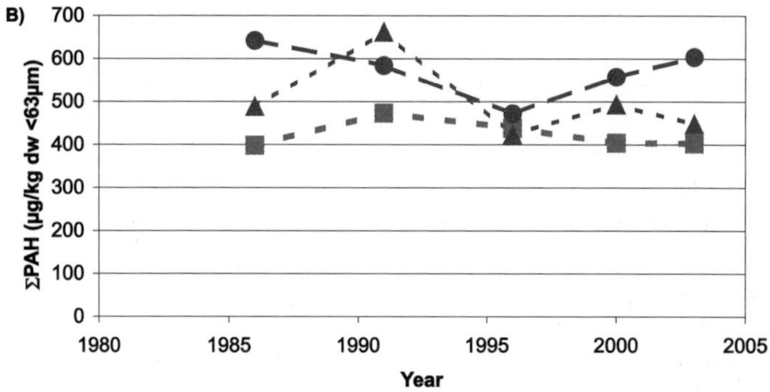

Fig. 14 (continued)

known quantified sources are described: The (il)legal discharges of oil (including PAHs) from ships have not yet been included, but could constitute a rather large source [116].

PAHs, azaarenes and oxygenated azaarenes were measured in the Dutch coastal zone in a monitoring programme in 2000. Spatial trends could be established for PAH from this and previous monitoring studies [112]. Azaarenes as well as their primary metabolic products are present in the marine environment in sediments, whereas dissolved concentrations in coastal surface waters generally are below detection limits [117]. In sediments the cumulative concentrations of azaarene transformation products amount to about four times the cumulative concentrations of the azaarenes (see Fig. 14). No significant decline in concentrations was found along a South to North (i.e. parallel to the Dutch North Sea coast) gradient. Levels of azaarenes amount to nanograms to several tens of nanograms per g of dry sediment. Levels of hydroxyquinolines (notably 2-OH and 5-OH quinoline) in marine sediments range from several tens to hundreds of ppb. Concentrations of azaarenes and oxygenated azaarenes in marine sediments tend to decrease with increasing distance from the coast, but beyond 30 km off the coast increase again.

4.2.4 Tributyltin Compounds

Tributyltin (TBT) compounds find their major applications in ship hull antifouling. When a relationship was discovered between TBT concentrations in aquatic systems and biological effects in molluscs, their application in yachting and other small vessels was banned or limited in many countries. The first studies in the Netherlands showed relatively high concentrations of TBT in the River Rhine, estuary and coastal zone and some harbours [84]. Since then the concentration of TBT has decreased substantially in different harbours along the Dutch coast [118, 119]. In the harbour of Scheveningen the concentration dropped significantly from more than 200 ng Sn L^{-1} (1990) to less than

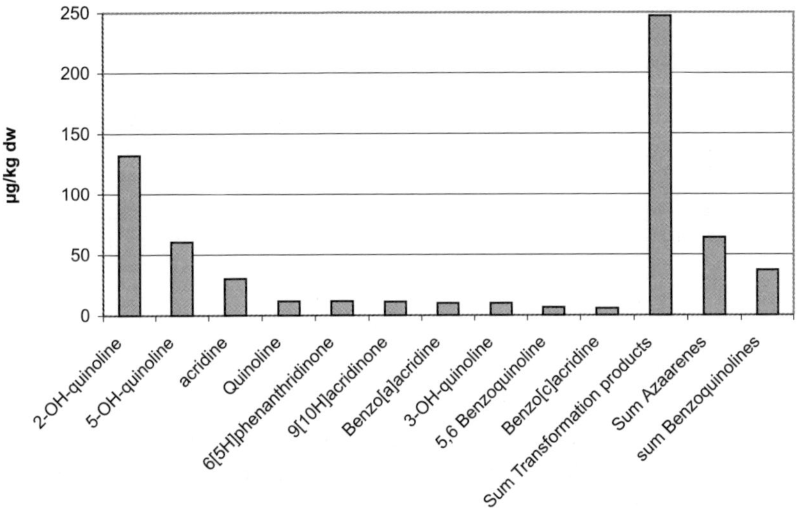

Fig. 15 Concentrations of azaarenes and oxygenated azaarenes in sediments from the Dutch coastal zone (From de Voogt et al. [144])

20 ng Sn L^{-1} [119]. In the Dutch coastal zone three out of nine organotin compounds measured, viz. MBT, DBT and TBT, were present in concentrations above the corresponding limits of detection.

The concentration of TBT in suspended matter is relatively high in the Rhine estuary (Maassluis) compared with the TBT concentrations at the locations where the Rivers Rhine and Meuse enter the Netherlands (Fig. 15).

This indicates that the shipping activities in and around the Rotterdam harbours are important sources of TBT. Although shipping on the North Sea has been established as the major source of TBT in the North Sea [120], the contribution of dredged material from the harbours of Rotterdam to the TBT concentration in the surface sediments of the Dutch coastal zone is also relatively high [121, 122]. The ratio of MBT : DBT : TBT changes from 24 : 29 : 47 in 1996 to 61 : 27 : 12 in 2000, and total butyltin concentration levels in the surface sediments of the Dutch coastal zone decreased from 55 µg Sn kg^{-1} d.w. in the fraction < 63 µm in 1996 to 15 µg Sn kg^{-1} d.w. in 2000 [123].

The substituent of TBT compounds, Irgarol 1051, could not be found in concentrations higher than the detection limit (0.2–0.6 g kg^{-1}) [124]. At various harbour locations and in the Scheldt estuary the concentration of dissolved Irgarol 1051 exceeded 24 ng L^{-1} many times [125].

4.2.5 Surfactants

Surfactants (surface active agents) are produced in many forms and find their application in e.g., detergents, cleaning agents and emulsifiers. Among the non-ionic surfactants the alkylphenol ethoxylates (APEO) have received particular attention because of the weak endocrine activity of some of their degradation products, notably nonylphenol (NP), and nonylphenol mono- and diethoxylate (A_9PEO_1 and A_9PEO_2, respectively).

Jonkers and coworkers [126, 127] investigated the fate of APEO and their transformation products in 1999 and 2000 during a field study in the Dutch coastal zone of the North Sea. A_9PEO, NP and the carboxylated metabolites (A_9PEC) were determined in surface waters, sediments and suspended matter, using Soxhlet extraction, solid phase extraction and liquid chromatography-electrospray ionisation mass spectrometry (LC-ESI-MS).

Concentrations of A_9PEO in estuarine sediments of the River Rhine decrease going downstream and this trend proceeds in the marine area (lower concentrations with greater distance from the shore) in 1999. In marine sediments, maximum concentrations of 277 and 86 ng g^{-1} (d.w.) for A_9PEO and NP were observed, whereas A_9PEC were only found sporadically. Relatively high concentrations of A_9PEO and NP were observed in marine sediments near the shore, which decreased until below 50 ng g^{-1} d.w. at about 10 km offshore. In the sediments, mainly the short chain A_9PEO_{1-3} was detected, indicating that extensive biodegradation has occurred.

Both the oxidative and non-oxidative hydrolytic biodegradation mechanisms were shown to occur in the Dutch coastal zone, although the oxidative route is less important than in the adjacent Scheldt and Rhine estuaries.

In these estuaries dissolved APEO are (bio)degraded into APEC, and dissolved concentrations of APEO decrease substantially due to dilution and degradation, depending on the residence time of the water in the estuary [128]. Thus in the Rhine estuary, where water residence times are short (typically several days), APEO concentrations decrease much slower than in e.g., the Scheldt estuary, with its residence times of several months. Dissolved concentrations of A_9PEO at the water surface are roughly one order of magnitude higher than several meters below the surface, both in freshwaters and marine waters.

Dissolved concentrations of A_9PEO in the coastal zone are low, but unusually high dissolved concentrations of A_9PEO were observed in a number of offshore samples with relatively high concentrations near or in shipping routes [126] (Fig. 16). Maximum dissolved concentrations of 36, 1.7 and 0.63 $\mu g\,L^{-1}$ were observed, respectively, for A_9PEO, NP and A_9PEC.

Jonkers and coworkers concluded that spatial profiles of A_9PEO in sediments and water in Dutch estuaries both exhibit declining trends going downstream, and that in the Dutch coastal zone this trend continues in sediments, whereas an opposite trend is observed in the dissolved phase.

Table 1 Estimated contributions of the freshwater sources of A_9PEO to the Dutch coastal zone (From Jonkers [129], with permission)

Source	Freshwater discharge[a] $10^9\ m^3\ y^{-1}$	Dissolved concentration $ng\ L^{-1}$	Concentration in SPM $ng\ g^{-1}$	Concentration SPM in water $mg\ L^{-1}$	Dissolved load $kg\ y^{-1}$	Load sorbed to SPM $kg\ y^{-1}$	Total load $kg\ y^{-1}$
Wadden Sea	10	66[b]	3228	17	660	549	1209
North Sea Canal	2.7	80[c]	6355	3	216	51	267
New Waterway[d]	44	94[e]	5356	13	4108	3043	7151
Haringvliet	22	50[f]	10 437	4	1105	923	2028
Scheldt estuary	4.3	88[g]	3023	14	378	182	560
Deposition (rain)	8.0[h]	91[i]	–	–	727	–	727
Total							12×10^3

[a] discharge data taken from [183]
[b] median concentration of 3 locations in the Wadden Sea
[c] concentration at location IJmuiden (freshwater side of the lock)
[d] this is the principal branch of the Rhine estuary
[e] median concentration in the Rhine estuary, taken from Jonkers et al. [128]
[f] concentration at location Haringvliet (freshwater side of the lock)
[g] concentration in Scheldt estuary near Vlissingen
[h] deposition is calculated for an area of 150×70 km (average rainfall in the Dutch coastal zone is 751 mm $year^{-1}$)
[i] data taken from Peters [184]

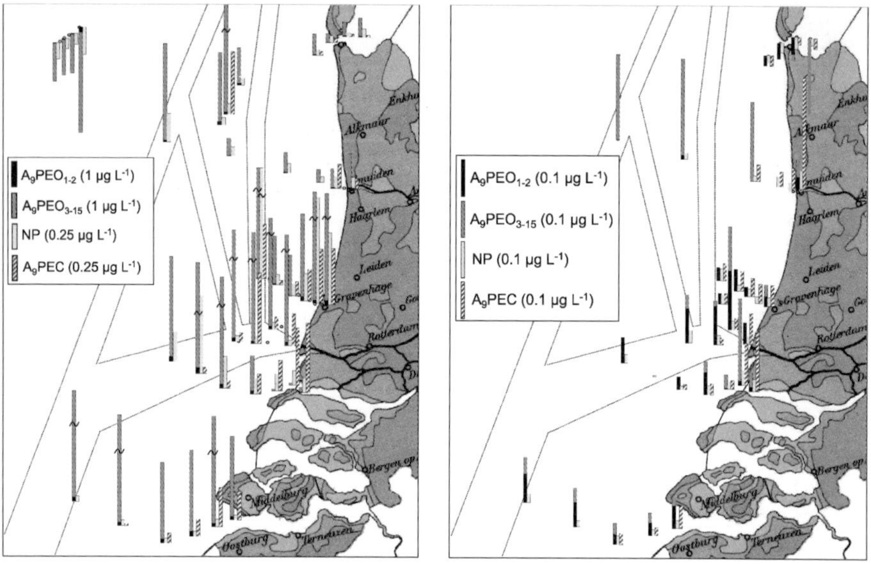

Fig. 16 Distribution of dissolved nonylphenol ethoxylates in the Dutch coastal zone and the River Rhine in 1999 and 2000 (From Jonkers et al. [126, 128])

Suspended matter generally contains much higher concentrations than sediments from the same location. This suggests that before settling of the particles degradation is likely to occur, and that degradation within the sediment is unimportant. More than 25% of the A_9PEO present in marine water is sorbed to suspended matter.

The main sources of A_9PEO in sediments in the Dutch coastal zone were identified as the Rhine and Scheldt estuaries, dump sites for harbour dredge and in some cases production platforms. For the water phase, the most likely sources of A_9PEO seem to be discharges from ships at open sea [126].

Jonkers and coworkers [127, 129] estimated the contribution of different freshwater sources of A_9PEO_n to the Dutch coastal zone from concentration and river discharge data. Table 1 shows the results of these calculations. The Rhine estuary (New Waterway and North Sea Canal) contributes approximately 60% to the estimated total load of A_9PEO_n in the coastal zone.

4.2.6 Phthalates

Phthalates are used as plasticisers to increase the flexibility of plastics and other polymers. Other applications include emulsifiers and carriers (of e.g., ink and paint). Some plastics may contain as much as 50% of phthalates by weight. Diethylhexyl phthalate (DEHP) accounts for about 25% of the total amount of plasticisers produced.

In a large national survey on estrogenic compounds carried out in 1999 in Dutch aquatic systems phthalates were measured in surface waters, suspended particulates and sediments [130]. The sample locations included marine and estuarine locations. It was concluded that the observed occurrence of phthalates in the Netherlands is mainly related to diffuse sources, because dissolved concentrations in rainwater were similar to those in surface waters. Dissolved concentrations of phthalates in marine waters were similar to those in freshwaters, with median levels generally below 0.35 $\mu g\,L^{-1}$.

All phthalates analysed were found to be present in one or more samples of suspended matter and sediment [130, 131]. In suspended matter and sediments levels of DEP, DMP, DBP, BBP, DEHP and DOP (Diethyl phthalate, dimethyl phthalate, di-*n*-butyl phthalate, butylbenzyl phthalate, di(2-ethylhexyl) phthalate and di-*n*-octyl phthalate respectively) were between the limit of detection and 4.1 $\mu g\,g^{-1}$ d.w. DEHP was the most abundant and reached the the highest concentrations. In 1999 the maximum concentrations of DEHP were 19 $\mu g\,g^{-1}$ in the North Sea Canal near IJmuiden and 14.5 $\mu g\,g^{-1}$ at Eijsden where the River Meuse enters The Netherlands. In general, levels in suspended matter in seawater were of the same order of magnitude as in freshwater bodies. Comparison of median levels in the Rivers Rhine, Meuse and Scheldt revealed a remarkable pattern: while DEHP levels in the suspended matter were high in all rivers (Rhine median 3 $\mu g\,g^{-1}$, Meuse 10 $\mu g\,g^{-1}$ and Scheldt 8 $\mu g\,g^{-1}$), in the Scheldt also the levels of the more polar phthalates DMP, DEP and DPP were relatively high. It is surprising that these more polar components were not found in the dissolved phase of the corresponding surface waters.

Klamer [124] analysed phthalates in sediments from the Dutch coastal zone. The phthalates (DMP, DEP, DBP) were below detection limits (0.5 $mg\,kg^{-1}$) in the surface sediments (fraction smaller than 63 μm) of the Dutch coastal zone in spring 2000 [124]. BBP and DOP were close to the detection limit in some samples. However, the concentration of DEPH varied between 0.28–5.5 $mg\,kg^{-1}$ (standardised on 10% organic matter).

4.2.7 Estrogenic Compounds

Several xeno-estrogens have been discussed in the previous sections. In this section the concentrations of some additional xeno-estrogens, viz. brominated flame retardants, bisphenol A, and estrogenic hormones in the Rhine estuary are described.

Estrogenic Hormones

Estrogenic hormones control the estrogen homeostasis in organisms. The compounds 17α-estradiol, 17β-estradiol and estrone are natural female sex hormones produced by humans, mammals and other vertebrates. They play an important and essential role in reproduction, development and behavior. 17α-ethynylestradiol is the active ingredient of the contraceptive pill. Estrogenic hormones are released into the environment through sewer systems and from cattle dung and manure.

In the Dutch national survey on estrogens and xeno-estrogens in the aquatic environment [130] estrogenic hormones were determined in the Rhine delta. Dissolved concentrations of 17α-estradiol (E2-17α), 17β-estradiol (E2) and 17α-ethynylestradiol (EE2) were invariably below the limit of detection (0.3 ng L^{-1} for both E2-17α and EE2, and 0.8 ng L^{-1} for E2). Estrone (E1) was observed in the New Waterway in 1999 (locations Maassluis and Splitsingsdam) with concentrations ranging from < 0.3 to 0.6 ng L^{-1} (3 out of 6 samples had levels above the l.o.d. of 0.3 ng L^{-1}). In the Meuse delta at Keizersveer concentrations varied from 0.4 to 1.0 ng L^{-1} ($n = 3$). Concentrations of estrone in the coastal zone were all below the l.o.d. of 0.3 ng L^{-1}.

Polybrominated Diphenyl Ethers

Flame retardants are used extensively in textiles and electronic equipment. Among the flame retardants applied, from an ecotoxicological point of view particular interest has been given to the polybrominated diphenyl ethers (PBDE) because they have been found widespread in the abiotic and biotic environment and their concentrations appear to increase steadily.

De Boer and coworkers [132] determined concentrations of polybrominated diphenylethers (PBDE) in sediments and particulate matter collected in Dutch estuaries and the Dutch coastal zone. The highest concentrations of BDE47 and 99 were found in suspended particulate matter from the Haringvliet (River Rhine delta) amounting to 5.2–9 and 4–12 ng g^{-1} d.w., respectively. The BDE47 and 99 concentrations at two other locations in the Rhine delta, the New Waterway and Rotterdam harbour were also relatively

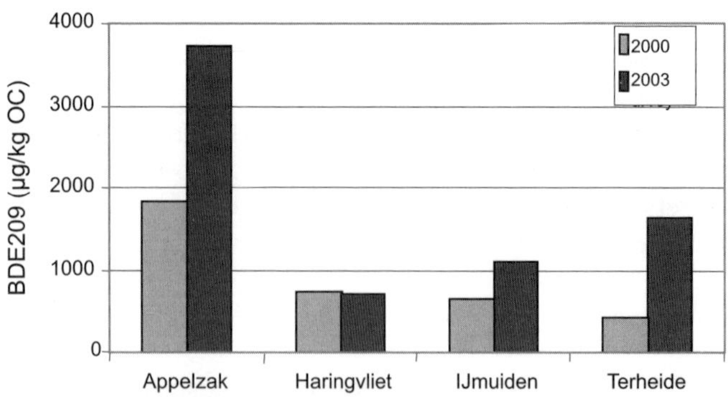

Fig. 17 Comparison of concentrations of the flame retardant BDE 209 in coastal sediments collected from four different locations in 2000 and 2003 (From Åkerman et al. [134])

high. Concentrations of decaBDE (BDE209) amounted to around 100 ng g^{-1} in samples from these locations. Given the extremely low solubility of higher brominated PBDEs (BDE209 < 0.1 μg L^{-1}) in water and the high PBDE concentrations found in these suspended matter samples, suspended particulate matter was identified as an important carrier for higher brominated diphenyl ethers in the aquatic environment.

Concentrations of BDEs 49, 77 and 209 in sediments from the Rhine estuary amounted to 3, 2 and 30 ng g^{-1}, respectively, in the Rotterdam harbour, and to 2, 1 and 20 ng g^{-1} at the mouth of the estuary.

The mean concentration of BDE209 in North Sea sediment off the Western Scheldt was 32 μg kg^{-1} dry weight (< 63 μm fraction, [133]). BDE concentrations in North Sea sediments elsewhere along the Dutch coast were slightly lower, and decreased further as sampling locations were further away from the coast [133]. Åkerman and coworkers [134] reported concentrations of PBDE in sediments collected from the same sites and noted an average increase of BDE 209 by a factor of 2 (see Fig. 17).

Bisphenol-A

Bisphenol-A is an intermediate product used in the chemical industry in the production of epoxy resins, polycarbonate plastics and flame retardants. Concentrations of Bisphenol-A (BpA) in the Rhine/Meuse delta have been reported by Vethaak and coworkers [130]. In the New Waterway in the dissolved phase, a mean ($n = 3$) concentration of 48 ng L^{-1} was found, whereas in the adjacent coastal zone concentrations were invariably below the limit of detection of 15 ng L^{-1}.

4.2.8
Pharmaceuticals

A special aspect of pharmaceuticals, compared with for instance pesticides, is that they are used the whole year around. Consequently, water organisms will be subjected to a mixture of pharmaceuticals and their metabolites during their whole life cycle. Little is known about the possible impact of these relatively low concentrations on organisms and population over the longer term [135].

In the Netherlands about 12 000 human and about 2500 veterinary pharmaceuticals are permitted. Beside the active substance, pharmaceuticals contain various other compounds to make them convenient to take in and administer the dosage [136]. About 850 active compounds are present in human pharmaceuticals and 200 active compounds in veterinary pharmaceuticals. In various literature studies the available information about possible sources and the possible risks were assessed for various pharmaceuticals [137–139]. The first measurements in Dutch surface waters were performed by Mons and coworkers [140].

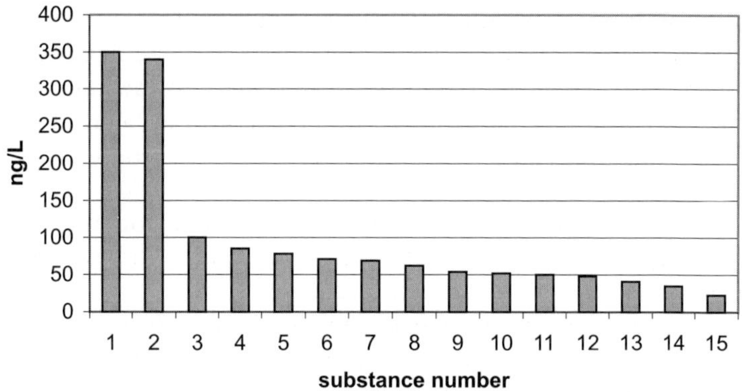

Fig. 18 Average dissolved concentrations of 15 pharmaceuticals, ranked from the highest to the lowest concentration, in river water collected near the Rhine estuary (Maassluis). Main area of application is given in brackets. (1) jopromide (X-ray); (2) carbamazepine (anti-epileptic); (3) amidotrizoinic acid (X-ray); (4) sulfomethoxazol (antibiotic); (5) jomeprol (X-ray); (6) ibuprofen (analgetic); (7) sotalol (β–blocker); (8) johexol (X-ray); (9) dimetridazol (coccidiostatic); (10) acetylsalicylic acid (analgetica); (11) diclofenac (analgetic); (12) incomycine (antibiotic); (13) bezafibrate (antilipaemic); (14) anhydro-erythromycin (antibiotic); (15) fenoprofen (analgetic). (From Schrap et al. [136])

Schrap and coworkers [136] selected different categories of pharmaceuticals, including antibiotics for a baseline study in 2002: coccidiostatics, analgesics, X-ray contrast media, drugs used to treat coronary vascular diseases, antineoplastic drugs and antiepileptics. No effort was made to detect metabolites in this study.

The concentrations observed for the various pharmaceuticals in Dutch surface waters were in agreement with those found abroad. In Fig. 18 the top 15 of the highest concentrations found in Maassluis and Keizersveer are given.

4.2.9
Perfluorinated Alkylated Compounds

Perfluorinated alkylated compounds (PFC) consist of fully fluorinated alkyl chains with a polar head group that can be a sulfonate or carboxylate anion, or a neutral moiety such as a hydroxyl or sulfonamide group. Since the 1950s PFC have been used extensively in textiles, carpets, in the manufacturing of fluorinated polymers, in paper and cardboard, and in fire fighting foams [97]. Their main applications include stain, moist and oil repellency, surfactant and as a polymerisation aid. The group of PFC includes several highly persistent chemicals, e.g., perfluorooctanesulfonate (PFOS) and perfluorooctanoic acid (PFOA). The PFC have recently emerged as new persistent organic pollutants with the discovery of their presence in biota from remote areas such as the Arctic [141–143].

Recently, the Dutch Ministry of Transport, Public Works and Water Management initiated a survey in order to map the presence of PFOA and PFOS in the aquatic environment in 2002 [144, 145]. Samples of sediments ($n = 11$) and suspended particulate matter ($n = 18$) were collected from rivers and the coastal zone and analysed by LC-MS methods.

In general PFOA was not found at levels above the limit of detection of 0.4 ng g^{-1} d.w. in suspended particulate matter. The exceptions included the border locations in the Rivers Meuse, Scheldt and Rhine where these rivers enter The Netherlands, where levels of 24, 3.5 and 2 ng g^{-1} were found, respectively, and two locations in the coastal zone, viz. IJmuiden outside the locks (18 ng g^{-1}), and De Panne offshore from the Belgian/French coast (2 ng g^{-1}). In coastal sediments PFOA levels ranged from < 0.4 to 3 ng g^{-1} with the highest levels (2–3 ng g^{-1}) offshore from Noordwijk, IJmuiden and Callantsoog, and decreasing further northeast to levels below 0.4 ng g^{-1} at Terschelling and in the Ems estuary.

PFOS levels were generally above the limit of detection of 0.4 ng g^{-1} in both suspended particulate matter and sediments. In suspended particulate matter from the River Rhine, PFOS levels varied from 31 ng g^{-1} at Lobith to 10 ng g^{-1} in the New Waterway at Maassluis, 5 ng g^{-1} in the Haringvliet, and 3 ng g^{-1} in the Rhine tributary Oude Maas. In suspended particulate matter ($n = 3$) from the coastal zone, levels varied between < 0.4 and 9 ng g^{-1}. In 9 out of 11 marine sediments from the coastal zone PFOS levels were above the limit of detection. Levels ranged from of < 0.4 to 5 ng g^{-1}.

The results of this screening survey indicated that the PFC are found widespread in the Dutch freshwater and marine aquatic environment.

4.3 Bioassays and Biological Effects

In general water quality is usually characterised by assessing the presence of individual chemical compounds. The pollution pressure is thus evaluated mainly through chemical analysis. In this way about 200–300 chemical compounds are analysed regularly in The Netherlands. However, little to nothing is known about the possible concentrations and impact of the other 100 000 odd chemical compounds. In addition, the presence of a certain compound does not necessarily mean that there will be an adverse impact. Moreover, it is currently recognised that the present chemical extraction techniques quite often do not take into account the bioavailable fraction of the chemical compounds.

Hendriks [146] showed that with the chemical compounds analysed in Rhine sediments only 30% of the negative effects observed in an in vitro bioassay exposed to sediment extracts could be explained. Hence, 70% of the observed effects were due to other, non-identified, chemicals and/or synergy between different compounds. Cornelissen [147] and Jonker [148] showed that only minor parts of the total concentrations of PCBs and PAHs in sediments are bioavailable. Finally, in the derivation of the targets for each individual chemical, the synergy of different compounds is currently taken into account by applying a safety factor. The present targets are for individual chemical compounds and indicate the level where there is a chance of a chronic effect.

Bioassays may provide a picture of the overall toxicity, and include the possible effects of the non-identified (groups of) compounds and their metabolites. With bioassays it is also possible to detect special biological effects, such as mutagenicity or estrogenicity.

Different bioassays can be distinguished: in vivo and in vitro bioassays. In vivo bioassays are experiments where living organisms are exposed to water, sediments, effluents, dredged material or extracts. They can be performed both in the laboratory and in the field. Acute effects (e.g., mortality) are measured during short-time period experiments, whereas the testing of chronic effects involves longer (weeks to months) lasting experiments.

In vitro bioassays are laboratory experiments where repaired cells or sub cellular systems are isolated from organisms or modified bacteria. They usually identify a specific mechanism of action.

Several bioassays have been tested and used for the monitoring of the quality of harbour sediments from Rotterdam and the coastal zone in the period 1999–2002 [122, 149]. Three bioassays passed the criteria for application to harbour sediments, and their application was subsequently standardised: (i) the survival rate of *Corophium volutator*; (ii) Microtox Solid Phase test (a measure of the impact on metabolic processes); and (iii) the DR-CALUX assay (luciferase expression as a measure of dioxin-like toxicity [150]).

In the sediments collected from the Rotterdam/Rhine estuary area and the coastal harbours in the period 1999–2001 no significant temporal or spatial differences were observed when using the *C. volutator* bioassay. In 2002, a significant decrease compared to the preceding period appeared to have occurred [149].

The response of the Microtox SP test did not show any temporal differences either. However, a geographical difference was observed when applying this bioassay. In the Wadden Sea (1999) and coastal zone (2000/2001) a significantly higher response compared to the Rhine estuary was reported. In 2001 the DR-CALUX responses of extracts from sediments from the Rotterdam harbours appeared to be elevated compared to the preceding years, when almost invariably levels below 50 ng TEQ kg^{-1} d.w. were recorded. The reason remains unknown and in 2002 normal values were measured again [149].

Four different types of bioassays were performed to identify the negative effects of chemical compounds in dredged material from the Rhine estuary dumped in the Dutch coastal zone in 1996. Up to 1996 the dredged material was dumped at a location, somewhat north of the mouth of the Rhine and within the 20 m depth line [26]. From then on the dumping site moved more to the northeast. In this way, it was calculated that the return flow of dredged material after dumping to the harbours of Rotterdam decreased considerably [26].

The bioassays involved in vivo tests with *Echinocardium* and *C. volutator* exposed to sediment extracts, and with *Bathyporeia sarsi* and *Brachionus plicatils* exposed to interstitial water. Other bioassays used included fish diseases (liver tumours and skin diseases) and several biomarkers (Cytochrome P450, DR-CALUX, DNA integrity, acetylcholine esterase and benzo(*a*)pyrene hydroxylase [26]).

None of these in vivo and in vitro bioassays elicited a significantly different response when compared to the responses registered for sediments or interstitial water from a reference location. The only difference between the locations was that the bioaccumulation of lead, zinc, mercury and PCBs was significantly higher at the dumping ground.

Up to 4 km from the dumping location, negative effects on the bottom fauna have been registered [26]. These effects could be attributed to physical and general chemical changes due to the dumping of dredged material. The silt content of the surface sediment at the dumping site (17%) was ten-times higher than in the surrounding sediments and the amount of organic matter was nearly two-times less, i.e. 2% instead of 5%. However, 78% of the silt fraction in the dredged material is removed within a short time by tidal currents and wave action. The dumping until 1996 resulted in a strong decline in numbers and biomass of bottom fauna (> 1 mm). The same effects were found at the new dumping site, located more seawards, after one year. The bottom fauna > 10 mm had been reduced considerably during dumping of dredged material. After one year the number of species had increased from 4 to 12 but still had not reached the number of 18 species that were recorded at the reference site. After 4 years the situation had not changed.

At the former dumping site, the recuperation of organisms was rather fast. In 1998 the number of species, density and biomass were even higher than at the reference location. After 4 years the bottom fauna had returned nearly completely, although the number of species was not as high as at the reference site [26]. Remarkably, only the mobile macrofauna species such as crabs and fishes had returned, whereas the sediment burying species had not.

In vitro bioassays were applied more recently to identify toxicity in the Dutch coastal zone. Åkerman [151] prepared hexane-acetone (3 : 1) extracts from the fraction < 63 μm of 10 samples taken from the surface sediments in the Dutch coastal zone in 2000. The activity of the extracts was tested on acute toxicity (Microtox) and genotoxicity (Mutatox). All extracts responded positively in the Microtox test: EC_{50} ranging from 276.4–1422 mg L^{-1} after 15 min. The highest toxicity was found at a location 10 km offshore Noordwijk. In general the toxicity of the coastal samples (< 20 km offshore) was higher than the toxicity of the offshore samples (> 70 km offshore). The acute toxicity values in the Dutch coastal zone appeared to be the same as in extracts from the surface sediments of the western Wadden Sea [152].

All extracts of the samples from the Dutch coastal zone in 2000 appeared to elicit low genotoxic responses (LOEC between 3×10^3 and 10^4 mg L^{-1}). The highest values were found at Noordwijk 10 and at the oyster grounds. The values were in the same order of magnitude as the extracts from surface sediments from the western Wadden Sea [152].

Dioxin-type toxicity was determined with the DR-CALUX assay in extracts (hexane-acetone; 3 : 1) of the fraction smaller than 63 μm from the surface sediments of the Dutch coastal zone. The toxicity could be explained in part by the presence of brominated flame retardants and PCBs [133] (see Sect. 4.2.7). Maximum responses were found close to the coast and near the outflow of the Rhine and North Sea Canal. Klamer and coworkers [133] concluded that unidentified compounds, like PAH-type or natural compounds, with a dioxin-like mode of action have to be identified before the response from the assay could be linked to compounds.

Bioassays were used again in the Dutch coastal zone in 2003 [134]. Three in vitro bioassays were used: Mutatox, ER-CALUX (estrogenic activity) and DR-CALUX. Three complementary in vivo assays were also used for the evaluation: Microtox SP, *Nereis diversicolor* and *C. volutator*.

Extracts made from sediments and suspended matter collected in the Dutch coastal zone in 2003 did not elicit a response in the Mutatox assay or in any of the in vivo bioassays [134]. In the ER- and DR-CALUX assays differences between areas and locations could be identified. In the Dutch coastal zone the response of the DR-CALUX in suspended matter and surface sediments was between 12 and 49 pg TCDD – TEQ g^{-1} d.w. Remarkably, the extracts that had been prepared with gel permeation chromatography (GPC) had a more than 300 times higher, unexplained, response than the extracts from the same samples prepared on a silica column. Estrogenic activity (measured with the ER-CALUX) ranged between 0.1 and 0.42 pM EEQ g^{-1} d.w. in suspended matter. In the surface sediments of the Dutch coastal zone invariably values were reported below 5 pM EEQ g^{-1} d.w. [134].

The response of the DR-CALUX assay (to sediment extracts prepared with GPC) was more than 100 times higher in the Dutch coastal zone in 2003 than in 2000 [134]. In that period the concentration of the brominated flame retardants (PBDE 209) also increased in the Dutch coastal zone, but a direct relationship with the strong increase in DR-CALUX could not be shown [134].

5 Assessment

As outlined above, different methods have been applied to describe the presence and impact of chemical compounds. In this section the focus will be on the comparison of concentrations with targets set for chemical compounds.

Also attention will be paid to the EU priority setting method COMMPS and to the comparison of bioassay responses with targets.

5.1 Chemical Targets

To protect the aquatic environment from the negative effects of chemical compounds, international policymakers focussed initially on emission reduction and agreed upon a reduction of the emissions by at least 50% [52]. Presently, a almost zero emission has been agreed upon for man-made compounds, while for natural compounds emissions up to or below background concentrations should be targeted [52, 153].

Beside these emission targets, emission targets were also set for the water quality policy. In the Netherlands the Maximum Permissible Risk concentration (MPRC) and the Negligible Risk concentration (NRC) have been derived for more than 100 compounds [154]. The MPRC presents the minimum level that has to be reached within a (hitherto non-specified) period of time (if possible by 2010). The NRC is the concentration that should be reached in the long term. The MPRC is based on scientific and policy-related considerations. "Reliable" NOEC (no effect concentrations) for various organisms from different taxa were used in a model to produce a log-normal chance distribution curve of the NOEC values [155, 156]. The MPRC is scientifically defined as the value where a randomly selected group of organisms has, with a certainty of 95%, a higher NOEC value [37]. The NRC is defined as the concentration corresponding to one tenth of the MPRC. If this concentration is lower than the natural background concentration, the latter is taken as the NRC [154]. The MPRC for suspended matter and for sediments is calculated from the MPRC value (valid for the dissolved concentration) using an average partition coefficient [37]. The targets for chemical compounds in sediments are calculated for a standardised sediment, thus accounting for the large variation in silt content [154].

Chapmann and coworkers [157] and Laane and coworkers [158] showed that for metals and nutrients respectively, the targets in different countries are not the same. The basis, i.e., the toxicological data, is the same. The calculation method, the uncertainty factors and the political choices differ, however, resulting in differences between the minimum and maximum values that amount to sometimes more than a factor of 100.

Presently, the targets from the European WFD [8] are setting the scene instead of the national targets. The European assessment stretches out to the first 11 miles offshore. In the more seaward part of the Dutch coastal zone the OSPAR targets ought to be used. Currently, there exists a difference between the OSPAR and the EU targets. Recently, however, a start has been made to harmonize the OSPAR methods and their values with the European targets [159].

The first time that the chemical status of the Dutch coastal zone was assessed broadly and systematically was in 1991 [160]. This assessment covered the period 1980–1990. The main focus was on nutrients and metals, and some organic compounds were also incorporated. Two important conclusions were drawn, viz. that the riverine and atmospheric input led to enhanced concentrations in the Dutch coastal zone, and that the concentrations of the chemical compounds studied started to decrease within the period 1980–1990.

Van Eck and coworkers [161] were the first to make a risk assessment of the Dutch coastal zone. The assessment method they used was based on the comparison of the 90 percentile of a chemical substance in the Dutch coastal zone with the chemical target [162]. The concentration of three metals (cadmium, copper and mercury), five pesticides, tributyltin compounds, lindane, dichlorvos, mevinfos and parathion-ethyl, PCB153 and the PAHs fluoranthene, benzo(*b*)fluoroanthene, benzo(*a*)pyrene, benzo(*ghi*)perylene and indenopyrene were above the MPRC. The general conclusion was that very often there were insufficient data available to make a proper assessment. Augustijn and coworkers [163] showed that the uncertainty in the 90 percentile concentration and the uncertainty in the chemical target are rather large and that the classification of a chemical compound is strongly dependent on the uncertainty in the ratio between the concentration and the target. Later, Swertz [164] also showed that the result of the assessment is strongly dependent on the compartment studied. It appeared that the ratio was below the target for the dissolved phase, above the target for suspended matter and below or above the target for sediments. This is mainly caused by the average partition coefficient that is used in the calculation of the target value in the suspended matter and sediment from the dissolved phase (see also [165]). Stolwijk and coworkers [166] showed, for instance, that the partition coefficient of cadmium derived from field data deviates tremendously from the mean partition coefficient used for the assessment.

In the Aquatic Outlook project [167] the assessment of the chemical status of the River Rhine, the Rhine estuary and the Dutch coastal zone was performed by comparing the chemical targets with the 90 percentile of the concentrations in a certain water system. Laane and Groeneveld [168] showed that this method is too drastic: if the concentrations at a few locations (often close to the outlet of rivers) are well above the target value then the whole water system can become identified as a problem area. Instead, Laane and Groeneveld [168], therefore, proposed to use the percentage of the sampling locations where the concentration is below the target. In this way a more realistic picture of the water quality of the whole system is presented (Fig. 19). Later, this method was applied for all sediment quality monitoring data in the Dutch coastal zone collected in the period between 1986 and 2003 [113]. As Fig. 19 demonstrates, the concentration of PAHs decreases with time, and the percentage of the sampling locations where the concentration is below the tar-

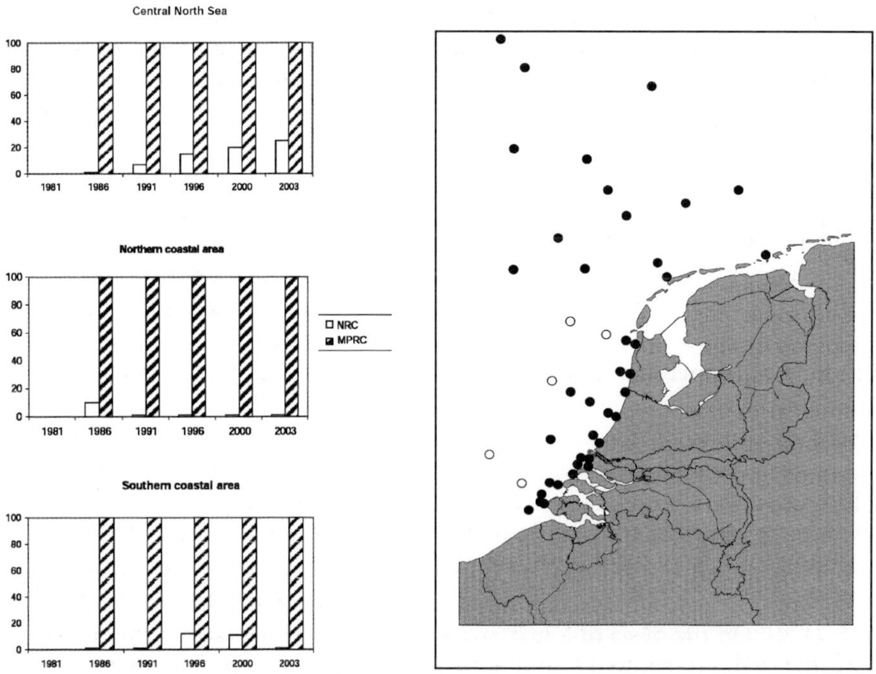

Fig. 19 Percentage of the sampling locations in three regions of the Dutch coastal zone, where the concentration of PAHs (six of Borneff) in the surface sediments (< 63 μm) is below the maximum permissible risk concentration (MPRC, *striped bars*) or below the negligible risk concentration (NRC, *open bars*) in the period 1986–2003. *Left hand panels*: grouping according to region. *Right hand panel*: sampling locations with concentrations below or above NRC; *filled circles*: ≥ NRC; *open circles*: < NRC (From Hegeman and Laane [113])

get (MPRC and NRC) increases with time. This water system was assessed as polluted if the 90-percentile method was applied.

Recently, the water quality of the Dutch coastal zone was assessed again [123, 169]. From these assessments the concentration of PAHs and TBT compounds appeared to exceed permissible levels in the Dutch coastal zone.

An overall picture of the assessment of the chemical compounds described above is presented in Table 2.

Targets have been set for the dissolved and total concentration of α- and γ-HCH in the Netherlands. The MPRC was set to 3300 ng L^{-1} for total and dissolved α-HCH, and to 860 and 800 ng L^{-1} for dissolved and total γ-HCH, respectively. The NRC has been set to 33 ng L^{-1} [99]. Although HCH can still be detected in the River Rhine and its estuary, its present concentrations are below the targets set. Hence, the chance of a possible negative effect of HCH on organisms occurring in the River Rhine and its estuary is nearly zero although after 2000 concentrations have exceeded the NRC level (see Fig. 9).

Table 2 Assessment of chemical compounds by comparing their environmental concentrations with their target values. MPRC = (Dutch) maximum permissible risk concentration, and FHI = preliminary targets derived for the European Commission by the Fraunhofer Institute [8]. *Black* = above target, *grey* = below target, white= no targets available

Substance	MPRC	FHI
HCHs (α, γ)	[grey]	[grey]
TBT	[black]	[black]
PAHs (six of Borneff)	[black]	[black]
Azaarenes		
Nonylphenol ethoxylates	[black]	
Nonylphenol	[black]	
Phthalates	[black]	
Estrogenic hormones		
Pharmaceuticals		

Target setting for TBT appeared to be a scientific and political problem [122]. On the basis of a few toxicological experiments, and the observed imposex effects, the MPRC and NRC targets for the dissolved concentration were set to 1 and 0.01 ng L^{-1} in marine waters, respectively. For fresh water the TBT targets were set to about 10 times higher levels. Recently, it has been demonstrated that the dissolved concentration of TBT has dropped significantly in the Dutch marinas and coastal zone; probably due to the ban on ships shorter than 25 m [119, 121]. Dissolved concentrations of TBT are presently around or below the MPRC.

However, the concentrations in the surface sediments of the River Rhine, the Rotterdam harbours and the associated Rhine estuary are far above the target (10 μg kg^{-1}) set for TBT compounds. Policymakers decided to increase this target for TBT in surface sediments by at least a factor of 10 [122]. This was the only possibility to continue the discharge of dredged material for the Rotterdam harbours in the Dutch coastal zone.

The presence of azaarenes and their oxygenated metabolites in the Dutch coastal zone has been measured and evaluated by De Voogt and coworkers [144] and Wiegman [170]. For these compounds it appeared that the concentrations of the parent compounds were relatively low and below the chemical targets, whereas the oxidised metabolites were present in higher concentrations than the parent compounds. The toxicity of the metabolites is also higher than their parent compound. Wiegman [170] argued that the toxicity of azaarenes to aquatic ecosystems increases due to the formation of transformation products, and therefore questioned if the existing targets provide sufficient protection against the demonstrated photoenhanced toxicity of azaarenes.

The presence of nonylphenol ethoxylates and nonylphenol in the Rhine estuary and Dutch coastal zone was evaluated by Jonkers [129]. He used indicative MPRC values for these compounds that were proposed recently [171]. For NP, MPRC values of 0.33 $\mu g\,L^{-1}$ in water, and 0.105 $\mu g\,g^{-1}$ d.w. in sediment were proposed, and for A_9PEO 14 $\mu g\,L^{-1}$ in water and 8.7 $\mu g\,g^{-1}$ d.w. in sediment. On the basis of these indicative target values Jonkers [129] concluded that at some locations in the Rhine and Scheldt estuaries the target values were exceeded in both water and sediments. In the coastal zone, in particular in shipping routes, indicative MPRC levels were exceeded by a factor up to 35 times for A_9PEO and a factor of almost 2 for NP, however. All levels were below the experimentally derived threshold (no effect) values for endocrine disruption. Since the latter mechanism of action is considered to be additive, the no effect values may be exceeded because of the presence of other endocrine disrupting compounds.

The indicative MPRC estimated for phthalates has been set to 1 mg DEPH kg^{-1} of standardised sediment [172]. The calculated NRC is set to 0.0085 $mg\,kg^{-1}$ of standardised sediment. The maximum value reported of DEPH close to the dumping area of dredged material exceeded the indicative NRC by a factor of 1258. On average the concentration of DEPH exceeded the indicative NRC by a factor of 660. Klamer [121] suggested that these rather high concentrations of DEPH could be partly caused by contamination during sieving of the samples.

For EE2 an ad hoc MPRC has been proposed of 1.0 $\mu g\,L^{-1}$ [173]. The dissolved concentrations of EE2 reported thus far in surface waters in The Netherlands are below 1 $ng\,L^{-1}$, and hence far below the MPRC level. Neither for the natural estrogenic hormones nor for pharmaceuticals are ecotoxicological target values currently available. Therefore, it is not possible to provide an assessment for these compounds.

For BpA an ad hoc MPRC has been proposed corresponding to 64 $\mu g\,L^{-1}$ in the dissolved phase and 22.9 $\mu g\,g^{-1}$ d.w. in sediments [173]. None of the BpA levels reported in the large survey on estrogenic compounds in the Dutch aquatic environment [130] exceeded these MPRC values.

As already discussed, specific chemical targets have been derived in the Netherlands for harbour sediments [149]. Compared to the Dutch objectives for sediments, the concentrations of different compounds in the dredged material from the Rotterdam harbour are relatively high. So, to allow dumping in the Dutch coastal zone, higher target values for dredged material have been set: the so-called product values [122]. Yland and coworkers [174] evaluated the chemical assessment of the dredged material from Rotterdam, against these product values. To take into account the variation in silt content (< 63 μm) the concentrations of the different compounds in the sediments were standardised to 25% lutum and 10% organic matter. Yland and coworkers concluded that in more than 25% of all samples the average concentrations of chrysene, phenanthrene, benzo(*a*)pyrene, benz(*a*)anthracene,

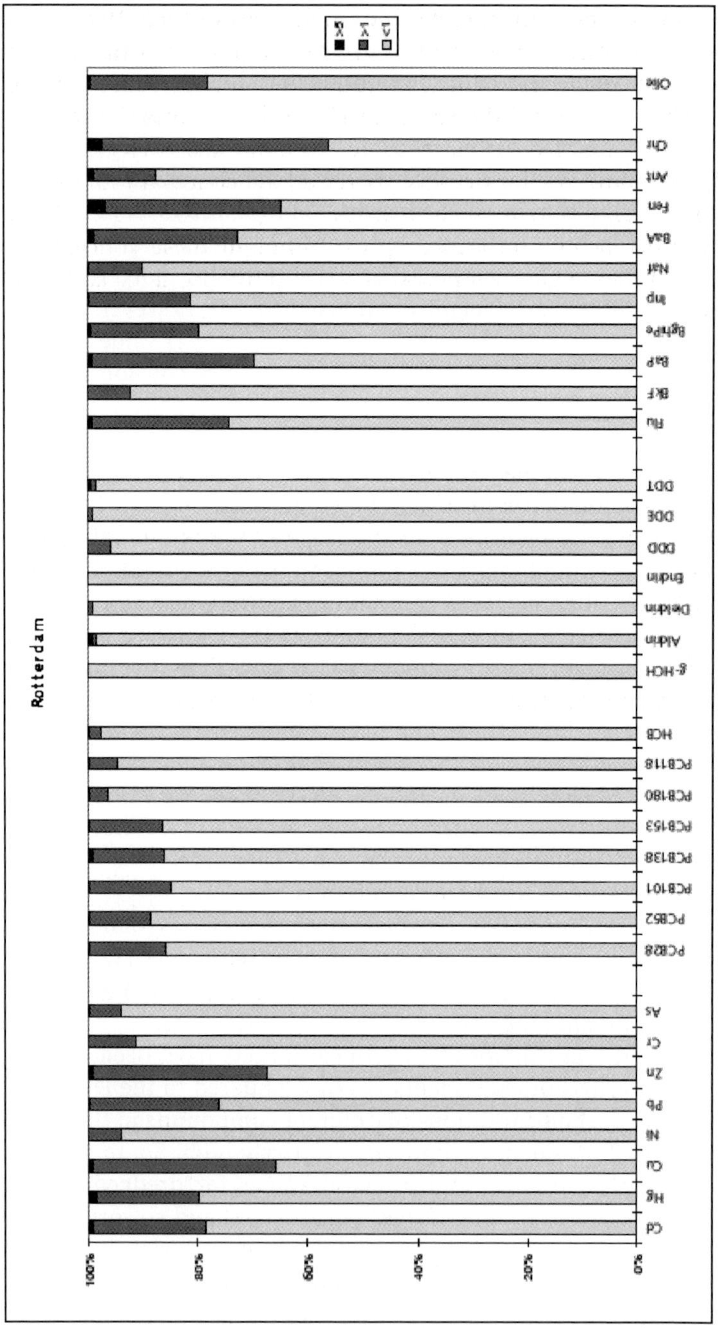

Fig. 20 Percentage of the sediment samples from the Rotterdam harbour area in which the concentration is less than the objective, between 1 and 5 times the objective, and 5 times above the objective, respectively, in the period 1986–1997

Cu and Zn were above the target values in the period 1986–1997. In 3% of the samples the concentrations of 13 of the 34 compounds measured was more than five times the target value (Fig. 20). This is mainly caused by the metals (Cd, Hg, Cu and Zn) and the PAHs. In 1% of the samples the concentrations of endrin also exceeded the target values by more than a factor of five.

At the dumping site for dredged material, the concentrations of the heavy metals, PCBs and PAH in surface sediments (standardised on the fraction < 63 um) were below the MPRCs [113] although these concentrations are two to five times higher immediately after dumping than at the reference location [26].

5.2 COMMPS

The COMMPS (Combined Monitoring-based and Modeling-based Priority Setting [99]) method has been developed for deriving a priority index for chemical compounds identified in screening surveys (see [101]). This method takes into account the bioaccumulation and toxicity potential of the various compounds. The index is based on multiplying the exposure index with an effect index. The determination of the exposure index is based on a logarithmic scaling of normalised concentrations using a scale from 1 to 10. It is assumed that the degradation of compounds identified is negligible. The effect index

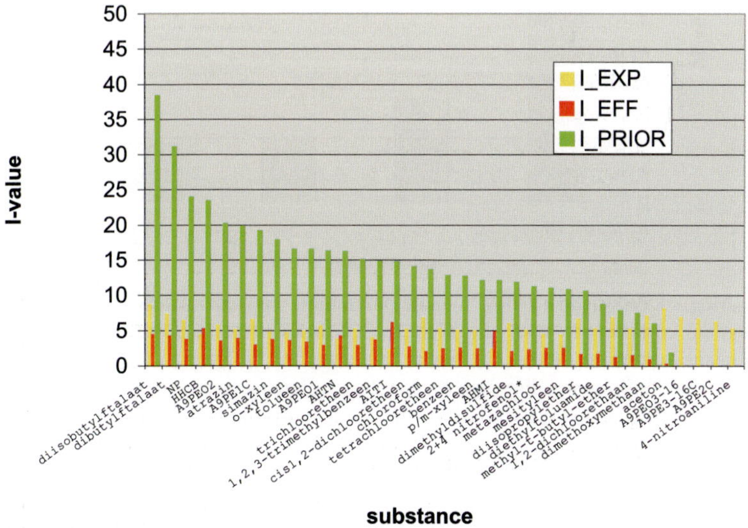

Fig. 21 Exposure (I-Exp), effect (I-Eff) and priority (I-Prior) index (according to the COMMPS method) of all organic compounds which have been found more than once in the dissolved phase of the Rivers Rhine and Meuse (From Geerdink and Schrap [175], with permission)

is based on aquatic toxicity data, bioaccumulation and information available from human toxicity. Here also the scale ranges between 1 and 10 [99].

Barreveld and coworkers [101] applied the COMMPS method for priority setting of the screening compounds identified in the Rhine estuary. At the Maassluis location in the New Waterway the most important priority

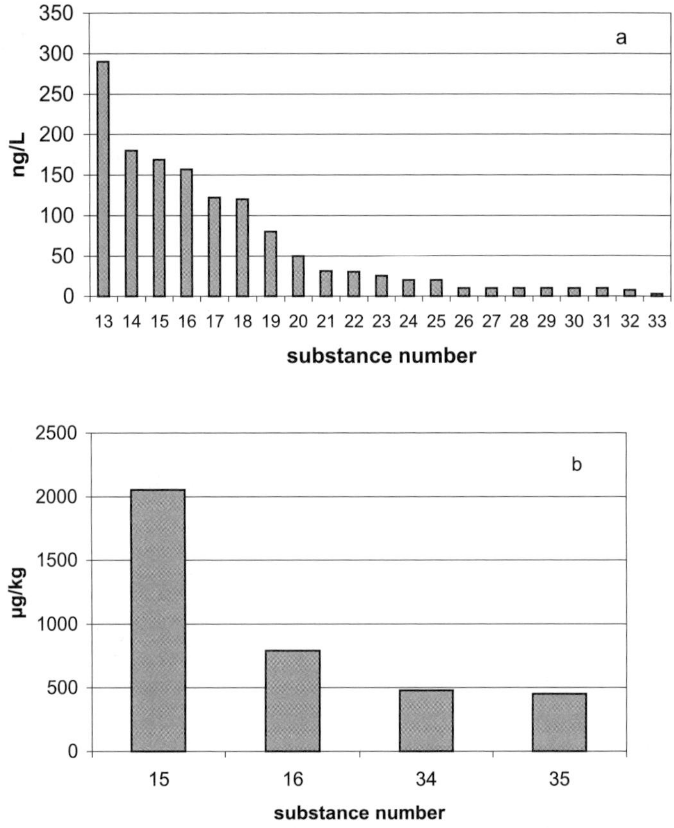

Fig. 22 Median dissolved (**a** ng L^{-1}) and particulate concentrations (**b** $\mu g\, kg^{-1}$) of various organic compounds with the highest COMMPS priority in the Rivers Rhine and Meuse near Rotterdam 2001. (13) methyl-*t*-butylether (solvent); (14) acetone (solvent); (15) nonylphenol (intermediate); (16) A_9PEO_{3-16} (surfactant); (17) A_9PE_1C (oxidation product of surfactant); (18) dimethoxymethane (solvent); (19) dimethyldisulfide (solvent); (20) diethyltoluamide (insect repellent); (21) A_9PEO_2 (surfactant); (22) atrazin (pesticide); (23) 2+4-nitrophenol (synthesis of drugs, pesticides); (24) simazin (pesticide); (25) tetrachloroethene (solvent); (26) metazachlor (pesticide); (27) *cis*-1,2-dichloroethene; (28) diisopropylether (solvent); (29) chloroform (solvent); (30) 1,2-dichloroethane (solvent); (31) trichloorethene (solvent); (32) HHCB (synthetic musk); (33) AHTN (1,1,3,4,4,6-hexamethyl-7-acetyltertaline) (synthetic musk, odour); (34) 4-chloro-3,5-xylenol (cosmetic); (35) brominated compounds (sum of BDE 28, 47, 66, 99, 100, 153, 209; BB 15, TBBP-A and me-TBBP-A). (From Geerdink and Schrap [175])

compounds thus identified were: TMPDD, musks (HHCB, AHTN = 7-acetyl-1,1,3,4,4,6-hexamethyl-1,2,3,4-tetrahydronaphthalene), 2,6-di-*t*-butyl-4-hydroxytoluene, octamethyltetrasilicone, and diisobutylphthalate.

Berbee and coworkers [102] also applied the COMMPS method to the chemical compounds identified in the screening study of the Rhine estuary. The most important priority compounds according to the COMMPS method were: HHCB (odour), di(2-ethylhexyladipate (softener), acridine (colouring agent), (+,–*meso*)2,3,7,9-tetramethyl-5-decyn-4,7-diol (anti-corrosion, foam inhibitor), benzophenone (additive to soaps), tris(2-chloroethyl)phosphate (flame retardant), tris(2-chloro-1-propyl)phosphate (flame retardant) and 2,6-di-*t*-butyl-4-methylphenol (BHT) as an anti-oxidant. Later, the chemical compounds identified in the Rhine estuary in the dissolved (Fig. 21) and particulate (Fig. 22) phases by Geerdink and Schrap [175] were also classified according the COMMPS method [99].

5.3 Bioassays

Schipper and Schout [149] and Stronkhorst [122] have derived targets for the allowed response of different bioassays to dredged material from the harbours in the Rhine estuary and dumping areas in the Dutch coastal zone. Sandeh [26], and Maas and coworkers [176] discussed the application of these targets to Dutch coastal waters.

The target for survival of *C. volutator* was set to 76% survival. For the microtox SP the target value was set to 48 toxic units, and the target for the DR-CALUX bioassay was set to 25 ng toxic equivalents kg^{-1} d.w. [149].

The target of the DR-CALUX assay was exceeded 3, 4, 6 and 4 times in the harbour sediments from the Rhine estuary in 1999, 2000, 2001, and 2002, respectively. The *C. volutator* response was three-times higher than the target value in the Rhine harbours between 1999 and 2001, and below the target in 2002. All responses of the microtox SP bioassay were below the target value in the period 1999–2002.

The responses of the bioassays in the dumping areas of dredged material from the Rhine harbours and in the Dutch coastal zone, measured by Åkerman and coworkers [134, 151] and Sandeh [26] were all below the target values.

6 Conclusions and Recommendations

The geographical situation of the Rhine estuary has shifted in time. In earlier days, the estuarine zone was located more land inwards, whereas presently the estuary extends along the Dutch coastal zone to the north. Due to these morphological and hydrological changes the transport and transfer of fluviatile compounds including anthropogenic compounds from the estuary to the Dutch coastal zone and Wadden Sea has increased.

In earlier days in was not possible to analyse and identify many different organic compounds. However, the extraction, isolation and detection methods including instrumental equipment, have improved considerably and today it is possible to detect almost any organic compound on request. Research has become more focussed on groups of chemical compounds with a certain toxicological syndrome or mode of action, such as dioxin and estrogen-like toxicity, and mutagenicity. In addition, the more water-soluble compounds have gained much interest. The compounds found in the Dutch coastal zone presently include many common household products. Widely used pharmaceuticals, personal care products and hormones, which are flushed down the toilet, are found in small amounts in surface waters of the Rhine estuary and the Dutch coastal zone. Concurrent with these technological and social developments a shift in research focus can be observed. Whereas in earlier days human health was the focus of concern associated with exposure to chemicals, presently the (aquatic) environment has become the focus of primary interest.

The organic compounds identified chronologically in the Rhine estuary start with the organochlorine pesticides, such as the drins and HCH, in the beginning of the 1970s, followed by PCBs and PAHs. Presently, attention is focussed on the more water-soluble chemical compounds, e.g. surfactants. The presence of about 250–300 organic compounds in the Rhine estuary has been confirmed.

Since the 1970s the concentrations of most of the organic compounds identified in the Rhine estuary and the Dutch coastal zone have decreased. The exceptions include the concentration of the PAHs (six of Borneff), which remained constant in the period 1986–2003, and the concentration of polybrominated flame retardants (PBDEs), which increased in the Dutch coastal zone in the beginning of the twenty-first century.

In general it is difficult to attribute changes in organisms, populations and ecosystems to the impact of organic compounds. Natural stress and the impact of for instance fishery activities confound the possible impact of organic chemicals. Yet, a few examples of convincing links between toxic effects on organisms in the field and the presence of organic compounds in the Dutch coastal zone and the western Wadden Sea have been demonstrated; these include the effects of drins on birds, of PAHs on flounders, of PCBs on seals and of TBT on (dog)whelks.

One of the main technical difficulties in the study of chronic effects of chemical compounds in the field is inherent in the field studies themselves. The area in question, i.e. the Dutch coastal zone, is about 16 times as large as that of the Netherlands territory. Furthermore, only a few aquatic toxicologists have the North Sea and the Rhine estuary as their working field. A few more marine biologists are studying organisms but very often they are not trained in observing the chronic effects of chemical compounds. Finally, too little is known up to now about the impact of mixtures of chemical substances on organisms. Hence, it is not yet possible to use this type of information for focusing field studies. Future research programs will also have to provide space for serendipity and accidental discoveries to identify chronic effects of substances in the field. The existing or proposed ecotoxicological target values can be used to obtain indications in this respect.

To assess the negative impact of organic compounds in the Rhine and Dutch coastal zone, targets have been derived scientifically from toxicological data and subsequently implemented by policy makers. Because it is not possible to analyse all organic compounds, priority selections out of the tremendous number of organic compounds have been proposed by different (inter)national organisations, using various methodologies. The main criteria for the selection of priority organic compounds are persistency, bioaccumulation and toxicity. Often a lack of toxicological and reliable physical chemical data occurs, in which case safety factors are applied. At the end of the selection procedure also expert judgment plays an important role to finalise the ultimate list. Various international selection processes for chemical substances have led to different lists. Quite remarkably, when the lists of OSPAR, WFD, HELCOM and UNEP for persistent organic compounds are compared, there is not a single compound that appears on all lists. Contrary to the other lists compounds can be added to or removed from the WFD list (currently containing 33 compounds) after a certain period of time if new scientific evidence becomes available.

The priority setting approach mentioned in the preceding paragraph often leads to relatively low targets for the individual organic compounds that currently appear on the several priority lists. Until now several compound characteristics such as bioavailability have not been taken into account, however, in the derivation of the targets. Moreover, the approach calculates the concentration at which a possible effect of one individual compound can occur. The possible combined effect of some chemical compounds together is accounted for by application of a safety factor that may overestimate the actual effects.

In earlier days the Dutch target values were defined as objectives that were considered as realistically achievable. If concentrations exceeded target values, this would not immediately lead to abatement/mitigation actions. Usually, further research was first conducted to identify if genuine effects were observed indeed before further actions were taken. The outcome of such work could lead and has led to modification of the target values. However, presently the legal WFD targets of the EU are more stringent. This implies that if these targets are exceeded, the sources and causes have to be identified and mitigated within 6 years.

One can expect that scientists will find "new" chemical compounds in the future in the Rhine estuary. It will be very difficult to derive targets for all individual chemical compounds because there will be a huge lack of toxicological and physical chemical data. An internationally accepted standardised method, like the COMMPS, could be used to estimate the chance of a possible effect to occur. If so, this should trigger more dedicated research before including the pertinent compound on the priority list.

Beside the classic chemical assessment approach of different countries and the EU, more and more attention is paid by researchers to biological assessment methods. These methods take into account the impact of mixtures of chemical compounds, which corresponds better to the actual field situation. For the Rhine estuary and coastal zone, it can be concluded that in vivo bioassays have not shown any responses in the field, i.e. at the dumping sites and surrounding areas, and slight responses in experimental studies with dredged material. In vitro assays elicited an almost similar picture, with an occasionally high DR-CALUX response observed in the field. Here, it must be recognised that the response of in vitro bioassays is measured with an extract in which the total concentration present is available to the exposed cells. The test system does not yet take into account the bioavailable fraction, which can be rather small compared to the total amount.

For instance, the chemical assessment of TBT and PAHs indicates that the concentrations are above the chemical targets set, both in the Rhine estuary and the Dutch coastal zone. In the field, imposex can still be observed, and the incidence of fish diseases has decreased to natural levels. The bioassays applied to samples from these areas, however, as noted above, gave either no responses or responses below the bioassay target values set.

As argued above, it is and will be impossible to continue the analyses and assessment of individual chemical compounds on a routine basis. Incorporation of more biological methods, that also take into account the bioavailable fractions, is mandatory for assessing the water quality of the River Rhine and the Dutch coastal zone. The bioassay battery should preferably cover a very broad spectrum of relevant toxicological syndromes. If a significant response of a bioassay is observed, then subsequently the chemical compounds that cause the response need be identified. To this end bioassay-directed chemical fractionation, isolation and characterisation could be used. This method is called TIE (Toxicity Identification Evaluation) [177]. There is an urgent need for international standardisation of TIE methodologies, as suggested by Cizmas and coworkers [178] and Hutchings and coworkers [179].

A further complication in the assessment of chemical compounds in the aquatic environment is that the management and politics, necessary to reduce the input of chemicals, are more complex today than 50 years ago. In the 1980s it was agreed (inter)nationally to reduce the input of certain black and grey list chemical compounds by at least 50%. In general, for most chemical compounds this policy measure has been successful because each national ministry could take measures within its own legal responsibility. Few authorities looked across borders at that time. The international Rhine Commission already in 1987 [180] recognised the need for the various Rhine countries to co-operate in order to reduce the burden of chemical compounds in the Rhine. Today, this international approach is one of the key principles behind the WFD of the EU: the so-called catchment approach. However, to realise a further step in the reduction of the emission of chemicals to the aquatic environment, different national ministries also have to work together.

As generic approaches (e.g., an overall emission reduction of 50%) will cost too much, tailor-made solutions have to be developed. An example of a tailor-made solution is the approach used to combat eutrophication in the Baltic Sea [181]. Unfortunately, in many countries it is not a single ministry or a single legislation that describes and regulates the production, use and emission of chemical substances. Instead, usually various ministries are involved each with different objectives. To give an example, the Ministry of Transport and Public Works in the Netherlands is responsible for the water quality, and they have adopted the European strategy, i.e., adopting near-background concentrations as target values, and "close-to-zero" emissions. On the other hand, the Ministry of Public Health, responsible for the atmospheric emissions has adopted goals and targets based on the best available technology, and "as-low-as-reasonably-achievable" principles.

Last but not least, a better compliance between source-oriented and quality-oriented targets ought to be realised. The so-called driver, pressure, state, impact and response (DPSIR) approach proposed by the OECD is a promising vehicle for a combined assessment of social, economical, political and scientific objectives.

Acknowledgements R. Bovelander (RWS-RIKZ) is acknowledged for supplying the River Rhine data. H. Barreveld (RWS-RIZA) adapted and provided the gas chromatogram of chemical compounds in the Rhine estuary.

References

1. Musters CJM, Weinrich JA (1991) Documentatie verandering Flora en Fauna. In: Hekstra GP, van Linden FJM (eds) Flora en Fauna chemisch onder druk. Verslag van nationalsymsoium Oecologische Kring. Pudoc, Wageningen, The Netherlands, p 41–47
2. Dijkzeul A (1982) De waterkwaliteit van de Rijn in Nederland in de periode 1970–1981. Rijkswaterstaat, Lelystad, The Netherlands, RIZA report 82(061):112
3. Kramer KJM, Duinker JC (1988) The Rhine/Meuse Estuary. In: Salomons W, Bayne BL, Duursma EK, Förstner U (eds) Pollution of the North Sea; an assessment. Springer, Berlin Heidelberg New York, p 194–212
4. Dijkzeul A (1981) De Waterkwaliteit van de Maas in Nederland in de periode 1953–1980. Rijkswaterstaat, Lelystad, The Netherlands. RIZA report 81(048):96
5. Ven GP (1993) Leefbaar Laagland. Uitgeverij Matrijs, Utrecht, The Netherlands
6. Van der Kooij LA (1982) De waterkwaliteite van de Westerschelde in de periode 1964-1981. Rijkswaterstaat, The Hague, The Netherlands, report 82(063):110

7. Wollast R (1988) The Scheldt estuary. In: Salomons W, Bayne BL, Duursma EK, Förstner U (eds) Pollution of the North Sea; an assessment. Springer, Berlin Heidelberg New York, p 183–193
8. Anonymous (2002) Towards the Derivation of Quality Standards for Priority Compounds in the Context of the Water Framework Directive. Draft final report of the study: Identification of quality standards for priority compounds in the field of water policy. Fraunhofer-Institut, Neuherberg, Germany, EAF(3)-06/06/FHI
9. Van Alphen JSLJ, Ruijter WPM, Borst JC (1988) Outflow and three-dimensional spreading of river Rhine water in the Netherlands coastal zone. In: Dronkers J, van Leussen W (eds) Physical processes in estuaries. Springer, Berlin Heidelberg New York, p 70–92
10. Van der Giessen A, de Ruijter WPM, Borst JC (1990) Three-dimensional current structure in the Dutch coastal zone. Netherlands J Sea Res 25(1/2):45–55
11. Reid PC, Taylor AH, Stephens JA (1988) The Hydrography and Hydrographic Balances of the North Sea. In: Salomons W, Bayne BL, Duursma EK, Förstner U (eds) Pollution of the North Sea; an assessment. Springer, Berlin Heidelberg New York, p 3–19
12. Verlaan PAJ, Spanhoff R (2000) Massive Sedimentation Events at the Mouth of the Rottterdam Waterway. J Coastal Res 16(20):458–469
13. Van Bennekom AJ, Wetstijn FJ (1990) The winter distribution of nutrients in the Southern Bight of the North Sea (1961–1978) and in the estuaries of the Scheldt and the Rhine/Meuse. Netherlands Journal of Sea Research 25(1/2):75–87
14. Eisma D, Cadée GC, Laane RWPM (1982) Supply of Suspended Matter and Particulate and Dissolved Organic Carbon from the Rhine to the Coastal North Sea. Mitteilungen Geologisch-Paläontologisch Institute Universität Hamburg, SCOPE/UNEP Sonderband 52:483–505
15. Van Alphen JSLJ (1990) A Mud Balance for Belgian Dutch coastal waters between 1969–1986. Netherlands J Sea Res 25(1/2):19–30
16. De Kok M, Lourens JM, de Ruig JHM (1992) Baggerspecie, van Waterweg tot Waddenzee: fysisch en milieukundig onderzoek naar alternatieven voor Loswal Noord. Rijkswaterstaat, The Hague, The Netherlands, report DGW-92(030)
17. Van Leussen W, Van Velzen E (1989) High concentration suspended matter: origin and importance in Dutch estuaries and coastal zone. J Coastal Res 5:1–22
18. Wollast R, Peters JJ (1980) Transfer of materials in estuarine zones. Review of water quality and transport of materials in coastal and estuarine waters. Copenhagen, Denmark, 200–216
19. Salomons W, Eysink WD (1981) Pathways of mud and particulate metals from rivers to the southern North Sea. In: Nio SD, Schuttenhelm RTE, van Weering TCE (eds) Holocene Marine Sediments in the North Sea. Blackwell, Oxford, p 492–450
20. Steyaert F, Van Maldegem D (1987) Research of mud transport Westerschelde. Poster presented at the International Conference on the Scheldt Estuary, Terneuzen, The Netherlands
21. Salden RM (1998) Prognose voor de ontwikkeling van de slibbalans in de beneden Zeeschelde bij continuering en beëindiging van de slibverwijdering bij Kallo. Report RIKZ/OS/1998.128X001, The Hague, The Netherlands
22. Mulder HPJ, Uddink C (1991) Modelling of cohesive sediment transport. A case study: the Western Scheldt Estuary. In: Edge BL (ed) 22nd Coastal Eng Conf Proc Int Conf, Delft, The Netherlands, p 3012–3023

23. Van Maldegem DC (1993) The mud balance of the Scheldt Estuary. Rijkswaterstaat, The Hague, The Netherlands, report GWAO-93(081)
24. Martin JM, Wollast R, Loijens M, Thomas A, Mouchel JM, Nieuwenhuize J (1994) Origin and fate of artificial radionuclides in the Scheldt Estuary. Marine Chem 46:189–202
25. Laane RWPM, personal communication
26. Stutterheim S (2002) Van Noord tot Noordwest: een studie naar de berging van baggerspecie op loswallen. Ministerie van Verkeer en Waterstaat, Rijkswaterstaat, The Hague, The Netherlands. RIKZ report 2002(047):310
27. Van Straaten LMJU, Kuenen PhH (1957) Accumulation of fine-grained sediments in the Dutch Wadden Sea. Geologie en Mijnbouw 19:3290–354
28. Postma H (1961) Susupended matter and Secchi disc visibility in coastal waters. Netherlands J Sea Res 1:359–390
29. Eisma D (1993) Suspended matter in the aquatic environment. Springer, Berlin Heidelberg New York, p 299
30. Molegraaf P (1987) Schatting van de aanvoer van aan slib gebonden metalen vanuit de Noordzee naar de Waddenzee voor het jaar 1985. Rijkswaterstaat, The Hague, The Netherlands, report GWWS-87(235)
31. De Jong F, Bakker J, van Berkel C, Dahl K, Dankers N, Gätje C, Marencie H, Potel P (eds) (1999) 1999 Wadden Sea Quality Status Report. Common Wadden Sea Secretariat, Wilhelmshafen, Germany
32. Lozán JL, Rachor E, Reise K, Sündermann J, Von Westenhagen H (2002) Warnsignale aus Nordsee und Wattenmeer. Wissemschaftlich Auswertungen, Hamburg, Germany
33. Otto F, Honkoop J (1998) Een Beeld van de Noordzee: economische gegevens van de belangrijkste gebruikfuncties. Rijkswaterstaat, The Hague, The Netherlands. RIKZ report 98(032):85
34. Laane RWPM (1992) Background concentrations of natural compounds in river, sea, water, atmosphere and mussels. Rijkswaterstaat, The Hague, The Netherlands, report 92(033)
35. Heessen HJL (1988) Fishery effects. In: Salomons W, Bayne BL, Duursma EK, Förstner U (eds) Pollution of the North Sea; an assessment. Springer, Berlin Heidelberg New York , 538–550
36. Zijlstra JJ, De Wolf P (1988) Natural effects. In: Salomons W, Bayne BL, Duursma EK, Förstner U (eds) Pollution of the North Sea; an assessment. Springer, Berlin Heidelberg New York, Germany, p 164–182
37. Ragas AMJ, Leuven RSEW, Sloof DJW (1994) Milieukwaliteit en normstelling. Handboek Milieukunde 1. Uitgeverij Boom, Meppel, The Netherlands
38. George SG (1989) Cadmium effects on plaice liver xenobiotic and metal detoxification systems: dose-response. Aquatic Toxicology 15:303–310
39. Eggens ML (1996) Cytochrome P450 1A induction in North Sea flatfish as a biomarker of exposure to dioxin-type compounds. Thesis, University of Utrecht, Utrecht, The Netherlands
40. Bourne WRP, G Vauk (1988) Accumulation in birds. In: Salomons W, Bayne BL, Duursma EK, Förstner U (eds) Pollution of the North Sea; an assessment. Springer, Berlin Heidelberg New York, p 579–595
41. Bucke D, Watermann B (1988) Effects of pollutants in fish. In: Salomons W, Bayne BL, Duursma EK, Förstner U (eds) Pollution of the North Sea; an assessment. Springer, Berlin Heidelberg New York, p 612–623

42. Cooper VA (1988) Effects on invertebrates. In: Salomons W, Bayne BL, Duursma EK, Förstner U (eds) Pollution of the North Sea; an assessment. Springer, Berlin Heidelberg New York, p 604–611
43. Ernst W, Boon JP, Weber K (1988) Occurrence and Fate of Organic Micropollutants in the North Sea. In: Salomons W, Bayne BL, Duursma EK, Förstner U (eds) Pollution of the North Sea; an assessment. Springer, Berlin Heidelberg New York, p 425–440
44. Harms U, Kerkhoff MAT (1988) Accumulation in fish. In: Salomons W, Bayne BL, Duursma EK, Förstner U (eds) Pollution of the North Sea; an assessment. Springer, Berlin Heidelberg New York, p 567–778
45. Reijnders PJH (1988) Accumulation and Body distribution of Xenobiotics in Marine Mammals. In: Salomons W, Bayne BL, Duursma EK, Förstner U (eds) Pollution of the North Sea; an assessment. Springer, Berlin Heidelberg New York, p 596–603
46. Becker PH, Bruhn R (2002) Schadstoffbelastung der Organismen im Küstenbereich. In: Lozán JL, Rachor E, Reise K, Sündermann J, von Westerhagen H (eds) Warnsignale aus Nordsee & Wattenmeer. Wissenschaftliche Auswertungen, Hamburg, Germany, p 204–210
47. Haarich M, Harms U (2002) Aktuelle Belastung der Organismen in der offenen See am Beispiel der Fische. In: Lozán JL, Rachor E, Reise K, Sündermann J, von Westerhagen H (eds) Warnsignale aus Nordsee & Wattenmeer. Wissenschaftliche Auswertungen, Hamburg, Germany, p 211–218
48. Semple KT, Doick KJ, Jones KC, Burauel P, Craven A, Harms H (2004) Defining Bioavailability and Bioaccessibility of Contaminated Soil and Sediments is Complicated. Environ Sci Tech 38:228–231A
49. Vijver MG, van Gestel CAM, Lanno RP, Van Straalen N, Pijnenburg WJGM (2004) Internal Metal Sequestration and its Ecotoxicological Relevance: A review. Environ Sci Tech 38(18):4705–4712
50. Leslie HA, ter Laak TL, Busser FJM, Kraak MHS, Hermens JLM (2002) Bioconcentration of organic chemicals: is a solid-phase microextraction fiber a good surrogate for biota? Environ Sci Technol 36:5399–5404
51. Ten Hulscher ThEM, Postma J, den Besten PJ, Stroomberg GJ, Belfroid A, Wegener JW, Faber JH, van der Pol JJC, Hendriks AJ, van Noort PCM (2003) Tenax extraction mimics benthic and terrestrial bioavailability of organic compounds. Environ Toxicol Chem 22:2258–2265
52. OSPAR (2002) Provisional Instruction Manual for the Dynamic Selection and Prioritization Mechanism for Hazardous Substances (DYNAMEC), DYNAMEC. OSPAR, London
53. Camphuysen CJ (1998) Olievervuiling en olieslachtoffers in Nederland, 1997–1998. Sula 12:41–72
54. Koeman JH (1971) Het voorkomen en de toxicologische betekenis van enkele chloorkoolwaterstoffen aan de Nederlandse kust in de periode 1965 tot 1970. PhD Thesis, University of Utrecht, 136
55. Koeman JH, van Genderen H (1972) Tissue levels in animals and effects caused by chlorinated hydrocarbon insecticides, chlorinated biphenyls and mercury in the marine environment along the Netherlands coast. In: Marine Pollution and Sea Life Fishing News (book) Survey, p 1–8
56. Swennen C (1972) Chlorinated hydrocarbons attacked the Eider population in the Netherlands. TNO nieuws 27:556–560

57. De Groot AJ (1963) Mangaantoestand van de Nederlandse en Duitse Holocene sedimenten in verband met slibtransport en bodemgenese. Verslag Landbouwkundig Onderzoek Wageningen 697:1–164
58. Van Bennekom AJ, Gieskes WWC, Tijssen SB (1975) Eutrophication of Dutch coastal waters. Proc Royal Soc London B 189:359–374
59. Zimmermann JTF (1976) Mixing and flushing of tidal embayment in the western Dutch Waden Sea I and II. Netherlands J Sea Res 10:149–191;397–439
60. Drescher HE, Harms U, Huschenbeth E (1977) Organochlorines and heavy metals in the harbour seal (*Phoca vitulina*) from the German North Sea coast. Marine Biol 41:99–106
61. Drescher HE (1978) Hautkrankheiten beim Seehund, *Phoca vitulina* Linné 1958, in der Nordsee. Säugetierk Mitteilungen 26:50–59
62. Reijnders PJH (1980) Organochlorine and heavy metal residues in harbour seals from the Wadden Sea and their possible effects on reproduction. Netherlands J Sea Res 14:30–65
63. Reijnders PJH (1980) On the causes of the decrease in the harbour seal (*Phoca vitulina*) population in the Dutch Wadden Sea: population dynamics and tissue levels of heavy metal and organochlorine compounds. PhD Thesis, Agricultural University Wageningen, The Netherlands
64. Reijnders PJH (1982) On the ecology of the harbour seal (*Phoca vitulina*) in the Wadden Sea: population dynamics, residue levels and management. Veterinary Quarterly 4:36–42
65. Klamer HC, Laane RWPM, Marquenie J (1990) Sources and fate of PCBs in the North Sea. In: North Sea Pollution, Proc Int IAWPRC/ECPCA Conf, Amsterdam, The Netherlands, p 179–194
66. Van de Kamp G (1977) Vertebral deformaties of herring around the British Isles and their usefulness for a pollution monitoring programme. ICES Copenhagen, Denmark. CM/E:5:9
67. Möller H (1979) Geographical distribution of fish diseases in the NE Atlantic. Meeresforschung 27:217–235
68. Dethlefsen V (1980) Observations on fish diseases in the German Bight and their possible relation to pollution. Rapport P-v Réun Cons Perm Int Explor Mer 179:110–117
69. Dethlefsen V, Watermann B, Hoppenheit M (1987) Diseases of the North Sea dab (*Limanda limanda L.*) in relation to biological and chemical parameters. Archive Fisch Wissenschaften 37:107–237
70. Dethlefsen V (1995) Biological changes in the German Bight of the North Sea as indicators of ecosystem health. In: Rapport DJ, Gaudet CL, Callow P (eds) Evaluating and monitoring the health of large scale ecosystems. NATO series I, 28. Springer, Berlin Heidelberg New York
71. Dethlefsen V, von Westernhagen H, Cameron P (1996) Malformation in North Sea fish embryos during the period 1984–1995. ICES J Marine Sci 53:1024–1035
72. Van Banning P (1987) Long-term recording of some fish diseases using general fishery research surveys in the south-east part of the North Sea. Dis Aquat Org 3:1–11
73. Vethaak AD (1985) Prevalence of fish diseases with reference to pollution of Dutch coastal waters. Netherlands Institute of Fisheries Investigations, IJmuiden, The Netherlands. Report CA 85(01):70

74. Vethaak AD (1987) Fish diseases, signals for a diseased environment? In: Peet G (ed) Reasons for Concern, Proc 2nd North Sea Seminar 1986, vol 2. Werkgroep Noordzee, Amsterdam, The Netherlands
75. Vethaak AD, Van der Meer J (1991) Fish disease monitoring in the Dutch part of the North Sea in relation to dumping of waste from titanium dioxide production. Chemical Ecol 5:149–170
76. Vethaak AD (1993) Fish Disease and Marine Pollution. PhD Thesis, University of Amsterdam, The Netherlands
77. Marquenie JM, Vethaak AD (1989) Effecten van verontreinigingen in zoutwater ecosystemen. In: Stortelder PMB, Koops WF (eds) Effecten van verontreinigingen op aquatische ecosystemen. Vereniging voor Milieuwetenschappen, Vught, The Netherlands, 91–103(78)
78. Cameron P, Berg J (1992) Morphological and chromosomal aberrations during embryonic development in dab *Limanda limanda*. Marine Ecol-progress Series 91(1–3):163–169
79. Parrett A (1998) Pollution impacts on North Sea fish stocks. WWF report to the EC, DG XIV. WWF, Godalming, UK
80. Evers EHG, van Meerendonk JH, Ritsema R, Pijnenburg J, Lourens JM (1995) Butyltinverbindingen: een analyse van de problematiek in aquatisch milieu. Rijkswaterstaat, The Hague, The Netherlands, RIKZ report 95(007):141
81. Ten Hallers-Tjabbes CC, Kemp JF, Boon JP (1994) Imposex in whelks *(Buccinum undatum)* from the open sea: relation to shipping activities. Marine Pollution Bull 28:311–313
82. Ten Hallers-Tjabbes CC, Wegener J-W, van Hattum B, Kemp JF, Ten Hallers E, Reitsema TJ, Boon JP (2003) Imposex and organotin concentrations in *Buccinum undatum* and *Neptunea antique* from the North Sea: relationship to shipping density and hydrographical conditions. Marine Environ Res 55:2032–233
83. Fent K (1996) Ecotoxicology of organotin compounds. Critical Reviews in Toxicology 26:1–117
84. Quevauviller P (1991) Speciation de l'etain dans les milieux estuariens et cotiers. PhD Thesis, University of Bordeaux, Bordeaux, France
85. Mertens O, Van Zwol C (1988) Purperslakken en organotin, een onderzoek naar effecten in de Oosterschelde en Grevelingen. Rijkswaterstaat, The Hague, The Netherlands, report GWAO-88(039)
86. Ritsema R, Laane RWPM (1991) Dissolved butyltins in fresh and marine waters in The Netherlands in 1989. Sci Total Environ 105:149–156
87. Ritsema R, Laane RWPM, Donard OFX (1991) Butyltins in marine waters in The Netherlands in 1988 and 1989. Concentrations and effects. Marine Environ Res 32:243–260
88. Harding MJC, Bailey SK, Davies IM (1992) TBT imposex survey of the North Sea. UK department of the Environment. Scottish Fisheries Working Paper 14/92, Aberdeen, UK
89. Mensink BP (1999) Imposex in the common dogwhelk, *Buccinum undatum*. PhD Thesis, University of Wageningen, The Netherlands
90. Mensink BP, van Hattum B, Ten Hallers-Tjabbes CC, Everaats JM, Kralt H, Vethaak AD, Boon JP (1997) Tributyltin causes imposex in the common whelk *Buccinum undatum*: mechanisms and occurrence. NIOZ, Den Burg, The Netherlands 1997–6:56
91. Laane RWPM, Yland E, Pijnenburg J, Groeneveld G, De Vries A (2001) Selectie potentiele probleemstoffen voor de Noordzee: stand van zaken en analyse, maart 2001. Rijskwaterstaat, The Hague, The Netherlands. Working document RIKZ 2000(034)

92. Carson R (1963) Silent spring. Hougton Mifflin Company, New York
93. Anonymous (1966) Report of a new chemical hazard. New Scientist 32:162
94. Bernes C (1998) Persitent Organic Pollutants: A Swedish View of an International Problem. Swedish Environmental Protection Agency, Sweden
95. Jensen S (1972) The PCB story. Ambio 1:123
96. De Voogt P, Brinkman UAT (1989) Production, properties and usage of PCBs. In: Kimbrough RA, Jensen AA (eds) Halogenated biphenyls, terphenyls, naphthalenes, dibenzodioxins and related products (2nd ed). Elsevier, Amsterdam, The Netherlands p 3–45
97. Hekster F, Laane RWPM, de Voogt P (2003) Environmental and toxicity effects of perfluoroalkylated compounds. Rev Environ Contamin Toxicol 179:99–121
98. OSPAR (2002) Provisional Instruction Manual for the Dynamic Selection and Prioritization Mechanism for Hazardous Substances (DYNAMEC), DYNAMEC.
99. European Commission (1999) Study on the prioritisation of substances dangerous to the aquatic environment. Revised proposal for a list of priority substances in the context of the Water Framework Directive (COMMPS procedure), Brussels, Belgium
100. CIW (2000) Stoffen overzicht internationale kaders. CIW notitie 8, Lelystad, The Netherlands
101. Barreveld HL, Berbee RPM, Ferdinandy MMA (2001) Vergeten stoffen in Nederlands oppervlaktewater. Rijkswaterstaat, Lelystad, The Netherlands, RIZA report 2001(020X)
102. Berbee RPM, Kalf D, van Duin P, Beek M (2004) Vergeten stoffen in RWZI-effluenten in het Maasstroomgebied. RIZA-report 2004. 018, Lelystad, The Netherlands
103. Geerdink RB, Schrap SM (2004) Vergeten stoffen. In: de Rijn-Maas monding. Rijkswaterstaat, Lelystad, The Netherlands, RIZA report 2004(015):82
104. POR II (2000) Dredged Material in the Port of Rotterdam-Interface between Rhine catchment Area and North Sea. Extended Summary, GKSS Research Center, Max Planck Strasse, Geesthacht, Germany
105. Duinker JC, Hillebrand MTJ (1979) Behaviour of PCB, pentachlorobenzene, alpha-HCH, gamma-HCH, beta-HCH, dieldrin, endrin and p,p′′-DDD in the Rhine-Meuse estuary and adjacent coastal area. Netherlands J Sea Res 13:256–281
106. Weber K, Ernst W (1983) Vorkommen und Fluktuation von Organischen Umweltchemikalien in deutschen Ästuarien. Wasser 61:111–123
107. Anonymous (1983) De Waterkwaliteit van de Noordzee 1975–1982. Resultaten van het routinematig meetnet Rijkswaterstaat, RIZA, report 83(084):94
108. Kammann U, Haarich M, Heyer K, Hühnerfuss H, Karbe L, Kersten M, Schmidt D, Steinhart H (1994) Large scale distribution of contaminants in spring 1986 and winter 1986/87. In: Sündermann J (ed) Circulation and Contaminants Fluxes in the North Sea. Springer, Berlin Heidelberg New York, p 90–138
109. Lohse JH (1988) Herkunft, Ausbreitung und Verteilung von schwerflüchtigen Chlorkohlenwasserstoffen in sediment der Nordsee. PhD Thesis, University of Hamburg, Germany
110. Lohse JH (1988) Distribution of Organochlorine Pollutants in North Sea Sediment. Mitteilungen Geologische-Palaontologisch Institute, Universität Hamburg 65:345–365
111. Lohse JH (1990) Chlorierte organische Verbindungen in Wasser und Sediment. In: Lozán JL, Lens W, Rachor E, Watermann B, von Westernhagen H (eds) Warnsignale aus der Nordsee. Verlag Paul Parey, Berlin, Germany p 75–87

112. Laane RWPM, Sonneveldt HLA, Van der Weyden AJ, Loch JP, Groeneveld G (1999) Trends in the spatial and temporal distribution of metals (Cd, Cu, Zn, and Pb) and organic compounds (PCBs and PAHs) in the Dutch coastal zone sediments from 1981 to 1996: a model case study for Cd and PCBs. Netherlands J Sea Res 41:1–17
113. Hegeman WJH, Laane RWPM (2004) Concentratie en normtoetsing van stoffen in het oppervlakte sediment van het Nederlandse Continentale Plat (1981-2003): Cd, Cu, Zn, Pb, Hg, Ni and Ag and organic compounds: PCB, PAH, HCB en TBT. Rijkswaterstaat, The Hague, The Netherlands, RIKZ report 2004(138):72
114. Sonneveldt HLA, Laane RWPM (2001) Prediction of Sediment Quality in the Dutch Coastal Zone: Model Validation and Uncertainty Analysis for Cd, Cu, Pb, Zn, PCBs and PAHs. In: Rainbow PS, Hopkin SP, Crane M (eds) Forecasting the environmental fate and effects of chemical. Wiley, Chichester, p 177–189
115. Bleeker A, Duyzer JH (2004) Belasting van de grote Nederlandse oppervlakte wateren en de Noordzee door de atmosferische depositie met PCBs, PAKs, pesticiden en stikstof. TNO, Apeldoorn, The Netherlands, report 2004/506
116. Compaan H, Laane RWPM (1992) Polycyclic Aromatic Hydrocarbons (PAH) In: The North Sea: an inventory. TNO, Apeldoorn, the Netherlands, report IMW-R 92/392:130
117. De Voogt P, Wiegman S, van Beusekom B, Serné P, Slot P, van der Wielen F, Aardenburg L, Admiraal W (2001) Occurrence and Toxicity of Azaarenes in the North Sea. Report VOTON, IBED-MTC. University of Amsterdam, Amsterdam, The Netherlands
118. Swertz OC (1999) Butyltingehalten in de zoute wateren in 1998: resultaten van de monitoring (MWTL). Rijkswaterstaat, The Hague, The Netherlands, Report RIKZ-99(021):19
119. Bovelander R (2005) 13 jaar TBT-monitoring. Rijkswaterstaat, The Hague, The Netherlands, report MI 2004 (in press)
120. CIW (2004) Water in cijfers 2004. Commissie Intergraal Waterbeheer, Lelystad, The Netherlands
121. Laane RWPM, Groeneveld G (1999) Butyltin-verbindingen in sediment van de Noordzee, Delta en Waddenzee. Rijkswaterstaat, The Hague, The Netherlands, report 99(019):18
122. Stronkhorst J (2003) Contaminants of Concern in Dutch Marine Harbors. PhD Thesis, Free University, Amsterdam, The Netherlands
123. Anonymous (2004) Van Inzicht naar Doorzicht. Beleidsmonitor Water (BMW), chemische kwaliteit oppervlakte water. RIVM, Mileiu- Natuurplan Bureau, Bilthoven, The Netherlands, RIVM report 500799004: Chapter Organotinverbindingen 200–207
124. Klamer H (2001) Potentiële probleemstoffen in oppervlaktesedimenten van de Noordzee: gebromeerde vlamvertragers, phthalaten en Irgarol 1051. Rijkswaterstaat, The Hague, The Netherlands, working document RIKZ/OS/2001.609X, 10
125. Steen RJCA (2002) Fluxes of pesticides into the Marine Environment. PhD Thesis, Free University, Amsterdam, The Netherlands
126. Jonkers N, Laane RWPM, de Voogt P (2005) Sources and fate of nonylphenol ethoxylates and their metabolites in the Dutch coastal zone of the North Sea. Marine Chemistry 96:115–135
127. Jonkers N, Laane RWPM, de Graaf C, de Voogt P (2005) Fate modelling of nonylphenol ethoxylates and their metabolites in the Dutch estuary Scheldt and Rhone estuaries: validation with new field data. Estuarine, Coastal and Shelf Science 62:141–160

128. Jonkers N, Laane RWPM, de Voogt P (2003) Fate of nonylphenol ethoxylates and their metabolites in two Dutch estuaries: evidence of biodegradation in the field. Environ Sci Tech 37:321–327
129. Jonkers N (2004) Environmental behaviour of nonylphenol ethoxylkates in coastal waters. PhD Thesis, University of Amsterdam, The Netherlands
130. Vethaak AD, Rijs GBJ, Schrap SM, Ruiter H, Gerritsen A, Lahr J (2002) Estrogens and xeno-estrogens in the aquatic environment of the Netherlands. RIZA/RIKZ report 2002(001), The Hague, Netherlands, ISBN 9036954010
131. Belfroid A, de Voogt P, Schrap SM (2005) Occurrence of estrogenic hormones, bisphenol-A, and phthalates in the aquatic environment of The Netherlands. In: Vethaak AD, Schrap SM, de Voogt P (eds) SETAC Press, Pensacola, FL, USA, Chapter 3 (in press)
132. De Boer J, Wester PG, Van der Horst A, Leonards PEG (2005) Polybrominated diphenyl ethers in The Netherlands—occurrence and indicative estrogenic potency. In: Vethaak AD, Schrap SM, de Voogt P (eds) SETAC Press, Pensacola, FL, USA, Chapter 5 (in press)
133. Klamer JCH, Leonards PEG, Lamoree M, Bakker JF (2002) Chemical and toxicological risk assessment of North Sea surface sediments. Brominated flame retardants and dioxine type toxicity. Organohalogen Compounds 59:11–114
134. Åkerman J, Klamer H, Schipper C, Bakker J, Bellert B, Pijnenburg J (2004) Stoffen in de Noordzee en de Nederlandse kustzone in 2003: ftalaten, vlamvertragers, organotin- en geperfluoreerde verbindingen en effectgerichte metingen. Rijkswaterstaat, The Hague, The Netherlands, report RIKZ 2004(040)
135. Kümmerer K (ed) (2001) Pharmaceuticals in the Environment. Sources, Fate, Effect and Risks. Universitatsklinikum Freiburg, Berlin, Germany
136. Schrap SM, Rijs GBJ, Beek MA, Maaskant JFM, Staeb J, Stroomberg G, Tiesnitsch J (2003) Humane en veterinaire geneesmiddelen in Nederlands oppervlaktewater en afvalwater. Rijkswaterstaat, Lelystad, The Netherlands, report RIZA 2003(023)
137. Derksen JGM, van Eijnatten GM, Lahr J, van der Linde P, Kroon AGM (2000) Milieueffecten van humane geneesmiddelen. Aanwezigheid en risico's. RIWA/RIZA Amsterdam/Lelystad, The Netherlands, report 2000(051)
138. Jongbloed RH, Blankendaal VG, Kan CA, van Dokkum HP, Bernhard R, Rijs GBJ (2001) Milieurisico's van diergeneesmiddelen en veevoederadditieven in oppervlaktewater: een verkennende studie. Rijkswaterstaat, Lelystad, The Netherlands, RIZA report 2001(035)
139. De Knecht JA, Schefferlie GJ, Janssen PAH, Monfort MHMNM, Baars AJ (2001) Risico's voor de volksgezondheid en het milieu door het gebruik van geneesmiddelen in de kweek van paling en meerval. RIVM-CSR report 08323AOO, Bilthoven, The Netherlands
140. Mons MN, Hoogenboom AC, Noij THM (2003) Pharmaceuticals and drinking water supply in the Netherlands. KIWA, Nieuwegein, The Netherlands, report BTO 2003(040)
141. Kannan K, Franson JC, Bowerman WW, Hansen KJ, Jones PD, Giesy JP (2001) Perfluorooctane sulfonate in fish-eating water birds including bald eagles and albatrosses. Environ Sci Technol 35:3065–3070
142. Martin JW, Smithwick MM, Braune BG, Hoekstra PF, Muir DCG, Mabury SA (2004) Identification of long-chain perfluorinated acids in biota from the Canadian arctic. Environ Sci Technol 38:373–380
143. Martin JW, Kannan K, Berger U, De Voogt P, Field J, Franklin J, Giesy JP, Harner T, Muir DCG, Scott B, Kaiser M, Järnberg U, Jones KC, Mabury SA, Schroeder H,

Simcik M, Sottani C, Van Bavel B, Kärrman A, Lindström G, Van Leeuwen S (2004) Analytical Challanges Hamper Perfluoroalkyl Research. Environ Sci Technol 38(13):249A–255A

144. De Voogt P, Slot P, Serné P, van der Ven I, van Leeuwen S (2003) Optimalisatie analysemethode en bepaling van PFOS en PFOA in zoetwater—en mariene organismen en—zwevend stof/sediment. Report to RIKZ. Universiteit van Amsterdam, Amsterdam, The Netherlands
145. Schrap SM, Pijnenburg JAMCM, Geerdink RD (2004) Geperfluoreerde verbindingen in Nederlands oppervlaktewater. Rijkswaterstaat, Lelystad, The Netherlands, RIZA-report 2004(037)
146. Hendriks J (1995) Concentration of microcontaminants and response of organisms in laboratory experiments and Rhine delta field survey. PhD Thesis, University of Utrecht, The Netherlands
147. Cornelissen G (1999) Mechanism and consequences of slow desorption of organic compounds from sediments. PhD Thesis, University of Amsterdam, The Netherlands
148. Jonker MTO (2004) Black Magic in the Aquatic environment. PhD Thesis, University of Wageningen, The Netherlands
149. Schipper CA, Schout P (2003) De weg naar implementatie van de Chemie-Toxicitet-Toets. Rijkswaterstaat, The Hague, The Netherlands, RIKZ report 2003(036):84
150. Murk AJ, Legler J, Denison MS, Giesy JP, van der Guchte C, Brouwer A (1996) Chemical activated luciferase gene expression (CALUX); a novel in vitro bioassay for Ah receptor active compounds in sediments and pore waters. Fundamental and Applied Toxicology 33:149–160
151. Åkerman J (2001) Acute—en genotoxiciteit van oppervlaktesedimenten uit de Noordzee. Rijkswaterstaat, The Hague, The Netherlands, working document RIKZ/OS/2001(610X):9
152. Aquasense (2001) Microtox and Mutatox testen met sediment (extracten) afkomstig uit de Waddenzee. Aquasense, Amsterdam, The Netherlands
153. EU (2001) Beschikking 2455/2001/EG van het Europese parlement en de Raad tot vaststelling van de lijst prioritaire stoffen op het gebied van het waterbeleid en tot de wijziging van Richtlijn 2000/06/EG. Brussels, Belgium
154. Van de Guchte C, Beek M, Tuinstra J, van Rossenberg M (2000) Normen voor Waterbeheeer; achtergronddocument NW4. Commissie Integraal Waterbeheer (CIW), Lelystand, The Netherlands, 184
155. Van Straalen NM (1987) Stofgehalten in de bodem—(geen) effecten op bodemdieren. In: symposium Bodemkwaliteit, Voorlopige Commissie Bodembescherming Leidschendam, The Netherlands
156. Aldenberg T, Slob W (1991) Confidence Limits for Hazardous Concentrations Based on Logistically Distributed NOEC data. Rijksinitituut voor Volksgezondheid en Milieuhydiene, RIVM, Bilthoven, The Netherlands. Report 719102002
157. Chapmann PM, Wang F, Adams WJ, Green A (1999) Appropriate Application of Sediment Quality Values for Metals and Metalloids. Environ Sci Technol 33(22):3937–3941
158. Laane RWPM, Brockmann U, Van Lier L, Bovelander R (2004) Immission targets for nutrients (N and P) in catchments and coastal zones: a North Sea assessment. Estuarine Coastal and Shelf Science, to be published
159. OSPAR (2004) OSPAR/ICES workshop 9–13 February 2004, the Hague, The Netherlands. On the evaluation and update of background reference concentrations and ecotoxicological assessment criteria and how the assessment tools should be used for assessing contaminants in water, sediments and biota. OPSAR (CD), London, UK

160. Anonymous (1991) Trends en toestand zoute wateren 1980–1990. Rijkswaterstaat, The Hague, The Netherlands
161. Van Eck GTM, Groeneveld G, Burger G (1994) Risicobeoordeling Nederlandse zoute watersystemen voor de Watersysteemverkenningen. Rijkswaterstaat, The Hague, The Netherlands, report RIKZ/AB-94.865X, 72
162. Swertz OC (1993) Berekening van 90 percentiel. Rijkswaterstaat, The Hague, The Netherlands, working report GWIO 93.210X
163. Augustijn DCM, Baas A, Laane RWPM (2000) Uncertainties in the chemical yardstick for judging the quality of Dutch water systems. European Water Management 3(2):23–27
164. Swertz OC (2001) Metalen in water blijven meten? Rijkswaterstaat, The Hague, The Netherlands, working document 21-8-2001
165. Ragas AMJ (2000) Uncertainty in environmental quality standards. PhD Thesis, University of Nijmegen, The Netherlands
166. Stolwijk S, Ramaker CM, Laane RWPM (2000) Sensitivity of water quality model results for the Dutch coastal zone to the distribution coefficient of Cadmium. Spill Sci Technol Bull 6(2):153–158
167. De Jong J, van Buuren JT, Luiten JPA (1996) Systematic approaches in water management: aquatic outlook and decision support systems combining monitoring, research, policy analysis and information technology. Water Sci Technol 34:9–16
168. Laane RWPM, Groeneveld G (1999) Normtoetsing van stoffen in het sediment van het Nederlandse continentale Plat. Rijkswaterstaat, The Hague, The Netherlands, report RIKZ-99(027)
169. Zevenboom W, Bommelé M, Reuther C (2003) Signalen uit de Noordzee: Signaleren, beoordelen, evalueren en reageren. Rijkswaterstaat, Rijswijk, The Netherlands, report NZ-2003/10:38
170. Wiegman S (2002) Photoenhanced toxicity of azaarenes to marine phytoplankton. PhD Thesis, University of Amsterdam, The Netherlands
171. Van Vlaardingen P, Posthumus R, Traas TP (2003) Environmental risk limits for alkylphenols and alkylphenol. RIVM, Bilthoven, The Netherlands, report 601501019
172. Van Wezel AP, Posthumus R, van Vlaardingen P, Crommetuijn T, van de Plassche EJ (1999) Maximum permissible concentrations and negligible concentrations for phtalates (dibutylphtalate and di(2-ethyl-hexyl)phtalate) with special emphasis on endocrine disruptive properties. RIVM Bilthoven, The Netherlands, report 601501007, 61
173. RIZA (2004) RIZA Institute for Inland Water Management and Waste Water Treatment: Steunpunt wateremissies (www.steunpunt.wateremissies.nl), Lelystad, The Netherlands
174. Yland EML, Sonneveldt HLA, Stronkhorst JH (2000) Evaluatie Chemische Toetsing Zoute Baggerspecie; gegevensanalyse periode 1986–1997. Rijkswaterstaat, The Hague, The Netherlands, report RIKZ 2000(005):45
175. Geerdink RB, Schrap SM (2004) Vergeten stoffen. In: de Rijn-Maas monding. Rijkswaterstaat, Lelystad, The Netherlands, RIZA report 2004(015):82
176. Maas JL, Van De Plassche EJ, Straetmans A, Vethaak AD, Belfroid AC (2003) Normstelling voor bioassays. Rijkswaterstaat, The Hague, The Netherlands, report RIKZ 2003(007)
177. Rotteveel S, Berbee R, den Besten P, Tonkes M (1999) Als Sherlock Holmes op zoek naar oorzaken van toxiciteit. De toepasbaarheid van Toxicity Indentification Evaluation op afvalwater—en waterbodemmonsters. H_2O 32(5):35–38

178. Cizmas L, Mcdonald TJ, Phillips TD, Gillespie AM, Lingenfelter RA, Kubena LF, Donnelly KC (2004) Toxicity Characterization of Complex Mixtures Using Biological and Chemical Analysis in Preparation for Assessment of Mixture Similarity. Environ Sci Technol 38(19):5127–5133
179. Hutchings M, Johnson I, Elaine H, Andrew G, Thain J, Kevin T, Rachel B, Graham W, Jo W (2004) Ecotoxicol 13(5):475–484
180. IRC (1987) Rhein Aktionsprogramme (Rhine action plan). Arnhem, The Netherlands
181. Gren I-M, Turner K, Wulff F (1998) Managing a sea. Earthscan, UK
182. Suijlen J, Duin R (2001) Variability of near-surface total suspended matter concentrations in the Dutch coastal zone of the North Sea: climatological study on the suspended matter concentrations in the North Sea. Rijkswaterstaat, The Hague, The Netherlands, report 2001(150X)
183. De Voogt P, Lourens J (1992) Fate and effects of pesticides in the North Sea. University of Amsterdam/National Institute for Coastal and Marine Management, The Hague, The Netherlands
184. Peters RJB (2003) Hazardous chemicals in precipitation. TNO-report R2003/198, Apeldoorn, The Netherlands

Subject Index

Printing: Krips bv, Meppel
Binding: Stürtz, Würzburg